Contemporary Trends and Issues in Science Education

Volume 44

Series Editor
Dana Zeidler, *University of South Florida, Tampa, USA*

Founding Editor
Ken Tobin, *City University of New York, USA*

Editorial Board
Hsing Chi von Bergmann, *University of Calgary, Canada*
Michael P. Clough, *Iowa State University, Ames, IA, USA*
Fouad Abd El Khalick, *University of Illinois at Urbana-Champaign, USA*
Marissa Rollnick, *University of the Witwatersrand, Johannesburg, South Africa*
Troy D. Sadler, *University of Missouri, Columbia, USA*
Svein Sjøberg, *University of Oslo, Norway*
David Treagust, *Curtin University of Technology, Perth, Australia*
Larry Yore, *University of Victoria, British Columbia, Canada*

SCOPE

The book series Contemporary Trends and Issues in Science Education provides a forum for innovative trends and issues connected to science education. Scholarship that focuses on advancing new visions, understanding, and is at the forefront of the field is found in this series. Accordingly, authoritative works based on empirical research and writings from disciplines external to science education, including historical, philosophical, psychological and sociological traditions, are represented here.

More information about this series at http://www.springer.com/series/6512

Leonard A. Annetta • James Minogue
Editors

Connecting Science and Engineering Education Practices in Meaningful Ways

Building Bridges

Editors
Leonard A. Annetta
George Mason University
Fairfax, VA, USA

James Minogue
North Carolina State University
Raleigh, NC, USA

ISSN 1878-0482 ISSN 1878-0784 (electronic)
Contemporary Trends and Issues in Science Education
ISBN 978-3-319-16398-7 ISBN 978-3-319-16399-4 (eBook)
DOI 10.1007/978-3-319-16399-4

Library of Congress Control Number: 2015958086

Springer Cham Heidelberg New York Dordrecht London
© Springer International Publishing Switzerland 2016
This work is subject to copyright. All rights are reserved by the Publisher, whether the whole or part of the material is concerned, specifically the rights of translation, reprinting, reuse of illustrations, recitation, broadcasting, reproduction on microfilms or in any other physical way, and transmission or information storage and retrieval, electronic adaptation, computer software, or by similar or dissimilar methodology now known or hereafter developed.
The use of general descriptive names, registered names, trademarks, service marks, etc. in this publication does not imply, even in the absence of a specific statement, that such names are exempt from the relevant protective laws and regulations and therefore free for general use.
The publisher, the authors and the editors are safe to assume that the advice and information in this book are believed to be true and accurate at the date of publication. Neither the publisher nor the authors or the editors give a warranty, express or implied, with respect to the material contained herein or for any errors or omissions that may have been made.

Printed on acid-free paper

Springer International Publishing AG Switzerland is part of Springer Science+Business Media (www.springer.com)

Contents

Part I Setting the Stage…The Culture and the Challenges

1 **Creating Disruptive Innovators: Serious Educational Game Design on the Technology and Engineering Spectrum** 3
   David Nelson and Leonard A. Annetta

2 **Grand Challenges for Engineering Education** 19
   Cary Sneider

Part II Student-Centered Design…Exemplary Projects and Programs that Transfer Theory to Practice

3 **Museum Design Experiences That Recognize New Ways to Be Smart** 39
   Dorothy Bennett, Peggy Monahan, and Margaret Honey

4 **Five Principles for Supporting Design Activity** 59
   Catherine N. Langman, Judith S. Zawojewski, and Stephanie R. Whitney

5 **Studio STEM: A Model to Enhance Integrative STEM Literacy Through Engineering Design** 107
   Michael A. Evans, Christine Schnittka, Brett D. Jones, and Carol B. Brandt

6 **Instrumental STEM (iSTEM): An Integrated STEM Instructional Model** 139
   Daniel L. Dickerson, Diana V. Cantu, Stephanie J. Hathcock, William J. McConnell, and Doug R. Levin

7 **Robotics Education Done Right: Robotics Expansion™, A STEAM Based Curricula** 169
   Anthony J. Nunez

8   Designing Serious Educational Games (SEGs) for Learning
    Biology: Pre-service Teachers' Experiences and Reflections 187
    Meng-Tzu Cheng and Ying-Tien Wu

Part III   Preparing Teachers for the Grand Challenges…
           Exemplary Professional Development Practices

9   Language of Design Within Science and Engineering 217
    Nicole Weber and Kristina Lamour Sansone

10 Teaching with Design Thinking: Developing New Vision
    and Approaches to Twenty-First Century Learning 237
    Shelley Goldman and Molly B. Zielezinski

11 Elementary School Engineering for Fictional Clients
    in Children's Literature 263
    Elissa Milto, Kristen Wendell, Jessica Watkins, David Hammer,
    Kathleen Spencer, Merredith Portsmore, and Chris Rogers

12 Teaching Engineering Design in Elementary Science
    Methods Classes 293
    Christine D. Tippett

13 Infusing Engineering Concepts into High School Science:
    Opportunities and Challenges 317
    Rodney Custer, Arthur Eisenkraft, Kristen Wendell,
    Jenny Daugherty, and Julie Ross

14 How Do Secondary Level Biology Teachers Make Sense
    of Using Mathematics in Design-Based Lessons
    About a Biological Process? 339
    Charlie Cox, Birdy Reynolds, Anita Schuchardt,
    and Christian Schunn

15 Final Commentary: Connecting Science and Engineering
    Practices: A Cautionary Perspective 373
    Michael P. Clough and Joanne K. Olson

Introduction

The Next Great Debate?

By elevating engineering design (practices) to the same level as scientific inquiry (practices), the crafters of the *Framework for K-12 Science Education* (NRC 2012) and the subsequent *Next Generation Science Standards* (NGSS 2013) have caused some controversy and lively debate, some of which is captured in Clough and Olsen's final Commentary. represented what engineers do. Where do the true connections between engineering and science practices reside. To what extent can and should science and engineering practices co-exist in educational spaces?

In a recent issue of the *Journal of Science Teacher Education*, Cunningham and Carlsen (2014) offer a rather critical review of the way the nature and methods of engineering are portrayed in these reform documents (NGSS 2013; NRC 2012). As the editors of this volume, we acknowledge and accept (even embrace) the fact that the *eight practices* described in these reform documents look different across science and engineering. We think these differences should be explicitly addressed with students (as several of the contributing authors also suggest) but the mere existence of differences in these disciplines does not preclude their successful integration. We think these differences actually enrich and deepen their relationship. In this book we do not really enter the debate over the relative value, importance, or proper placement in the "standards" of science and engineering as separate disciplines. Rather we maintain that, while different in nature and methods, science and engineering are intimately intertwined and their thoughtful integration is essential to the development of a scientifically literate citizenry moving forward. This book aims to help researchers and practitioners better leverage power of these shared practices to promote science proficiency. This book is intended to help those looking for productive ways to harmonize science and engineering practices to propel STEM teaching and learning within a culture of innovation.

Building Bridges…

While the purposes of science and engineering may be different, their practices are parallel and often quite complementary. In this book, we move beyond the prototypical "bridge building" activity that one might envision when they hear the words "engineering design challenge". Here the "bridges" being built are novel and integrated approaches to teaching science and engineering practices that span diverse and traditionally isolated research communities to foster dialogue and fruitful synergies. In Chap. 1, Nelson and Annetta help set the stage by defining "design thinking" and remind us of the power of "contextualization". They underscore the value of productive failures and paint a picture of how to develop "disruptive innovators" to feed the next generation STEM workforce, a theme of this volume.

This volume also highlights the many ways in which the prudent integration of science and engineering practices can be used to create new and exciting opportunities to learn in K-16 educational spaces (both formal and informal). From Cox's and colleagues' (Chap. 14) forward-looking approach to curricular integration and cutting edge work with serious educational games (SEGs) and robotics by Cheng (Chap. 8) and Nunez (Chap. 7) to the foundational work by Goldman and Bullock (Chap. 10), viewed alone or collectively these efforts represent thoughtful and meaningful cross-cutting connections between research and practice within and across diverse communities.

Facing the Challenges

We borrow Sneider's *Grand Challenges for Engineering Education* (Chap. 2) to help orient the reader to the content of this book. The early Grand Challenges (*Explaining Technology* and *Explaining What Engineers Do*), while not the intended focus of this book, are foundational to any work in this area. Dickerson and colleagues present an interesting approach to the treatment of this inherent disciplinarity in their chapter about the *Instrumental STEM* (iSTEM) project. They put forward a novel instructional model that includes attention to the "nature of the domains" followed by "domain specific instruction". This explicit attention to difference in the domains is also highlighted in Tippett's chapter (Chap. 12) titled *Teaching Engineering Design in Elementary Science Methods Classes* where she examines the consequences of embedding engineering design in elementary science methods courses and the "trouble with terminology" she has experienced. *Project Infuse*, discussed in Chap. 13 by Custer and colleagues, attacks this issue head on by involving teachers in *concept-driven engineering*, used in contrast to simply "doing" engineering-type of activities without a significant understanding of what engineering is and of engineering practices and core concepts.

This book also addresses Grand Challenge *#3 Developing New Curriculum Materials* as numerous exemplary projects are showcased. In their chapter

(Chap. 4), Langman, Zawojewski, and Whitney describe the interdisciplinarity and portability of model-eliciting activities (MEAs). They go on to use representative MEAs to outline a set of *Implementation Design Principals*, the central focus of which is to maintain students' engagement in the foundational design process: cycles of expressing, testing, and revising the object under design.

The *Instrumental STEM* (iSTEM) project shared by Dickerson and colleagues (Chap. 6) serves as another example of new curricular materials being developed. Here students design and build the tools and instruments they need to do authentic scientific inquiry. They assert that this novel approach creates relevance for students by requiring the successful design and fabrication of tools and instruments necessary to answer questions that they have about things they care about.

In their chapter (Chap. 11) titled *Elementary School Engineering for Fictional Clients in Children's Literature*, Milto and team introduce us to the *Integrating Engineering and Literacy* (IEL) project and chronicle how engineering that is situated within the literature that students are reading in their class helps them to frame engineering problems and design solutions for the problems that the characters in the book are experiencing.

The authors in this volume also offer some really keen insights into and practical examples of ways to *teach the design process* to the K-12 students, their teachers, and even teacher educators (Sneider's Grand Challenge #4 and 6, respectively). Part II (*Student-Centered Design...Exemplary Projects and Programs that Transfer Theory to Practice*) is full of examples that engage K-12 students in innovative STEM programs that promote the development of science and engineering practices. Bennett, Monahan, and Honey's showcasing of *New York Hall of Science's* (NYSCI) *Design Lab* (Chap. 3) helps set the stage for the rest of the programs featured. Their focus on the "what" and "how" of children's experiences mirrors the NRC's new view of three-dimensional learning (NRC 2014) well, and their idea of helping children find a "new way to be smart" captures the spirit of this part nicely.

Evans and his team present *Studio-STEM* (Chap. 5), an engineering-based out of school program that engages learners in open-ended real-life problems around energy and sustainability. Their work looks closely at motivation and career intent but at its core maintains that learning is the result of social practices and communicative acts.

This vision is shared by Weber and Sansone (Chap. 9) in their description of the *Language of Design*. We placed this work at the front of *Part III: Preparing Teachers for the Grand Challenges...Exemplary Professional Development Practices* because the problem-based transdisciplinary teacher professional development experience they describe sets the tone for the rest of this part. Their efforts to "transform science teaching by engaging teachers experientially in local, inquiry-based research projects with the integration of science and engineering with the graphic design processes" is the sort of thoughtful teacher (both preservice and inservice) professional development that we need as a field.

Goldman and Bullock from Stanford University (Chap. 10) take us even further down a productive path forward with their sharing of the d.Loft STEM Learning project. Their extraordinary work helping teachers develop their own "design thinking" (abilities to find answers to complex problems that have multiple viable solutions) epitomizes the kinds of attitudes and "mindshifts" that are necessary for building of solid traversable bridges between science and engineering practices.

Unmet Challenges

To be frank, this volume does not give Sneider's Grand Challenge #7 *Balancing Technical and Academic Subjects* and Grand Challenge #8 *Engaging Technology and CTE Teachers* the attention they probably deserve. Sneider's call for the "nation's technology teachers and CTE teachers to join with science teachers to provide the kind of education that all students need to meet the global challenges" needs to be heard, but as he suggests this call must also be heard (and answered) by school administrators and community leaders…these bridges cannot be built by teachers and teacher educators alone. That being said, we feel that (as a diverse but integrated community of STEM teacher-scholars) perhaps the gravest challenge we face lies in Grand Challenge #5 *Developing Assessments*.

In a recent National Research Council (NRC 2014) report titled "Developing Assessments for the Next Generation Science Standards" the authors term the integration of content knowledge, crosscutting concepts, and science and engineering practices "three-dimensional learning." They go on to describe it as instruction that engages students with the practices (of science and engineering) in the context of a core idea and crosscutting concepts. They next suggest that practices (and crosscutting ideas) are at once tools for addressing problems and the topics for learning in and of themselves (NRC 2014). This notion of "three-dimensional learning" is exciting and many of the projects and programs featured in this volume surely capture the essence of this vision of teaching and learning, but the accurate and robust assessment of this sort of instruction simply does not exist as the committee notes by rather bluntly stating that:

> Developing new assessments to measure the kinds of learning the framework describes presents a significant challenge and will require a major change to the status quo. The framework calls for assessments that capture students' competencies in performing the practices of science and engineering by applying the knowledge and skills they have learned. The assessments that are now in wide use were not designed to meet this vision of science proficiency and cannot readily be retrofitted to do so. To address this disjuncture, the Committee on Developing Assessments of Science Proficiency in K-12 was asked to help guide the development of new science assessments. (NRC 2014, p. 12)

The situation seems even more dire when one reads that:

> Most National Research Council committees rely primarily on syntheses of the research literature in areas related to their charge as the basis for their conclusions and recommendations. However, the approach to instruction and assessment envisioned in the framework and the NGSS is new: thus, there is little research on which to base our recommendations for best strategies for assessment. (NRC 2014, p. 17)

The fact that these new standards (NGSS 2013) are in the form of *performance expectations*, specifying what students should know and *be able to do*, necessitates that future assessment tasks be designed and built to *capture evidence of students' ability to use the practices in situ* (as they apply their understanding of crosscutting concepts and disciplinary ideas) to address specific problems (NRC 2014, p. 32).

A Culture of Innovation?

We end this introduction by looking a bit at what Sneider called the greatest challenge of all, that of *Teaching the Teacher Educators*. We feel good about the potential of this book to engage university professors who prepare tomorrow's teachers in supporting the NGSS and cultivating its thoughtful implementation. We asked all of the contributing authors to consider the same question: *Given the rapidly changing landscape of science education, including the elevated status of engineering design, what are the best approaches to the effective integration of the science and engineering practices?*

They answered with rich descriptions of pioneering approaches, critical insights, and useful practical examples of how embodying a culture of interdisciplinarity and innovation can fuel the development of a scientifically literate citizenry. We are confident this collection of work builds ***traversable bridges*** across diverse research communities and begins to break down long-standing disciplinary silos that have historically often hamstrung well-meaning efforts to bring research and practice from science and engineering together in meaningful and lasting ways.

George Mason University   Leonard A. Annetta
Fairfax, VA, USA
North Carolina State University   James Minogue
Raleigh, NC, USA

References

Cunningham, C. M., & Carlsen, W. S. (2014). Teaching engineering practices. *Journal of Science Teacher Education, 25*, 197–210.

National Research Council. (2012). *A framework for K-12 science education: Practices, crosscutting concepts, and core ideas*. Committee on a Conceptual Framework for New K-12 Science Education Standards. Board on Science Education, Division of Behavioral and Social Sciences and Education. Washington, DC: The National Academies Press.

National Research Council. (2014). *Developing assessments for the next generation science standards*. Committee on Developing Assessments of Science Proficiency in K-12. Board on Testing and Assessment and Board on Science Education, In J. W. Pellegrino, M. R. Wilson, J. A. Koenig, & A. S. Beatty (Eds.), Division of Behavioral and Social Sciences and Education. Washington, DC: The National Academies Press.

NGSS Lead States. (2013). *Next generation science standards: For states, by state*s. Washington, DC: The National Academies Press.

Part I
Setting the Stage…The Culture and the Challenges

Chapter 1
Creating Disruptive Innovators: Serious Educational Game Design on the Technology and Engineering Spectrum

David Nelson and Leonard A. Annetta

By early 1892, the World's Columbian Exposition in Chicago was months behind schedule for its October dedication and official opening in May of 1893. After the Manufacturer's and Liberal Arts Building collapsed, the Exposition designers were acutely aware of the need to move even more quickly than before in completing the building structures as well as the aesthetics such as painting and landscaping. Daniel Burnham, the architect of the Exposition, knew that the laborious painting of the rebuilt Manufacturer's building—then the centerpiece of the Exposition and possessing the largest footprint of any building in the world—would be the most likely reason that the building would not be ready in time. As with any other design need, Burnham recognized a problem, and he needed a designed solution in order to achieve the anticipated October dedication.

The solution to Burnham's problem was soon found through the same process of design suggested by the Next Generation Science Standards in the United State, which is discussed in the next section below. In essence, the problem was resolved through a collaborative process among designers whereby alternative solutions were considered, evaluated and revised, all while considering the limitations of time and the constraints of available technology. For this vignette, the final solution was the invention of spray painting, which decreased the labor, time, and expense for Burnham and many others very soon after (Larson, 2003).

D. Nelson (✉) • L.A. Annetta
George Mason University, Fairfax, VA, USA
e-mail: dnelso16@masonlive.gmu.edu

Design Thinking

Although the concept is not necessarily new, the notion of design as a 'way of thinking' is a creative action that has application across numerous disciplinary fields. Design thinking is an approach to practical and creative solutions to problems or issues framed as a design question. Human desires are therefore expressed in the design question with the intent on creating solutions that ultimately impact humans. Human needs provide insights that help the designer refine the design question and form a goal to what is meant to be achieved. Instead of starting with a certain problem, design thinking begins with a question and the acknowledgement that we may not understand the problem. Then, by focusing on process and human needs, the parameters of the problem and the resolutions are concurrently explored.

Design thinking is a creative process that evolves as a building of ideas, where the solution is often actually the starting point. There are no judgments early on in design thinking. This eliminates the fear of failure and encourages maximum input and participation in the ideation and prototype phases. One can characterize the stages of the design thinking process as: *define, research, ideate, prototype, choose, implement*, and *learn*. Within these seven steps, problems can be framed, productive questions can be asked, more ideas can be created, and the best answers can be chosen. This is neither a linear process nor a process that cannot be repeated or occur simultaneously.

Design and the Next Generation Science Standards

Unlike previous science standards such as *Science for All Americans*, the Next Generation Science Standards explicitly place the engineering practices (i.e., *design*) within the context of the science framework by positioning design alongside the practice of science rather than positioned as solely an application of the science content. Importantly, the NGSS characterize engineering as "...[A]ny engagement in a systematic practice of design to achieve solutions to particular human problems" (NGSS Lead States, 2013). As such, design is a *process* through which discipline-specific theories, models, procedures and practices are used to create a useable and tangible solution; today, design is no longer viewed simply an applied science.

The progression of the sophistication in learning and using the design process in the NGSS has been carefully well-defined for practitioners, as well. In applying the NGSS to the classroom, curriculum will highlight the nature of design as defining the needs and limits of a problem that needs to be addressed, designing alternative solutions to a problem based on how well each meets the needs of a problem, and optimizing the final selected design to include the most important features essential to solving problems. Moreover, the expectations of design evaluation in the NGSS increase along the K-12 continuum as students increase the level of evaluation of different designs (for example, design failure in grades K-2; effects on the environment in grades 9–12) and build toward leaving high school with an ability for sophisticated examinations such as evaluating large-scale trade-offs for different design solutions.

Design in the NGSS has several notable features, which are not found in more generic frameworks focusing on design as a subtopic of science or as an applied science; specifically, design:

- Starts with a goal as its conclusion, which is a solution to a problem;
- Will require understanding of and utilize broad principles and concepts of literacy but will move toward narrow application of those principles and concepts;
- Emphasizes reevaluation and revision in an iterative process to develop an optimized solution;
- Focuses on the collaborative nature of finding solutions; and
- Considers carefully the important constraints and limitations provided for the problem for which a solution is sought.

The NRC (2012) report, *'Education of Life and Work,'* addresses both mainstream and pipeline issues of science, technology, engineering, and mathematics (STEM) literacy under the general rubric of attempting to define '21st Century' skill sets, 'deeper learning', and 'competencies' needed to meet future challenges and applied to various tasks—citizen, employee, entrepreneur, manager, parent, and volunteer. The report identified three clusters of cognitive competencies—processes and strategies, knowledge, and creativity that subsume critical thinking, information literacy, reasoning and argument, and innovation; three clusters of intrapersonal competencies—intellectual openness, work ethic and conscientiousness, and positive core evaluation that subsume flexibility, initiative, appreciation for diversity, and metacognition; and two clusters of interpersonal competencies—teamwork and collaboration and leadership that subsume communication, collaboration, responsibility, and conflict resolution.

The semi-model (Fig. 1.1) of the NGSS describes the general inquiry teaching strategies for scientists and engineers. NGSS refers to this as a semi-model because it was built from the frameworks and converted into standards the model framework was excluded. This decision resulted in an illustration of how the practices of scientists and engineers are integrated with both inquiry and design.

Design in the Classroom

In implementing these elements alongside the science curriculum in the classroom, teachers and students alike may struggle with the underlying foundation upon which design is built: the systematic and collaborative nature of the process. Students will approach new problem-based design scenarios with an untrained response in which many variables are altered at once, the process of design begins before considerations of constraints and limitations are considered, and the solution or outcome has

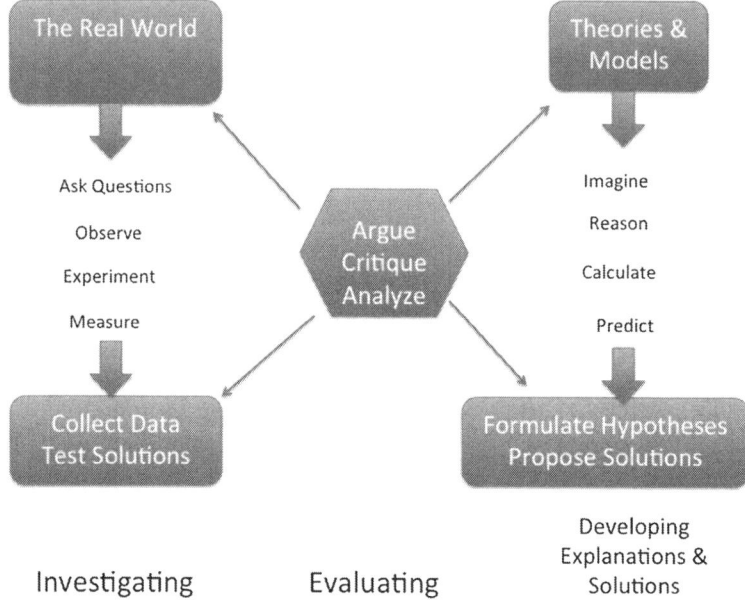

Fig. 1.1 NGSS semi-model of activities for scientists and engineers

not been well-defined. As students develop a stronger understanding of the process of design, they should be expected to exhibit deliberative, structured and justifiable analyses of initial and alternative solutions to the problems they have been asked to solve; trial-and-error, for example, will not support an iterative or evaluative approach to design strategy.

Like engineers in the field, students will learn to utilize standards of design and exhibit behaviors that lead to successfully arriving at desirable outcomes. The NGSS reflect this in the progressive nature of the standards through the K-12 grade bands. When fully implemented and aligned with the NGSS's framework, classrooms will feature engagement in collaborative workgroups with discourse and challenges among participants to justify and critique one's own and others' choices rather than working in isolation from peers in researching, designing, and evaluating design solutions. Exposure to design and the encouragement of a creative and innovative setting in the classroom will allow students to engage early in the practices and introduce careers and fields into which students might not have otherwise found a path. Students will recognize the social and environmental impacts of design options, and they will be able to consider how trade-offs in the design affect the desired outcome and optimize a final design. Application of design will occur alongside disciplinary-specific literacy and will not be subsumed as simply an application of the scientific disciplines; it will also not be reserved only for students in standalone engineering courses. Finally, students of diverse backgrounds will be afforded opportunities to use design as a conduit to seeing the relevance of science and its related fields in their lives while at the same time finding ways to explore areas that might have previously been less accessible to certain populations.

Significant, positive changes related to the move toward focusing on innovation as a collaborative team effort, encouraging broad student knowledge in the process of design, and finding ways for students to take risks by not considering the need for revision a failure can be realized with the NGSS. The NGSS provide a solid framework around which curriculum rooted in engineering design can be developed to foster the inquisitive nature of students in asking questions about why a particular solution is better than another as easily as to reinforce the more abstract and hypothetical capabilities of twelfth grade students.

Participatory Learning in NGSS Engineering Design

The processes of the engineering practices and the design aspect emphasized in the NGSS draw upon the principles of participatory learning, which involves aspects of learning fundamentally different than what are seen in many classrooms today. The behaviors and pedagogies that form the tenets of participatory learning are integral to the faithful implementation of the NGSS in the classroom. As design itself is a process rather than a single event, participatory learning involves similar reflective, critical and engaging social collaborations rather than a once-and-done activity or lesson.

Today, effective and efficient participatory learning for engineering and design education is often discussed in terms of three overarching ideas: using interactive technologies and environments to communicate and allow for simulation, ensuring real-world contextualization and situational frameworks, and providing settings in which failures are viewed as challenges for innovation.

Technology in Participatory Learning As discussed earlier, engaging and collaborative experiences underlie the NGSS expectations for the standards related to the engineering practices. Especially, the use of technology such as mobile devices, virtual reality simulations and Internet-connected "hubs" where data, information and ideas can be shared offer platforms upon which programs of participatory learning supporting the NGSS can be built.

Such technology-based experiences afford teachers and students several advantages over traditional technology and direct instruction-based methods. Such advantages are:

- Real-time data analysis and exchanges between students allow for quick changes to a design—that is, students can respond immediately to observations to adjust and fine-tune designs on-the-fly;
- Challenges to ideas can be easily communicated, evaluated and used to augment design plans and subsequent trials; and

- Concrete manifestations of abstract concepts can be made visible, and the effects of manipulations to variables, conditions or interactions can be assessed and revisited.

Contextualization in Participatory Learning Importantly, participatory learning provides students real-world and meaningful contexts in which they can visualize themselves as being important and relevant novice researchers who are creating as well as consuming information.

The contextualization of science and engineering learning has developed alongside the reform movements rather than in response to them. Rivet and Krajcik (2008) described contextualization in terms of their seminal research surrounding the method in middle schools as "[S]cience instruction [involving] utilizing students' prior knowledge and everyday experiences as a catalyst for understanding challenging science concepts" (p. 79). Not too much earlier, the concept of contextualization was described as "…a diverse family of instructional strategies…[that focus] teaching and learning squarely on concrete applications in a specific context that is of interest to the student" (Mazzeo, Rab & Alssid, 2003, p. 3). And, Stinner (1989, 2006), who was a pioneer in the more specific large context problem aspect of contextualized learning, has long defined the process in terms of "contexts of inquiry," (p. 19) which surrounds a five-pronged framework rooted in questions, methods, problems, experiments, and histories.

Contextualization can be realized by providing students opportunities to explore needs in their own communities or areas of interest, examples of which might be investigating solutions to water quality or land use problems or finding alternative energy sources where traditional delivery is unstable. At any rate, even as the essential processes of contextualization continue to be refined, well-developed descriptions of the methods associated with the strategy have been articulated, and the behaviors attendant with the strategies can be readily observed (e.g., Perin, 2011; Singer, Marx, Krajcik, & Clay-Chambers, 2000) to support participatory learning and its place in terms of engineering design education.

Innovation in Participatory Learning Participatory learning in the design classroom has at its core the notion that a student-driven environment results from the technology-enhanced interaction of actors in the classroom(s) resulting in an evaluative, contextualized atmosphere where intellectual risks and plan modifications are encouraged and supported. This leads to an innovative setting where students are supported and more-freely able to negotiate their own learning through collaboration, discussion, and engagement in finding answers to a common problem—even if the tangible result is a different product or solution. In a successfully innovative environment constructed around the principles of participatory learning (McLoughlin & Lee, 2007), content is learner-generated, authentic, and allows for varying perspectives; curriculum is dynamic rather than static and is scaffolded by a large network of players; and communication occurs thorough multiple media and is open and peer-to-peer.

Disruptive Classrooms

Most would argue that innovation is a disruptive process. A disruptive innovation is not necessarily an earth shattering improvement but rather disrupts the trajectory of the innovation to a social system. When this happens the innovation is often not as good as what is currently available but for some reason the target audience cannot consume the product. In education this is not uncommon, especially as it pertains to technology. Often there are better products than what schools can afford, maintain and/or network. Therefore, many schools and classrooms create disrupted innovations that are simpler and more affordable but still an improvement to what they currently have.

Infusing design into the science classroom will be a disruptive innovation in the near future. Teachers and their respective administrators need to understand and support disruptive innovation. Disruptive innovation can be confused with poor classroom management or just chaos. Unless first establishing a set of classroom rules, teachers will not find overall success in a disruptive environment. There could be communication problems due to the difficulty of getting students' attention; communicating and navigating the learning environment will be a challenge unto itself unless firm rules have already been established.

Large companies have research and development teams that understand and embrace the design thinking process. Research and development consists of creating and not consuming. Thinking outside the box is a term often used in the research and development process. It is a disruptive experience that often results in what can only be described as failure. However, learning through failure is something successful companies and individuals all possess, but failure in schools is a very bad word. When children fail in school, their attitudes toward that subject and efficacy drop considerably and sometimes to a point of diminishing return. Failure can be positive: it allows the designers to reformulate their question and enact a different path toward the end goal. Iteration is crucial to success. The more we fail, the more we learn; and, ultimately the end solution is the pinnacle of design.

Schools are asked to do more with less. The NGSS addition of design to the standards is a perfect example. With high stakes tests looming over teachers and their schools, how could one possible spend time iterating a design procedure while preparing for tests?

In his book "Creating Innovators: The Making of Young People Who Will Change the World," Wagner (2012) suggests five specific problems with our current educational system when it comes to creating innovators (p. 288):

1. Individual achievement is the focus: Students spend a bulk of their time focusing on improving their GPAs, but innovation is a team sport. Problems are too complex to innovate or solve by oneself.
2. Specialization is celebrated and rewarded: You can neither understand nor solve problems within the context and bright lines of subject content. Learning to be an innovator is about learning to cross-disciplinary boundaries and exploring problems and their solutions from multiple perspectives.

3. Risk aversion is the norm: We penalize mistakes. Innovation is grounded in taking risks and learning via trial and error. Educators could take a note from design firm IDEO with its mantra of "fail early, fail often."
4. Learning is profoundly passive: For 12–16 years, we learn to consume information while in school. Innovative learning cultures teach about creating, not consuming.
5. Extrinsic incentives drive learning: Young innovators are intrinsically motivated, he says. They aren't interested in grading scales and petty reward systems. Parents and teachers can encourage innovative thinking by nurturing the curiosity and inquisitiveness of young people. Parents of innovators encouraged their children to play in more exploratory ways.

Creating Disruptive Innovators through Design, Curiosity, and Creativity

If we are truly committed to creating the next generation STEM workforce, then it is most critical to create design thinkers that possess an ability to create networks between nodes that are seemingly unconnected. We need to teach students above and beyond the curriculum both in school and out of school. Disruptive innovators have the ability to probe deeply in their design thinking questions. For example, a student might ask, "What if we were able to capture all of the sun's energy each day and distribute it across the world at the speed of light?" Disruptive innovators build a network of people from varying backgrounds and intellect to gain access to new and different ways of thinking. We often envision some of the most famous innovators as being socially inept. Truthfully, they are extremely competent when it comes to their personal network established for the sake of their design passion. Finally, disruptive innovators have a knack for observing the world around them and questioning new ways to achieve desired ends. They are leaders, thick-skinned, daring, calculating, methodical, critical, observant, pattern recognizing, scenario planning, committed, and goal oriented.

Too often, students have been conditioned to believe that there is only one path to the right answer during their formal schooling. But we know that students learn through play (Vygotsky, 1963, 1978; Piaget, 1951; Jackson et al., 2012) and even more through design (Annetta et al., 2014). Creating a passion for science through design thinking can encumber much of the curricular goals in one process. We argue a model for:

$$\text{Prevail} = \text{Play} + \text{Passion} + \text{Purpose} + \text{Persistence}$$

Teachers must create a learning environment and interact with students in a manner that promotes playing with purpose, persistence toward excellence, and design questions that students find intrinsically compelling. We don't do this by rote memorizing facts for a standardized test but rather by cultivating the innate creativity in children. We often confuse creativity with trial-and-error learning, often counting this lower-order reasoning as creative on the same level as what Mozart or Einstein accomplished. The truth is, great innovations come from years of dedicated practice. Yes, trial-and-error can yield interesting and creative learning. But the remarkable creative and innovative breakthroughs we point to typically came from a disruptive innovator who understood how his/her breakthrough was important and how to explain the significance.

Design by Modding the Mod

Over the last 12 years, we have been refining a design-thinking model for both students and teachers through the National Science Foundation funded HI FIVES (Highly Interactive Fun Internet Virtual Environments in Science) and GRADUATE (Games Requiring Advanced Developmental Understanding and Achievement in Technological Endeavors) projects. Teachers and students in grades 5–12 have been exposed to Serious Educational Game (SEG) (Annetta, 2008) design and development through a proprietary platform built on top of a commercial game engine. This platform allows teachers and students to create SEGs that align with science and mathematics content standards.

A *mod* is a modification of a game engine and is used by many gamers as a way to create their own games through manipulating the source code of said engine. Many popular commercial game engines that tend to be part of the modder community are: Unreal, Half Life Source, Unity, and Never Winter Nights. In our projects, we effectively created a mod of a commercial game engine by layering libraries of model/objects, animations, and level environments with an object-oriented programming interface. The mod effectively became an authoring tool that allowed the user to create games of that engine and join the modder community without knowing or learning any of the myriad of technical skills most modders posses; skills such as programming, 3D art, animation, level design, etc.

The SEG created from the original mod is essentially a mod of the mod. To successfully mod a mod and create an SEG, one might first partake in the design process. We've learned over the course of the last 12 years (see Annetta publications) that SEG design and development is an effective learning tool in grades 5–16. Although game design may seemingly be an unusual topic to bring up in relation to education, it is not without reason. Games are very permissive, with the current statistics of United States gamers revealing that 65 % play video games, and of that, 23 % are youth under the age of 18.

Although the NGSS are driving today's science instruction, it was the National Science Education Standards (NSES) driving U.S. science education for the previous 17 years. The NSES posited, "children's abilities in technological problem solving can be developed by firsthand experience in tackling tasks with a technological purpose" (NRC, 1996, p. 135). Using technology without purpose does not provide meaningful learning experiences. The "T" in STEM continues to be in question, but for us that "T" means technological endeavors centered and used for development by students.

Design thinking can be situated within the context of authentic problems that allow students to address personal and societal needs (Atman, Kilgore, & McKenna, 2008). In particular, design thinking offers students an opportunity to experience the iterative nature of science as well as the meaning of testing alternative ideas in problem solving (Bers & Potsmore, 2005; Cunningham et al., 2005; Kahn & Bers, 2005), the soft failure that encourages deeper understanding and advanced discovery (Vallett & Annetta, 2014), and affords students the opportunity to begin to understand systems, a common theme in science education (Sullivan, 2008).

SEG Design and Development Strategy

We know the gamers have certain needs they generally are not getting in traditional schooling. Gamers are creative, but creativity is not always being cultivated in today's schools. Gamers crave constant feedback and constructive assessment. Teaching and learning in the K-12 is generally a process by which teachers are forced to follow a pacing guide so to cover a very crowded curriculum in the average 180-day school year. If students have not yet learned a given content or mastered given skills, they are generally left behind to catch up on their own. This sometimes happens through tutoring or as in Taiwan, cram school. Conversely, in games students get constant feedback and are assessed to the point of failure. If students fail in a game they are demoted to the beginning of that level until they master the skills necessary to be promoted to the next level. The game logic, and sometimes artificial intelligence, scaffolds learning for players so they can be "tutored" and learn the necessary and desired skills to gain promotion.

Our model promoted creativity and allows a safe soft failure environment. Most importantly, teachers are involved in the learning process throughout the entire SEG design mechanism but in a very different role from what they may have been trained to serve. Figure 1.2 is an illustration of the SEG Design Mechanics we have refined through the aforementioned National Science Foundation projects. This model is an amalgam of game design and instructional design. We superimposed these two strategies to help us create an environment in which we ask students to become the teacher as they design and develop their SEG. Many of us know through experience that one of the best ways to learn something is by being asked to teach it first, and we have learned over the years that students learned the targeted science topics more deeply when they have to design and develop and SEG. To this end we teach students, albeit at a very basic level, to become teachers while they infuse game design principles as well. What follows in a description of each step of the process and what we have learned through students creating SEGS through each step. It is important to note that all of the steps of the SEG design happen without technology (although you could invite technology into each piece). However, we have mostly used poster paper and white boards.

1 Creating Disruptive Innovators: Serious Educational Game Design on the Technology... 13

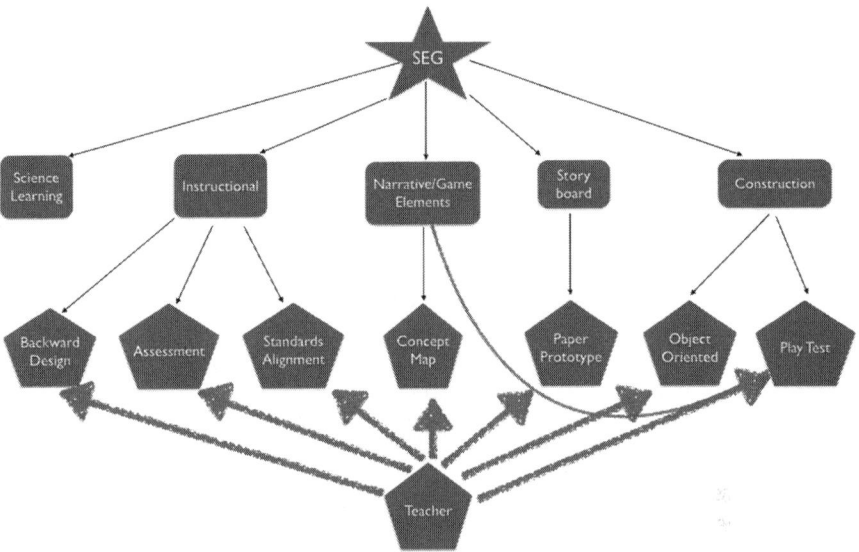

Fig. 1.2 SEG design mechanics model

The first, and most the important step is for the student to learn content (science in our cases). Like a good teacher, if you are not well versed in the content then it will become incredibly difficult to teach it. We have used varying strategies to teach students the content. Such strategies as:

1. Traditional classroom instruction where the student learns the content through a teachers' designed unit.
2. Through science research where the student conducts experiments not unlike they would in a science fair but in a much deeper fashion
3. Through mentors where students work with university faculty and/or community partners in the field that aligns most closely with the student SEG topic. For example, students designing an SEG about wind energy might partner with an employee from the local electric cooperative that is volunteering his/her time.
4. Some combination of the first three but invariably we see students going further and deeper in each of the strategies.

Peer pressure often comes into play and students who want to make the best game possible generally go online, read books, and ask experts questions so they understand the content well enough to drive the rest of the design mechanisms.

After students have learned the content, we then ask them to learn to teach. By using a backward design approach, we invite students to understand how teachers use learning objectives to drive their units and lessons while game designers have level objectives that drive the game play in said level. We also clue students to content standards and ask that they also align their SEG with state content standards much like their teacher does within the classroom. We ask students first what they want the learner to know at the end of the game or the end of the level and then ask them to think about what the learners/players would need to do to learn the content in the SEG. How do teachers know if their students are learning? They assess them of course. In SEG design, we instruct student designers to embed assessments in the game but also know that we can collect click stream data on the back end server so students are being assessed without knowing they are being assessed, something (Annetta et al., 2007) called virtual observations. Therefore, the assessment in the backward design will ultimately dictate the game logic.

Our SEGs are narrative driven, and we instruct student designers on how to create a story by including fundamental game elements. We aim for a cross-disciplinary approach to the design process by including the language arts to this model. We have students develop the essential story elements such as characters, plot, setting, theme, and conflict. This approach includes students who may not be scientifically inclined or science phobic but excels at the creative and open writing components of learning.

Concurrently, student designers embed common game elements into their narratives. Obviously, this is a very time consuming step in the design process but a very important step. Students need to generate interest, and more importantly, how to make people want things innately rather than by external motivation. We believe this adds something to traditional schooling that just isn't occurring in today's classrooms. Students seem to not be generally intrinsically motivated to learn, but when learning is stealthy enough to be hidden behind the cloak of SEG design, students assimilate content knowledge because they want to make a fun, accurate game.

Story needs to provide a sense of **identity** to the reader and subsequent game player. The player needs to have a sensation of being immersed and part of a community while providing the aesthetics to attract players to come back for more. This is often done through **fantasy**, which is the play of imagination and perspective of environment, which feeds the addictive experience seen in good commercial games. Much of our work has seen students take this element to a role-play scenario where the player empathizes with game characters and the surroundings and takes the perspectives of the main character-whether or not the player is the main character.

The narrative needs to be dramatic and emotionally progressive for the immersion to occur. This is where the **conflict** element is played out. Allowing players to have control of the environment is a big challenge to the student designers but they begin to understand this challenge later in the process. Practicing the art of decision-making has resulted in student designers being more engaged with critical thinking and creative skills.

These games need to also have an **economy**. Whether it is a score or artifacts collected, player engagement greatly hinges on this game element. Although commercial games are won and lost by the economy, SEGs, and more specifically the design and development of SEGs are won and lost by how well the economy is articulated within the narrative.

Once the narrative and game elements are written and fleshed out, student designers create a concept map/flow chart/decision tree. Call it what you will, but this component of the model is simply a graphical representation of the path players may take through the narrative. What decisions the player must make and the game logic that reacts to a player decision is mapped out here. It allows student designers to rethink their narrative and enhances the learning cycle by giving a visual of what was written.

The next step is the storyboard. Not unlike video/movie creation, the storyboard takes a scene-by-scene approach to the narrative. It allows designers to draw what objects need to be added to the game to make it aesthetically pleasing to the player to keep immersion and engagement at its highest. From there, student designers paper prototype their storyboard by creating a representation of the user interface. It allows for usability testing before getting into the game creation software so the narrative and game elements can be adjusted if needed. This is an important step so the construction component isn't wasted time.

The final step is the construction phase where student designers build out their design in a proprietary software we developed that is object-oriented in nature so students don't need to learn programming per se, or 3D art and animation, which of high order skills often not seen in K-12 students. Upon completing construction, games are play tested with peers in the intended audience and feedback from play tests allows student designers to go back to the narrative and game elements to make necessary adjustments and the cycle plays out again until the final SEG is ready to go live.

The most important piece to this model is the change in the teacher role. The teacher now becomes the facilitator of content and pedagogy through the design process. Since SEG are intended to be used to teach content and/or concepts, teachers are the masters of how best to teach and assess learning-even if they don't play or understand video games. As previously mentioned, students generally learn more content through this design model than do their peers who gets similar instruction from their teachers. This allows teachers to challenges students to think on a much higher order level than they normally get to do. Teachers seem to love being able to make students think about content more deeply.

Conclusion

A fundamental challenge in the classroom in the era of the NGSS will be making the necessary shift toward redefining the role of students as teacher-learners and training teachers to become facilitators in environments that support such a shift. The iterative approach of SEG design described here supports the scaffolding, feedback and assessment foundation needed in any good lesson, and it acts to counter Wagner's (2012) suggested problems with our education system. Emphasizing the often neglected but sophisticated ways of thinking that students bring to the classroom based on experiences (NRC, 2007), design thinking motivates students to position themselves as active participants in the learning process. Moreover, because it is a process rather than a singular event, design thinking manifested as we describe here encourages challenges, social interaction and peer feedback, all of which are powerful motivators to help students find intrinsic motivation in classroom (Psotka, 2013).

References

Annetta, L. A. (2008). *Serious educational games: From theory to practice*. Amsterdam: Sense.

Annetta, L. A., Cook, M. P., & Schultz, M. (2007). Video games and universal design: A vehicle for problem-based learning. *Journal of Instructional Science and Technology, 10*(1). http://www.usq.edu.au/electpub/e-jist/docs/vol10_no1/papers/current_practice/annetta_cook_schultz.htm

Annetta, L. A., Lamb, R., Vallett, D., & Cheng, R. (2014, March 30). *Changes in high school science student affect through serious educational game design and development*. Paper presented at the annual meeting of the National Association for the Research in Science Teaching (NARST). Pittsburgh, PA.

Atman, C., Kilgore, D., & McKenna, A. (2008). Characterizing design learning: A mixed-methods study of engineering designers' use of language. *Journal of Engineering Education, 97*, 309–326.

Bers, M. U., & Portsmore, M. (2005). Teaching partnership: Early childhood and engineering students teaching math and science through robotics. *Journal of Science and Technology, 14*, 59–73.

Cunningham, C. M., Lachapelle, C., & Lindgren-Streicher, A. (2005). Assessing elementary school students' conceptions of engineering and technology. In *Proceedings of the ASEE annual conference*.

Jackson, L. A., Witt, E. A., Games, A. I., Fitzgerald, H. E., von Eye, A., & Zhao, Y. (2012). Information technology use and creativity: Findings from the children and technology project. *Computers in Human Behavior, 28*, 370–376.

Kahn, J., & Bers, M. (2005). *An examination of early elementary students' approach to engineering*. Presented at the proceedings of the 2005 American Society for Engineering Annual Conference and Exposition.

Larson, E. (2003). *Devil in the white city*. New York: Vintage Books.

Lead States, N. G. S. S. (2013). *Next generation science standards: For states, by states*. Washington, DC: Achieve.

Mazzeo, C., Rab, S. Y., & Alssid, J. L. (2003). *Building bridges to college and careers: Contextualized basic skills programs at community colleges*. Brooklyn: Workforce Strategy Center.

McLoughlin, C., & Lee, M. (2007, December 5). *Social software and participatory learning: Pedagogical choices with technology affordances in the web 2.0 era*. Presented at the ICT: Providing choices for learners and learning, Singapore.

NRC. (1996). *National science education standards*. Washington, DC: National Academy Press.

NRC. (2007). *Taking science to school: Learning and teaching science in grades k-8*. Washington, DC: National Academies Press.

NRC. (2012). In J. Pellegrino & M. Hilton (Eds.), *Education for life and work: Developing transferable knowledge and skills in the 21st century*. Washington, DC: National Academies Press.

Perin, D. (2011). *Facilitating student learning through contextualization* (No. 29). New York: Community College Research Center, Columbia University.

Piaget, J. (1951). *Play, dreams, and imitation in childhood*. (G. Gattegno and F. M. Hodgson, transl.) New York: Norton.

Psotka, J. (2013). Educational games and virtual reality as disruptive technologies. *Educational Technology & Society, 16*, 69–80.

Rivet, A., & Krajcik, J. (2008). Contextualizing instruction: Leveraging students' prior knowledge and experiences to foster understanding in middle school science. *Journal of Research in Science Teaching, 45*, 79–100.

Singer, J., Marx, R. W., Krajcik, J. S., & Clay-Chambers, J. (2000). Constructing extended inquiry projects: Curriculum materials for science education reform. *Educational Psychologist, 35*, 165–178.

Stinner, A. (1989). The teaching of physics and the contexts of inquiry: From Aristotle to Einstein. *Science Education, 73*, 591–605.

Stinner, A. (2006). The large context problem. *Interchange, 37*, 19–30.

Sullivan, F. R. (2008). Robotics and science literacy: Thinking skills, science process skills and systems understanding. *Journal of Research in Science Teaching, 45*, 373–394.

Vallett, D., & Annetta, L. A. (2014). Re-visioning k-12 education: Learning through failure–not social promotion. *Psychology of Popular Media Culture, 3*(3), 174–188.

Vygotsky, L. S. (1963). Learning and mental development at school age (J. Simon, Trans.). In B. Simon & J. Simon (Eds.), *Educational psychology in the U.S.S.R* (pp. 21–34). London: Routledge & Kegan Paul.

Vygotsky, L. S. (1978). Interaction between learning and development (M. Lopez-Morillas, Trans.). In M. Cole, V. John-Steiner, S. Scribner, & E. Souberman (Eds.), *Mind in society: The development of higher psychological processes* (pp. 79–91). Cambridge, MA: Harvard University Press.

Wagner, T. (2012). *Creating innovators: The making of young people who will change the world.* New York: Simon & Schuster.

Chapter 2
Grand Challenges for Engineering Education

Cary Sneider

In 2013 the National Research Council released *A Framework for K-12 Science Education: Practices, Core Ideas, and Crosscutting Concepts* (NRC, 2012), which laid the groundwork for revising state science standards. Unlike previous documents that presented long lists of concepts and skills, the *Framework* specified just thirteen core ideas that all students should learn at increasing levels of sophistication from kindergarten through twelfth grade.

What is even more remarkable than agreement on a coherent set of core ideas was the vision of practices of science and engineering that all students should learn. It was a vision both inspirational and practical:

> *We anticipate that the insights gained and interests provoked from studying and engaging in the practices of science and engineering during their K-12 schooling should help students see how science and engineering are instrumental in addressing major challenges that confront society today, such as generating sufficient energy, preventing and treating diseases, maintaining supplies of clean water and food, and solving the problems of global environmental change. In addition, although not all students will choose to pursue careers in science, engineering, or technology, we hope that a science education based on the framework will motivate and inspire a greater number of people—and a better representation of the broad diversity of the American population—to follow these paths than is the case today.* (NRC, 2012, p. 9)

The *Framework* included "engineering" alongside "science," and declared that students should study major global problems that require at least equal measures of engineering know-how and scientific knowledge. The document also included explicit instructions for presenting to students the engineering design process as both core ideas (what students should know) and practice (what students should be able to do.) Also included were important ideas about the two-way relationship between science and engineering (that science helps engineering advance, and

---

C. Sneider (✉)
Portland State University, Portland, OR, USA
e-mail: csneider@pdx.edu

engineering drives science forward), and the influence of science, technology, and engineering on society and the natural environment.

Development of the *Framework* was just the first step in the most recent effort to remake our nation's science education infrastructure. A coalition of 26 states, working with the independent organization Achieve, Inc., used the *Framework* as the blueprint for *Next Generation Science Standards* (NGSS Lead States, 2013), which spells out, grade by grade for K-5, and in grade bands for 6–8 and 9–12, statements that translate the major ideas from the *Framework* into specific learning targets, or "performance expectations." Together, the *Framework* and NGSS project an entirely new vision of science education to guide the development of new curricula, new assessments, new methods of teacher education, and new goals for our students.

These documents have launched what is likely to be a long campaign to integrate engineering and technology into our nation's educational infrastructure. Although a similar goal was put forward in *Science for All Americans* (AAAS, 1989), and the *National Science Education Standards* (NRC, 1996), the immense inertia of our educational system has so far resisted any significant integration of engineering and technology into science education, let alone social studies, mathematics, or language arts (although there are clear connections to all of those curriculum areas).

These global problems mentioned in the paragraph quoted above—such *as generating sufficient energy, preventing and treating diseases, maintaining supplies of clean water and food, and solving the problems of global environmental change*— are among the grand challenges that engineers will face with increasing urgency in the decades ahead as the human population continues to grow. The thesis of this chapter is that realizing this vision also poses grand challenges for science and engineering teachers at the K-12 level, as well as for school principals, district and state educational leaders, and those of us who work at universities charged with preparing tomorrow's teachers. This chapter will describe the sources of that resistance with the aim of alerting readers to the nature and depth of the challenge ahead, and suggest new pathways forward.

Grand Challenge #1 Explaining Technology

According to the *Framework* and the NGSS, science, engineering, and technology are interrelated but distinct terms:

> *In the K–12 context, "science" is generally taken to mean the traditional natural sciences: physics, chemistry, biology, and (more recently) earth, space, and environmental sciences … We use the term "engineering" in a very broad sense to mean any engagement in a systematic practice of design to achieve solutions to particular human problems. Likewise, we broadly use the term "technology" to include all types of human-made systems and processes—not in the limited sense often used in schools that equates technology with modern computational and communications devices. Technologies result when engineers apply their understanding of the natural world and of human behavior to design ways to satisfy human needs and wants.* (NRC, 2012, p. 11–12)

According to this definition the earliest uses of rock, bone, and wood to make implements for hunting and preparing food were *technologies*, as were the invention of fire, woven fabrics, and the earliest forms of agriculture. Although the nameless inventors who created these technologies did not have degrees in engineering, there is no doubt that they created what they did to solve very real problems in their environment.

Our early human ancestors carried technologies with them, but for the most part they lived in a natural environment. Today we are surrounded by technologies and we experience very little of the natural world. To appreciate the extent to which we depend on them, imagine what would happen if all of our technologies disappeared. First, this book would dissolve. Whether it's electronic or made of paper, it's a product of human invention. Next the lights would go out, as would everything that runs on electricity, oil or gas, since these all depend on technologies to utilize Earth's resources for energy and power. If you are indoors the furniture, rugs, and walls would disappear, and soon the entire building would be gone. Say goodbye to your glasses, cosmetics, and every stitch of clothing. Without the comfort and support of the technological world, you would be standing naked in a field or forest.

Actually, the above scenario is optimistic. Chances are without technology very few of us would survive long at all. In 1900 people could expect to live about 47 years. The vastly extended life expectancy that we enjoy today is only partly due to advances in medicine and improved child mortality rates. The technologies involved in processing fresh potable water is largely responsible for our increased lifespan, just as the technologies involved in growing and processing food have greatly increased the carrying capacity of our planet.

Despite the wide diversity of technologies that we encounter daily, and their importance for our very existence, most people don't even think about them. And when they do, they use the term "technology" in a very limited sense. According to a pair of Gallup polls, for the great majority of people the word *technology* is "tied more to the modern apparatus, machines, and gadgets people have developed" (Rose et al. 2004, p. 1). In 2001, most people who were asked: "When you hear the word 'technology' what is the first thought that comes to mind?" the majority responded "computers" (67 %), while a few responded "electronics" (4 %). Those numbers were virtually unchanged in 2004 (68 % and 5 % respectively).

For the most part teachers of all subjects and grade levels also use the term "technology" in a limited sense, although in a way that is somewhat different from the general population. When teachers claim that their students "don't have access to technology" they are not saying that their students have no pencils and paper. Instead they usually mean that their school does not have sufficient computers or tablets for their students to use. And a classroom "equipped with technology" usually means a Smart Board, which offers the functions of a computer and projector rolled into one.

If we expect our students to understand what engineers do, an important step is coming to understand the products of engineering—the technologies that engineers design and modify to meet people's needs and wants.

To gain some insight into the nature of technology, pick up an object within reach. If you're sitting at a desk a pen will do, as will a piece of paper or more complex technology such as a calendar or cell phone. If you're reading in bed pick up a tissue or alarm clock, and ask yourself these questions:

- What was this technology designed to do?
- What did this particular piece of technology replace?
- How does this technology function better than what was used in the past? How is it worse?
- Where did the materials used to make this technology come from?
- What technologies were required to produce it, and transport it here?
- What will happen to this technology when I'm done with it?
- Could this technology be improved? If so, how?

Helping people realize that the vast number of products around them are technologies would be a step in the right direction; but only a step. People who do understand that technologies are all of the ways that people change the world to meet human needs and wants tend to think of products. But technologies also include processes and systems. A bus schedule is a technology. A recipe for baking a cake is a technology. Life insurance is a technology. Our nation's system of government is a technology. All of these have been created by people, and modified and improved over time. While the people who shaped these technologies may not have been licensed engineers, they were nonetheless "doing engineering." That is they were solving problems in a way that is systematic and iterative.

Why is it important for everyone to learn about technology? Isn't it enough for the professionals to understand it, since most people seem to do just fine with their limited understanding? A thoughtful answer to that question was provided by the National Academy of Engineering and the National Research Council in a short report entitled *Technically Speaking: Why All Americans Need to Know More About Technology*.

> *As far into the future as our imaginations can take us, we will face challenges that depend on the development and application of technology. Better health, more abundant food, more humane living and working conditions, cleaner air and water, more effective education, and scores of other improvements in the human condition are within our grasp. But none of these improvements is guaranteed, and many problems will arise that we cannot predict. To take full advantage of the benefits and to recognize, address, or even avoid the pitfalls of technology, Americans must become better stewards of technological change.* (Pearson et al. 2002, p. 12)

Technically Speaking points out that it is not only our standard of living that is at stake. As the world's population grows, so does our impact on the environment. While developing nations mechanize agriculture, produce more energy, goods, and services, and turn more arable land into cities, the impact on the environment grows at an ever faster rate. To counter these trends we need to be both leaders and collaborators in finding new solutions to the unanticipated effects of yesterday's technologies, such as our changing atmosphere due to the burning of fossil fuels, the impact of pesticides on amphibians and other fragile species, and industrial wastes from thousands of sources. In other words, we need a strong, creative, and flexible technical workforce *and* a technologically literate populace to solve these global challenges. Given how little people's understanding of technology has changed in recent years, that is a grand challenge indeed.

A pathway forward proposed in *Technologically Speaking* consists of 11 recommendations that include incorporating technology into state standards, curriculum, and assessment, as well as the preparation of teachers. The recommendations call upon the National Science Foundation and other federal agencies to support research in how people learn about technology. Museums, private industry, and engineering societies are asked to educate the public, and especially journalists about the nature and importance of technology. The eleventh recommendation is for the White House to add a Presidential Award for Excellence in Technology Teaching to those it currently offers for mathematics and science teaching.

To some extent these recommendations foreshadowed the rise of STEM education as a new national goal, and the *Framework* and *Next Generation Science Standards*. Nonetheless, we have a long way to go before we begin to turn the tide, so that a majority of people have a broad and deep understanding of the "T" in STEM.

Grand Challenge #2: Explaining What Engineers Do

You've checked into a hotel room only to find that the toilet does not flush. You call the front desk, and after apologizing for the inconvenience the clerk promises to notify "Engineering" right away. Does that sound familiar? Perhaps you've also noticed that many public buildings have a room where janitorial supplies are kept that is labeled "Engineering." A somewhat more elevated vision of engineering is portrayed in Star Trek, where unsung heroes in "Engineering" often save the day by fixing the warp drive just in time to fend off a Klingon attack.

The common conception of engineers as the people who repair and maintain modern conveniences is widespread, and presents one of the greatest challenges to implementing new educational standards related to engineering. Why, after all, would a parent want their child to spend valuable hours in school learning the skills needed for menial jobs? A reflection of this view has been a policy of the National Collegiate Athletic Association (NCAA) that established a Clearinghouse for reviewing every high school course in the country to ensure that college athletes were prepared to meet the academic rigors of college. When Massachusetts adopted engineering as a part of its science standards in 2001, a number of high schools developed rigorous engineering courses. The NCAA Clearinghouse rejected all of these courses as "vocational" subjects—that is, not a college preparatory course. A letter from the Commissioner of Education in Massachusetts to the President of the NCAA was required to reverse the policy—but only for schools in Massachusetts.

School guidance counselors, who presumably have their fingers on the pulse of the nation's job markets, have a more nuanced view of engineering. The Museum of Science in Boston investigated conceptions of engineering among school guidance counselor and found two prevailing viewpoints: One view was that engineering referred to trades such as plumbing, sanitation, or similar vocations. The other view was that engineers were brilliant people to whom science and mathematics came easily. Consequently, in some schools the only students who were counseled to consider engineering were those who struggled with academic work, while at other schools only the top students were counseled to apply to top engineering schools such as MIT. To counter these narrow views the Museum of Science developed a daylong program that brought guidance counselors together with engineers and engineering graduate students. Many of the guidance counselors were surprised at the wide variety of engineering specialties, and the number of educational institutions that offered various levels of engineering degrees.

Increasing the public's understanding of the engineering profession to the extent that they encourage their children to consider engineering as a career is grand challenge #2. To meet the challenge it will be important to enlist the help of museum educators, journalists, and other thought leaders to help public audiences understand the essential role of engineers in modern society.

Grand Challenge #3 Developing New Curriculum Materials

Since Massachusetts was one of the first states to include a very strong engineering thread in its science standards, the Museum of Science in Boston undertook a major project to develop curriculum materials that teachers could use to teach children and youth about the world of technology and engineering. The Museum developed curricula at the elementary, middle, and high school levels. The best known of these is an elementary program called *Engineering is Elementary* (Cunningham & Hester, 2007).

Engineering is Elementary introduces children to engineering through a series of stories about children who live in different countries. Each story features a technology that is important in that country. Career awareness is built by including a different type of professional engineer in each story—usually a parent, aunt, or uncle of the story's main character. The story sets the context for a design challenge that the children will do in class, using simple materials. All EiE units emphasize connections among science, language arts, and social studies, so teachers will not see this effort as "something else they have to add." Instead, the EiE units illustrate the connections among the different school subjects. For example:

Materials Engineering and the Great Wall of China tells the story of Yi Min. Students learn how materials engineers investigate the properties of earth materials like pebbles, soil, sand, and silt, and how different materials were combined to create the Great Wall of China. They then investigate on their own to determine which earth materials would make the strongest, sturdiest wall. For the design challenge, students construct their own "mini Wall of China."

Environmental Engineering and Drinking Water for India centers on the story of Salila, a girl in India whose family cannot just tap a faucet to get a drink of fresh water. In this book students learn about the human requirement for clean and safe drinking water and the consequential need for environmental engineers to ensure water quality. This unit addresses the increasingly important issue of water quality through lessons that teach students about water contamination and the ways that people ensure the quality of their drinking water. Students plan, construct, test, and improve their own water filters.

Mechanical Engineering and Denmark's Windmills explains how engineers design machines to capture wind energy as told by a young boy named Leif. The story includes the science concepts of air resistance, air pressure, and air as wind, and a description of Denmark's extensive wind turbines, which provide a renewable energy source. Students explore different materials and shapes conducive to catching the wind. For the design challenge, students create their own windmills that can lift a small weight.

These instructional materials aim to do much more than explain what technology is and what engineers do. The goal is to teach student to *think* like engineers. For students at the elementary level, that means identifying a situation that they want to change as a problem to be solved, and to approach the problem with a systematic design process involving five phases—asking pertinent questions, brainstorming ideas, planning, creating, and improving the design. A number of evaluation studies have shown the curriculum to be highly effective (Lachapelle, Phadnis, Jocz, & Cunningham, 2012).

The Museum of Science also developed a middle school mathematics curriculum called *Building Math*, in which students learn mathematics concepts and skills in the context of engineering design challenges, and a high school course entitled *Engineering the Future: Science, Technology, and the Design Process*. The latest curriculum, *Engineering Today*, provides enrichment units to complement existing science materials. Although these materials were developed before the *Framework* and NGSS, they can easily be adapted to align with the new standards.

At the high school level teachers need to decide if they will teach engineering design in a course that focuses on engineering and uses science to support the engineering concepts; or a course that primarily focuses on the science and uses engineering to help students better learn the science. Both approaches are valid. The science first perspective is that science concepts and processes are more fundamental than practical applications. The engineering first perspective is that students are likely to be more motivated by applying science in the real world.

An Investigation of the Impact of Strengthening the "T" and "E" Components of STEM in High School Biology and Chemistry Courses is an NSF project led by Debra Brockway at Stevens Institute of Technology in New Jersey, to develop and evaluate engineering units that would be integrated and taught in the context of high school chemistry and biology courses. The rationale for that project is that today, if engineering is taught at all, it is typically part of a physics course. However, only about a third of all high school students take physics. That's up from about 18 % in the 1970s and 1980s (Neuschatz, McFarling, & White, 2005; Tesfaye & White, 2010), but it means that most students would miss engineering entirely if it is just taught in the context of physics. However, most students take biology or chemistry, so if engineering is built into these courses most students will have an opportunity to learn what engineering is all about.

Luckily, there are a substantial number of curriculum materials that combine science and engineering. The *Go-To Guide for Engineering Curricula* is a three volume series that describes 40 curriculum programs, ranging from pre-school to high school seniors (Sneider, 2015). The curricula employ a wide variety of different methods. Although these materials are not fully "aligned" to the NGSS since they were developed before the standards were released, they have nonetheless been developed in the spirit of the new standards; and to some extent they helped to influence the standards since they provided an existence proof that curricula can be developed that blend science and engineering.

In summary, we do have some instructional materials that blend science and engineering; but none of these materials are a precise match for the NGSS. The grand challenge of developing instructional materials for teaching engineering in the context of science can be met—but as we show in subsequent sections, it's not an easy lift. Challenges include recognizing that designing and building things alone is not necessarily engineering, learning about the various dimensions of engineering design that students need to learn, and the common misconceptions and difficulties that students encounter. In the next section we drill deeper, into what it means to teach the design process.

Grand Challenge #4 Teaching the Design Process

Today many teachers claim that they already teach engineering because they occasionally have their students build newspaper towers or bridges from cardboard or popsicle sticks and test them to failure. Another popular "engineering" activity is designing a holder for a raw egg that will keep the egg from breaking when it is dropped. None of these are in fact engineering if students are not being taught design principles. They also do not belong in the science curriculum if students are not encouraged to apply scientific ideas and mathematics when doing these activities.

Curriculum developers need to base their work on research showing which instructional methods represent best practice. Unfortunately, the body of research literature on how to accurately and effectively teach the design process is quite limited, particularly in contrast to the science-education research base.

Crismond and Adams (2012) found a way around the problem of too few studies of engineering in K-12 schools by casting the net wider to include *any* studies on the teaching of engineering design, including such related fields as industrial design and teaching engineering at the college level, based on the reasonable assumption that engineering design is a transferrable skill and that people of various ages in many different fields encounter similar problems when engaged in designing a product, process, or system to solve a problem. Their work is based on an analysis of more than 400 papers from 170 peer-reviewed journals concerning the cognitive aspects of design. The results are organized in a table that summarizes expert and novice strategies. Table 2.1 is an abbreviated version of the table published in *The Science Teacher* (Crismond, 2013). The descriptions in the table of how beginners vs. informed designers meet design challenges provide insight into what it means to teach design principles to students.

In its extended form the table provides suggestions for how teachers can help their students progress from "beginning" to "informed" designers. Let's look at an example. The first pattern—Problem Solving vs. Problem Framing—poses the challenge of helping students move from treating a design task as a well-defined, straightforward problem posed by the teacher, to a situation that needs further exploration and definition in terms of criteria and constraints. Instructional strategies that are recommended include having the students state the problem in their own words, explain how they think a good solution would function, and to restate the problem in a way that would allow them to begin investigating possible solutions.

While Crismond and Adam's (2012) contribution to engineering design education is helpful, moving students from beginning to informed designers is complex and a grand challenge for engineering education.

Grand Challenge #5 Developing Assessments

Grand Challenge number 5 has two parts: (1) to develop ways to assess large numbers of students in ways that tap their creative abilities to engineer solutions to problems as called for in the NGSS; and (2) to develop assessments that teachers can use to find out what their students have learned (or not) and how they think about engineering and technology, so they can adjust instructional appropriately.

Starting with large-scale assessments, it's important to keep in mind that the NGSS is an assessment framework. That is, the performance expectations that make up the heart of the NGSS are intended to be endpoints in instruction. They illustrate what students are expected to be able to do to demonstrate their understanding after instruction. In contrast, prior sets of standards were statements of facts. Consider, for example what a fifth grader should be expected to know and be able to do about the sun, according to the *Next Generation Science Standards* (NGSS Lead States 2013) and the *National Science Education Standards* (NRC, 1996), the most recent comparable document.

National Science Education Standards (p. 43)	*Next Generation Science Standards* (p. 49)
The sun, and average size star, is the central and largest body in the solar system	Support an argument that differences in the apparent brightness of the sun compared to other stars is due to their relative distances from Earth

Table 2.1 Characteristics of Beginning vs. Informed Designers

Design strategies	Beginning vs. informed designer patterns	
	What beginning designers do	What informed designers do
Understand the design challenge	**Pattern A. Problem solving vs. problem framing**	
	Treat design task as a well-defined, straightforward problem that they prematurely attempt to solve	Delay making design decisions in order to explore, comprehend and frame the problem better
Build knowledge, do research	**Pattern B. Skipping vs. doing research**	
	Skip doing research and instead pose or build solutions immediately	Do investigations and research to learn about the problem, and how the system works
Generate ideas	**Pattern C. Idea scarcity vs. idea fluency**	
	Work with few or just one idea, which they can get fixated or stuck on, and may not want to	Practice idea fluency in order to work with lots of ideas by doing divergent thinking, brainstorming, etc
Sketch and represent ideas	**Pattern D. Surface vs. deep drawing and modeling**	
	Propose superficial ideas that do not support deep inquiry of a system, and that would not work if built	Use multiple representations to explore and investigate design ideas and support deeper inquiry into how a system works
Weigh options and make decisions	**Pattern E. Ignore vs. balance benefits and tradeoffs**	
	Make design decisions without articulating reasoning, or attend only to pros of favored ideas and cons of lesser approaches	Use words and graphics to display and weigh both benefits and tradeoffs of all ideas before making a decision
Conduct tests and experiments	**Pattern F. Confounded vs. valid tests and experiments**	
	Do few or no tests on prototypes, or may run confounded experiments that cannot provide useful information	Conduct valid experiments to learn about materials, key design variables and how the system works
Troubleshoot prototypes	**Pattern G. Unfocused vs. diagnostic troubleshooting**	
	Use an unfocused, non-analytical way to view prototypes during testing and troubleshooting ideas	Focus attention on problematic areas and subsystems when troubleshooting devices and proposing ways to fix them
Revise and iterate	**Pattern H. Haphazard or linear vs. managed & iterative designing**	
	Design in haphazard ways, or do design steps once in linear order	Do design in a managed way, where ideas are improved iteratively via feedback, and strategies are used ultiple times as needed, in any order
Reflect on process	**Pattern I. Tacit vs. reflective design thinking**	
	Do tacit designing with little self-reflective or monitoring of actions taken	Practice reflective thinking by keeping tabs on design strategies and thinking while working and after finished

Table from Crismond and Adams (2012), with permission from the authors

Both of these statements include the idea that the sun is a star. However, they are vastly different from an assessment point of view. To assess the older statement all that is needed is a multiple-choice question or two, to find out if students know about the sun's position in the solar system, and how big it is compared with the planets. To assess whether or not a student meets the performance expectation from the NGSS, the student needs to have an opportunity to construct and articulate an argument (verbally or in writing) about why he or she believes the sun to be a star, even though it is much, much, brighter than the stars that can be seen in the sky.

The National Assessment of Educational Progress (NAEP), also known as "The Nation's Report Card" is not a high stakes test. Students do not receive individual scores. Instead, assessments are given to large samples of students to gauge the effectiveness of our nation's educational system, and to compare how well different states and 21 major cities prepare students in reading, writing, mathematics, social studies, science, and most recently, technology and engineering literacy. Many of the items ask students to perform challenging tasks like the one from the NGSS in which students are asked to support an argument. Students' papers are scanned and sent to hundreds of scorers across the country (many of whom are retired teachers) to score at home, using a rubric. The fact that hundreds of thousands of tests that involve constructed responses can be scored within a reasonable time demonstrates that it is possible to assess individual students' achievement of these new standards.

The second part of challenge number five concerns "formative" assessments—what teachers do every day to find out what their students have learned so that can better shape the learning experience. Some educators think of formative assessment only in terms of instruments or quizzes, while others think of formative assessment as a process that enables perceptive teachers to gain insight into student thinking. In fact, both are important, as illustrated in a recent series of studies to develop a new physics course (Osowiecki & Southwick, in press) that used several different methods of formative assessment keyed to traditional summative mid-term and final exams (Sneider & Wojnowski, 2013).

Assessment has received a bad reputation in recent years because of high stakes testing. Certainly we need to change the punishing tactics built into law concerning high stakes tests. However, when those laws are reformed we don't want to throw out the baby with the bathwater. Assessment is essential for teachers and students to measure progress and to plan instructional moves. We just need to replace the "sticks" with "carrots" and integrate assessment smoothly into our instructional programs. Without assessment there is no way to determine if our students are achieving the standards; and if we don't know what they know (or don't know) there is no way we can help them.

As curriculum developers and teachers begin using the NGSS both types of assessments should improve, since the NGSS clearly specifies not just what students should know, but also how they should demonstrate their abilities to use the knowledge. While that may not be easy to assess with multiple-choice tests, assessments like NAEP are demonstrating that it can be done, even with large numbers of students.

Grand Challenge #6 Teaching the Teachers

The greatest challenge is likely to be experienced by teachers. Preparing elementary teachers to teach science has always been difficult; adding engineering just increases the burden. At the high school level it will be challenging to figure out how to fit five subjects into 3 years. Why five subjects? First, physical science includes both physics and chemistry. That's two. Then there's Earth and space science, which includes more at the high school level than physics and chemistry combined. Life science also includes a lot of really big ideas that take some time to teach; so that cannot be done in less than two semesters. And finally there's engineering. That's five subjects!

A report from the National Research Council (2015) recommends that educators at all levels take some time to figure out how to implement the new standards, and not rush to buy new curriculum materials that say "NGSS Aligned" on the cover. Teachers at all levels will need experience, practice, and opportunities to collaborate in developing new skills including, but not limited to:

- Integrating engineering design into science in ways that help their students develop engineering design skills alongside science inquiry skills;
- Engaging their students in all eight practices of science and engineering and helping them become more skilled at using the practices;
- Helping their students see the deep connections among the different fields of science and engineering through crosscutting concepts;
- Using formative assessment to monitor student progress, and enabling their students to gauge their own progress;
- Teaching fewer topics in greater depth;
- Teaching their students not only to use new technologies, but also how to acquire new technical skills on their own; and
- Communicating not only the enjoyment of science and engineering as interesting and challenging activities in themselves, but also the importance of all four STEM fields in developing sustainable practices that will allow society to thrive while maintaining healthy natural environments.

There is an especially bright ray of hope from informal educators, including afterschool and summer programs as well as museums and science centers. For example, 4-H is a huge informal education program in this country, with clubs and summer camps and afterschool programs for six million children. In recent years 4-H has greatly expanded their science and technology offerings such as robotics (Baker, Nugent, & Hampton, 2008). Science centers have also taken leadership in engineering education, both through exhibits and programs on site, as well as outreach (Alpert, Isaacs, Barry, Miller, & Busmaina, 2005).

There is no silver bullet, no single approach to helping teachers acquire these skills. Many approaches will be needed, and they will certainly need help from their fellow teachers of all subject areas, principals and other administrators, parents, local businesses and industries. In short they will need the support of their entire communities to meet these formidable challenges.

Grand Challenge #7 Balancing Technical and Academic Subjects

The U.S. and Great Britain have had a long history of establishing educational programs aimed at teaching technical skills, then eliminating them in favor of more "academic" pursuits (Firth, 2005; Donnelly, 1989; Christiansen, 1975). For example, at one time Boston Technical High School was a leading institution for preparing students to enter technical fields. As late as the 1950s graduates would be admitted to MIT if they maintained all A's. However, during the 1960s many of the "shop" teachers retired and were not replaced, and the space that had been occupied by those shops were reallocated (Sneider & Moss, 2004). That story is being repeated today in most states, as technology programs are closed and teachers laid off. According to the California Industrial and Technology Education Association and Foundation (2007) in the 1980s, nearly every public high school in California had a technology education program. After years of budget shortfalls, today only 20 % of California schools have such programs.

The grand challenge is to reconcile two conflicting educational philosophies. One that values learning how to solve a problem and actually produce something that meets a societal need, and the other that values learning for its own sake, and disdains the time spent in "getting one's hands dirty."

In "A Turn to Engineering: The Continuing Struggle of Technology Education for Legitimization as a School Subject," Theodore Lewis presents his view that the new emphasis on "engineering" rather than "technology" is a strategy to paint the technical arts with a high status brush, making it more acceptable in the eyes of society. He acknowledges the success that this approach seems to be enjoying, but cautions that "we may take ourselves too seriously, throwing out those aspects of engineering that remind us of our humble practical traditions, and keeping only those aspects that resonate with the dominant academic ideology of schools" (Lewis, 2004).

Grand Challenge #8 Engaging Technology and CTE Teachers

Grand challenge #8 is to persuade the nation's technology teachers and CTE teachers to join with science teachers to provide the kind of education that all students need to meet the global challenges that will surely increase in their lifetimes. In order for that to happen it will be important for school administrators and community leaders to recognize the special skills of these educators and the value that they bring to the school overall.

Support for technical education in secondary schools dates from the 1917 Smith-Hughes Act, which provided funds for vocational education in agriculture and home economics, and had the effect of isolating vocational education from the other high school subjects, a legacy which is evident even today. Federal support of vocational education continued throughout the twentieth century and into the twenty-first, primarily as a result of legislation beginning with the 1973 Perkins-Morse bill, most recently revised as the Perkins Act of 2006, which provides approximately $1 billion per year for Career and Technology Education (CTE) in the United States (Bennett, 2009).

The profession of technology teachers has evolved along with changes in national educational goals and sources of funding. Happily, not all states have eliminated their CTE programs, and in many states CTE is thriving. According to the Association of Career and Technical Education (ACTE), the broad field of career and technical education prepares youth and adults for a wide range of high-wage, high-skill, high-demand careers, and 94 % of all high school students take advantage of some CTE courses, which prepare students for hundreds of jobs organized in 16 career clusters.

Some consider technology education (TE) to be a specialty within CTE. However, others advocate technology education as a core subject for all students, not just those who are focusing on course work for specific careers (Wright, Washer, Watkins, & Scott, 2008). With the rise in support of STEM for all students, and especially the inclusion of engineering in the NGSS, the argument today is clearly in favor of engineering and technology for all students.

The educators who are most knowledgeable and capable of providing technology and engineering education are today's technology teachers, many of whom belong to the International Technology and Engineering Education Association (ITEEA). The initial response of the ITEEA to the Framework's inclusion of engineering as a core subject for all students was negative. A letter from the ITEEA to the NRC committee that drafted the framework argued that "science teachers might not have sufficient background to teach the new material and, moreover, that there is currently no agreement in the field about what the core ideas in engineering and technology should be. The letter also pointed out that a corps of technology teachers at the secondary level already exists" (NRC, 2012, p. 337).

In Beverly, Massachusetts, where *Engineering the Future* was being piloted as a ninth grade course, a science teacher was not confident that she would be able to help her students build prototypes. So she talked with the technology teacher who had a fully-equipped wood shop. He was more than pleased to work with her since he liked to include relevant science content in his courses, and often had students design and build projects such as hovercraft. The two planned the curriculum together and worked out schedules that allowed the students to build their prototypes in the wood shop, where they were able to receive training in how to use power tools. The technology teacher was also actively involved in developing educational uses of a large photovoltaic array adjacent to the school, which would make an excellent enrichment to the course. Unfortunately, a year later the technology teacher's position was eliminated and a new school was planned and built without wood shop facilities.

The point of this story is to emphasize the importance of supporting technology teachers and CTE teachers as co-leaders with science, mathematics, and other "core subject" teachers in order to realize the tremendous potential of engineering education for all students. Given the emphasis in the NGSS on both engineering and science, such collaboration would appear to be a winning strategy.

Grand Challenge #9 Teaching the Teacher Educators

Perhaps the greatest challenge is engaging university professors who prepare tomorrow's teachers in supporting the NGSS. A colleague interviewed a number of college and university professors in engineering to see what they thought of the new plan for including engineering within the high school science curriculum. He was dismayed to find that the few who knew about it were unenthusiastic, preferring instead for their incoming students to have a rigorous background in traditional science and mathematics. While there are legitimate concerns about infusing engineering into the K-12 science curriculum, and a need for conversations about issues such as reducing attention to subjects long included in the curriculum to make room for engineering, it makes little sense to consider only the knowledge and skills needed to succeed in college engineering courses. Most students will not major in engineering. The purpose of K-12 engineering education is to educate all students about the designed world, and to help them develop broad skills, such as defining and solving problems, that will serve them well in whatever career they pursue.

The recognition that effective K-12 engineering education can be of service to college engineering departments is recognized at a few universities, such as Tufts and Olin College, in which professors place a high value on motivation, and engage incoming students in interesting engineering activities from the start. Even more important are the universities, such as Purdue, Texas A&M, and Virginia Tech, that have departments of engineering *education*, where PhD candidates are learning what it takes to develop curricula and assessments in support of the NGSS, and to lead STEM education reform at the district and state level.

Grand challenge #9 is to find ways to engage an increasing number of university professors responsible for educating teachers at the elementary, middle, and high school levels to learn about the NGSS, recognize and support its purpose and goals, and figure out what it means for their own practice. The pathway forward must be led by college professors who understand the importance of engaging students in interesting engineering activities early, and are willing to reach out to provide assistance and encouragement to their colleagues who teach at the K-12 level.

Conclusion

Education is a conservative endeavor. It has tremendous momentum, in part because it is deeply embedded in society. The first two challenges, helping our entire population understand technology and what engineers do, is vast in scope. Until these challenges are at least partially met, it is difficult to see how teachers will receive support from their students' parents and community stakeholders. The next set of challenges, involving curriculum, instruction, assessment, and professional development of teachers, involves transformation of a profession. The history of educational reform that has swung back and forth between the scholarly and practical arts suggest it may be difficult to find a balance. The last two challenges are equally daunting, engaging technology educators who may be threatened by science teachers "taking over" their profession, and college professors who may have a narrow focus on the preparation of their incoming students. These grand challenges involve everyone in our society—not just the science educators.

Creating the NGSS with a strong engineering component and getting states to adopt it is just the first step. We will not succeed in transforming our educational enterprise so that our students will have the tools they will need to meet the global challenges of the future, if we don't meet the grand challenges of engineering education today.

References

AAAS. (1989). *Science for all Americans*. American Association for the Advancement of Science. New York: Oxford University Press.

Alpert, C. L., Isaacs, J. A., Barry, C. M. F., Miller, G. P., & Busmaina, A. A. (2005). Nano's big bang: Transforming engineering education and outreach. In Proceedings of the 2005 American Society for engineering education annual conference & exposition, Portland, Oregon.

Baker, B. S., Nugent, G., & Hampton, A. (2008). Examining 4-H robotics in the learning of science, engineering, and technology topics and the related student attitudes. *Journal of Youth Development: Bridging Research and Practice, 2*(3), 7–18, Spring 2008. Article: 0803FA001.

Bennett, W. (2009). *Vocational education philosophy and historical development*. Online instructional material for EVOC 637: Foundations of Career & Technical Education, California State University, San Bernardino. Retrieved from: http://cis.msjc.edu/courses/evoc/637/Assignments/DL6-8.pdf

California Industrial and Technology Education Association and Foundation. (2007). *Career and technical education: Problems and solutions*. Position Paper. Fresno, CA: CITEA.

Christiansen, P. (1975). *Theory and practice in the formative years of American mechanical engineering education: A cultural and historical analysis*. Doctoral dissertation. City College of New York.

Crismond, D. (2013). Design practices and misconceptions: Helping beginners in engineering design. *The Science Teacher, 80*(1), 50–54.

Crismond, D., & Adams, R. (2012). The informed design teaching and learning matrix. *Journal of Engineering Education, 101*(4), 738–797.

Cunningham, C. M., & Hester, K. (2007). *Engineering is elementary: An engineering and technology curriculum for children*. In American Society for engineering education annual conference & exposition. Honolulu.

Donnelly, J. F. (1989). The origins of the technical curriculum in England during the nineteenth and early twentieth centuries. *Studies in Science Education, 16*, 123–161.

Firth, A. (2005). Culture and wealth creation: Mechanics' Institutes and the emergence of political economy in early nineteenth-century Britain. *History of Intellectual Culture, 5*(1). Retrieved from: http://www.ucalgary.ca/hic/issues/vol5/3

Lachapelle, C. P., Phadnis, P., Jocz, J., & Cunningham, C. M. (2012). *The impact of engineering curriculum units on students' interest in engineering and science.* Presented at the NARST annual international conference, Indianapolis, IN.

Lewis, T. (2004). A turn to engineering: The continuing struggle of technology education for legitimization as a school subject. *Journal of Technology Education, 16*(1), 21–39.

Neuschatz, M., McFarling, M., & White, S. (2005). Highlights from the 2005 high school physics teachers survey. *American Institute of Physics.* Retrieved from: http://www.aip.org/statistics/trends/highlite/hs2/hshigh.pdf

NGSS Lead States. (2013). *Next generation science standards: For states, by states* (Volume 1 the standards. Volume 2 appendices). Washington, DC: The National Academies Press.

NRC. (1996). National Science Education Standards. Board on Science Education, Division of Behavioral and Social Sciences and Education. National Research Council (NRC). Washington, DC: The National Academies Press.

NRC. (2012). *A framework for K-12 science education: Practices, crosscutting concepts, and core ideas* (Committee on a Conceptual Framework for New K-12 Science Education Standards. Board on Science Education, Division of Behavioral and Social Sciences and Education, National Research Council (NRC)). Washington, DC: The National Academies Press.

NRC. (2015). *Guide to implementing the next generation science standards.* Board on Science Education, Division of Behavioral and Social Sciences and Education, National Research Council. Washington, DC: The National Academies Press.

Osowiecki, A., & Southwick, J. (in press). *Energizing physics.* New York: Freeman. The course is also described at energizingphysics.com.

Pearson, G., & Young, T. A. (Eds.). (2002). *Technically speaking: Why all Americans need to know more about technology.* Washington, DC: The National Academies.

Rose, L. C., Gallup, A. M., Dugger, W. E., & Starkweather, K. N. (2004). The second installment of the ITEA/Gallup Poll and what it reveals as to how Americans think about technology. International Technology Education Association. *The Technology Teacher, 64*(1), 1–12.

Sneider, C. (Ed.). (2015). *The go-to guide for engineering curricula* (Three volumes: K-5, 6–8, and 9–12). Thousand Oaks: Corwin Press.

Sneider, C., & Moss, M. (2004). *John D. O'Bryant School for Mathematics and Science situation report.* Unpublished internal report to the Headmaster.

Sneider, C., & Wojnowski, B. (Eds.). (2013). *Opening the door to physics through formative assessment* [monograph]. Portland: Portland State University. Retrieved from National Science Education Leadership website: www.nsela.org/publications

Tesfaye, C. L., & White, S. (2010). High school physics courses and enrollments. *AIP Focus On,* August 1–6, 2010. Retrieved from: http://www.aip.org/statistics/trends/reports/highschool3.pdf

Wright, M. D., Washer, B. A., Watkins, L., & Scott, D. G. (2008). Have we made progress? Stakeholder perceptions of technology education in public secondary education in the United States. *Journal of Technology Education, 20*(1), 76–93.

Part II
Student-Centered Design…Exemplary Projects and Programs that Transfer Theory to Practice

Chapter 3
Museum Design Experiences That Recognize New Ways to Be Smart

Dorothy Bennett, Peggy Monahan, and Margaret Honey

Introduction

Invite deep participation
Make sure everyone feels welcome
Agency
Whimsy
NO BORED KIDS!

Selected entries on the Design Lab office's "Manifesto Wall" at NYSCI

In our office, we have an area called the "Manifesto Wall" where we tack up phrases that we feel are central to our work. Phrases such as "Access to and confidence in using tools," sit side-by-side with "Whimsy (…and gravitas), and "See design opportunities all around." They serve to remind us of our aspirations and ground us in the purpose and opportunity of the Design Lab project. NYSCI's Design Lab is an innovation laboratory for science, technology, engineering, and math learning through design. It aims to deeply engage *all* types of science learners in solving personally motivating problems via a creative design process. Through our project work, both in the unique museum environment and through collaboration with teachers, we've developed some principles to help us get ever closer to meeting this ideal. With this chapter, we'd like to share these principles, and suggest some ways that the insights borne of our project might help classroom teachers to consider new engagement strategies to effectively reach a broader range of students. Finally, we discuss directions for future work.

Developed for the general public, teachers, and schools, Design Lab includes a 9500 square foot museum exhibition with facilitated design activities (opened in the spring of 2014), a series of digital resources and mobile tools for teachers and

---

D. Bennett • P. Monahan • M. Honey (✉)
New York Hall of Science (NYSC), Corona, NY 11368, USA
e-mail: mhoney@nysci.org

classrooms, and ongoing professional development for teachers. From transforming a musical greeting card into an audio surprise for a friend, to designing working solar ovens from recycled materials, Design Lab is creating new possibilities for young people to identify design problems worth solving, notice design opportunities in the real world, and think creatively about the redesign and reuse of materials to solve everyday problems involving STEM concepts and skills.

At the heart of Design Lab's mission is to create museum experiences and resources for teaching and learning that engage students in design as a problem-solving process that motivates young people to explore and master these skills, and at the same time find new pathways into science content. National trends in the reform of science education recognize the strong link between learning and motivation. Stimulating students' interests, engaging them in problem solving, and demonstrating relevance are the recommended strategies for creating stronger attractions to STEM for diverse groups of students.

Even before the release of the Next Generation Science Standards (NGSS), NYSCI witnessed through its own programming focused on engineering that design-based approaches to teaching and learning afford many opportunities for young people to develop 21st Century skills such as critical thinking, creativity, entrepreneurial thinking, collaboration, communication, and innovative use of knowledge, information, and data to solve problems (Partnership for 21st Century Skills, 2008). The NGSS are now strongly advocating for engaging students in motivating, real world problems that blend these practices and the cross cutting concepts of engineering and science in meaningful ways.

The Next Generation Science Standards provide a unique opportunity to build strong partnerships between science museums and schools that take advantage of the respective strengths of informal and formal education. While the inclusion of engineering in the standards is not new, the new emphasis on the melding of concepts and cross cutting ideas with practices is. NGSS strongly states that, *"students cannot fully understand scientific and engineering ideas without engaging in the practices of inquiry and the discourses by which such ideas are developed and refined. At the same time, they cannot learn or show competence in practices except in the context of specific content (National Research Council, 2012, p. 218).*

By melding core disciplinary ideas with the practices, the NGSS recognize that it's not just *what* is getting taught that matters, but it is also the *how* of the science curriculum. Whether deliberate or incidental, content is always learned alongside skills, even if those skills are memorization and recitation. With its focus on practices, the NGSS encourage deliberately having students practice the skills of science and engineering while they learn the content, ensuring that the core ideas are learned along with a sense of efficacy.

In Design Lab, we've adopted a phrase from an essay called "The Genius of the Tinkerer" by Steven Johnson (2010) to help us focus on that sense of efficacy that comes from melding content and practices: "the adjacent possible."

> *The strange and beautiful truth about the adjacent possible is that its boundaries grow as you explore them. Each new combination opens up the possibility of other new combinations.*

The term originated in the field of biochemistry, where it referred to the ever-expanding complexities of chemical compounds that are possible starting with just the few simple organic compounds available in a primordial soup. Every new reaction expands the set of compounds possible in the next. Steven Johnson adopted that phrase to stand for the way innovation works to build the future from the present, and we adopted it as part of our purpose: One of our goals is to give students an ever-expanding sense of their own abilities and knowledge. The problem-solving nature of design is well-suited to providing students with this kind of opportunity, because it invites them to apply a wide variety of skills and knowledge to finding a possible solution. These activities can not only welcome current knowledge, but also prod students to build new knowledge through the creative application of their combined skills and ingenuity.

Since our aim is to build not only content knowledge, but also efficacy, we try to deliberately build bridges from kids' current knowledge base to new content and skills, and whenever possible, call students' attention to these newly adjacent possibilities.

Teaching content alongside practices, particularly in the area of engineering and design, is a challenge for schools. Typically, schools have taught science content and engineering process separately from the practices that lead to further understanding of how our world works. As a result, students often develop a static view of these fields and fail to see science and engineering as the creative endeavors that they are, and could fail to engage with these fields as a result.

Museums like NYSCI are in a great position to help educators take an inspired approach to bringing the standards to life. Science centers are particularly adept at meeting children where they are and at creating intrinsically motivating experiences that build on curiosity, confidence, challenge, and play (Perry, 1994). Many studies have shown that museums are uniquely successful in generating interest in science, personalizing science learning, and in engaging people in activities that help them realize and experience their own agency in activities associated with learning or doing science and engineering (Holland, Lachicotte, Skinner, & Cain, 1998; Hull & Greeno, 2006). Museums develop that expertise out of necessity, because engagement is crucial: Exhibitions are a free-choice environment, and if visitors aren't truly engaged, they'll vote with their feet. As organizations that are free of the requirements that schools have to adhere to, we are able to experiment with developing design experiences that are genuinely motivating and support greater degrees of self-guided learning.

For the past three years, our work on Design Lab has revealed ways that museums are in a unique position to help deliver on the promise of the NGSS by providing resources and expertise for supporting engineering practices that make content more relevant and widely accessible to diverse groups of learners. Through design activities that bridge the museum experience with the classroom, we have the opportunity to celebrate previously unrecognized strengths of students who might not otherwise participate deeply in science class. We are uncovering new ways to be smart.

Design at NYSCI: The Core Ingredients

NYSCI has been experimenting with ways our museum can serve as a lever for teachers and schools to bring compelling engineering and design experiences that meld content with practices into the classroom, bringing the NGSS to life. From the outset we conceptualized Design Lab as a place that would serve multiple audiences: family audiences, teachers who bring their classes to the museum, teachers who turn to NYSCI for professional development experiences, and others who could not engage in location-based activities, but might benefit from curricular and digital resources. We wanted to build exhibit experiences that would be irresistible for different audiences with different purposes, and we wanted to leverage these experiences to extend design-based learning beyond the walls of the museum. Our aim in all of this is to help maximize the science center's role in the educational ecosystem by creating experiences through which diverse groups of people could fall in love with science and engineering.

While we knew from prior work that design is a powerful way to engage children with a broad range of STEM content, we needed to define what design and engineering could be for these different audiences in our museum and beyond the walls. This would take a diverse team that could approach design from different vantage points—a creative mash-up of exhibit developers, science experts and developers, educational researchers, museum educators and facilitators, digital learning experts, and a steady stream of Teacher Design Fellows, K-12 teachers who could co-develop and test activities with us. We needed people who understood the museum setting and the classroom setting, who brought a love and deep knowledge of science and engineering alongside practical skills involving materials and classroom management techniques.

This has resulted in a multilayered project. For the general public, the Design Lab team has been prototyping physical spaces and hands-on activities that can invite visitors to see design possibilities in their everyday lives and to have a sense of agency to change their surroundings in small and big ways. For teachers, Design Lab has been creating formal professional development experiences along with an informal tinkering space where educators can find like-minded colleagues, stimulating conversation, and opportunities to be creative, playful, and inventive in coming up with relevant, interdisciplinary teaching approaches for their students. To extend our work beyond the walls, NYSCI Design Lab is developing a suite of digital learning resources that include videos, mobile apps, and other virtual resources to support design-based learning.

With this mix of people and experiences, we have gained important insights into core ingredients that make design a transformative experience for STEM teaching and learning. These core ingredients include: (1) problems worth solving to you (2) interest-driven iteration (3) materials literacy, (4) divergent solutions, (5) sharing and reflection.

Problems Worth Solving: To You

We have come to think of design primarily as a problem-solving methodology by which people create artifacts, systems, and tools intended to solve a broad range of problems. Through design, you learn how to identify a problem or need, how to consider design options and constraints, and how to plan, model, test, and iterate solutions to vexing problems, making higher-order thinking skills tangible and visible. Important in that process is the task of finding, defining, and truly understanding the problem itself. This process of problem definition and ideation is an important part of design, but often exhibit experiences and school-based design activities give students pre-defined problems with no opportunity to practice defining the problem themselves.

While children use all kinds of STEM content and skills when they solve design problems, the part that makes it relevant to them is the problem itself. It needs to be something that they are willing to invest themselves in. We knew early on that we needed to care deeply about the qualities of the problems themselves and the thinking behind the solutions as much as the specific science content that would be revealed.

Through extensive prototyping of design activities on the museum floor and with teachers in professional development workshops, we have come to realize that truly generative design-based activities are more open-ended and messier than standard museum and classroom activities. We decided to create activities that encourage students to work out solutions to design problems of their own choosing, enabling them to find relevance and pursue interests that are completely their own, to engage with "problems worth solving." What makes it a "problem worth solving" isn't necessarily that it's a large real-world problem. Instead, it needs to address a purpose that has relevance to the individual problem solver. This kind of personalization goes beyond embellishment or decoration, and is instead a personalization of purpose. Students define at least part of the problem for themselves, practicing this important part of the engineering process. Students frequently explore problems that incorporate significant humanitarian or community issues, but they can also choose to incorporate acts of whimsy, aesthetics or humor. In Design Lab, a problem worth solving is worth it to you. Case in point: the testing of our first exhibit activity, an exploration of conductivity and circuits that we call Happy City:

> *The program of the day in the Design Lab prototyping space is "Happy City"—a circuits activity that challenges visitors to build things with boxes, LEDs, and motors and then add them to an ever-evolving cityscape in order to make it a happier place. Currently, the city includes a playground, a science museum, and a profusion of pizza places, all built by previous visitors. A boy and a girl from a class on a field trip sit next to each other, freely sharing advice to each other and those around them as they get busy building their individual additions: the Happy City Police Station with red lights in front and on top, and a sparkly "BFF house". When museum staff come by, they share their creations, explaining each feature in great detail, dwelling especially on how they managed to make the LEDs*

light up. The teacher pulls the museum staff aside, explaining that she's excited by the activity and amazed at the work of the class, but of that little boy in particular. He's autistic, normally doesn't sit still, rarely speaks to anyone, and almost never engages in the classwork. In this setting, he's focused, engaged, and even sharing with others about his own unique addition to Happy City.

In this activity, the challenge invites idiosyncratic responses as each child is able to choose what they think would make the city happier. But the carefully-chosen materials promote the exploration of circuits and conductivity, and encourage students to wrestle with how to use circuits to solve their problem (Fig. 3.1). We've noticed that children often define problems for themselves that are harder than the ones we might assign to them. For example, a young boy who reportedly often struggled in his science class decided that he wanted to make a basketball hoop that lit up when someone scored. Given his age, it would have been impressive enough for him to simply create a circuit with conductive materials. But he set himself a more complicated goal, and needed to find a way to control that circuit. He invented his own pressure switch that turned on the light when a scoring basketball fell on it.

When creating design problems where everyone feels smart the problem context and invitation matters. Through its focus on solving problems that are personally relevant or purposeful, the design-based approach has been shown to be effective at

Fig. 3.1 Happy city design problems

reaching underrepresented groups who are not always motivated by more traditional STEM activities (Caleb, 2000; Davidson & Schofield, 2002; Margolis & Fisher, 2002; Rosser, 1990; Schofield, 1995; Turkle, 1984, 1988; Turkle & Papert, 1990, 1992). Since the definition of personally relevant varies from person to person, the nature of our invitation must be broad enough so that each learner finds his own skills and interests reflected in the activities. We opened up the design challenges enough to let learners define their own problems so that especially reluctant learners, found things that motivated them, and began to explore content they might not have encountered successfully before.

Involving children in activities like "Happy City," where the activity does not entail the traditional "rockets and robots" approach to science and engineering that appeals more to boys than to girls, we have opened up the world of circuits, switches, and conductivity to a new audience of young people who might otherwise shut down. In the conventional model used in most classrooms, the learning topics are pre-defined, and if they are not motivating, the students are out of luck. Opening up opportunities for personalization of purpose celebrates the interests of individual learners. It provides a space where anyone can learn.

Interest-Driven Iteration

One sign of a successful design experience is when children are motivated to keep working on the problem long after it is introduced, sometimes coming back to the activity area after wandering around the museum to add to the design they left behind or continuing to work on school design projects outside of class time and long after the project was intended to end.

> *A group of 20 elementary and middle school teachers gather together for our Thinkering session, a night for our design fellows to share insights gained from trying out the classroom design activities they developed with us in the summer design institute. One of Nyema's students had insisted on coming himself, and the 11 year-old young man in a suit speaks eloquently about the problem they were invited to solve: to redesign the drop-off experience at school which has led to big traffic jams with long-idling cars. He describes how they came before school to analyze the problem by videotaping traffic flow with their teacher, how they came up with a ticketing system for parents, and how they are pursuing a new public information campaign to get people to stop car idling with support from the district's superintendent. As he speaks, he holds up newspaper articles: "Do you know the amount of car emissions that results from car idling?" The student reports how he and a few others are now forming a club after school to think of new ways to address the car idling and traffic problem at the school. Nyema chimes in that the students don't want the project to end.*

Maria, a special education middle school teacher who was part of the same cohort, expressed some trepidation about implementing her design project in the classroom, one involving students designing diving crafts that achieve neutral buoyancy (a state of neither sinking nor rising in a liquid). She wasn't really sure how her students would react since they often got easily discouraged and disengaged when it came to challenging academic tasks. When we arrived in her class she bubbled with excitement: "I was surprised that they did it and they kept trying. Some of the students that struggle the most were the most successful, and that was really great; to offer my students a chance to have success, where sometimes they struggle in class."

Many teachers in our design fellowship who implemented design activities in their classrooms spoke about how their students researched problems before and after school and how they did not want to stop until they arrived at a solution they were satisfied with. Teachers were impressed at how often students would adopt new strategies when addressing problems they may have been initially frustrated with or felt they had done incorrectly. Students who often had the most difficulty with problem solving in class were in many instances the ones who gained confidence when applying design-learning principles.

This kind of iteration, the ability to persist in the face of failure, to recalibrate, redesign, and reevaluate are some of the requisite skills for engineering and science. It is the development of what is being referred to by many researchers as perseverance, tenacity, and grit (Duckworth, Grant, Loew, Oettingen, & Gollwitzer, 2011; Duckworth, Peterson, Matthews, & Kelly, 2007; Dweck, Walton, & Cohen, 2011). We believe this happens because children find these problems intrinsically motivating and sufficiently complex. Design projects that are not overly prescribed enable children to find their own ways into the engineering process and allow teachers to see new strengths of their students.

Materials Literacy

A big part of engineering involves seeing design opportunities in surprising places. One way to promote this is by providing opportunities for children to use familiar materials in unexpected applications. For example, when doing circuit-building activities in the Happy City activity with kids, we use folded strips of aluminum foil instead of wires. This use of the familiar has several qualities that help draw deliberate connections back to the everyday.

- Knowable—Students start with something they already know about. They know how to fold, tear, crumple, and smooth aluminum foil.
- Humble—Students can tell exactly what's connecting one part of the circuit to the next: just aluminum. There's nothing unseen beneath a rubber coating.
- Get-able—Students and teachers probably already have aluminum foil at home in the kitchen.

When children are able to find new uses for everyday materials they develop what we refer to as *materials literacy*, a potent skill that enables children to see possibilities in the world around them. In the Happy City activity, the deliberate use of familiar materials like aluminum foil facilitates this transformation of the everyday for children and adults working with them. Children were comfortable manipulating this decidedly non-precious material but were still surprised to see it used in this way (Fig. 3.2).

This idea of materials literacy has become a cornerstone of our work with schools and the exhibit. Through the use of everyday materials we are communicating the notion that materials can be reused and repurposed depending on needs at hand. When we work with teacher groups, they are especially surprised and appreciative of our use of readily available materials. In our professional development workshops, one teacher transformed dramatically from heavy dependence on prepackaged science kits to being confident that he can assemble his own materials largely by gathering things he already has.

Presenting familiar materials for use in new ways can also focus attention not just on their potential uses, but on their properties as well. It wasn't uncommon in our sessions to see children testing the other materials (pipe cleaners, paper clips, etc.) to see if they, too, would conduct electricity in the same way as the aluminum foil. The surprise of seeing the aluminum foil strips conduct electricity led to explorations of conductivity: What other materials are good for acting like wires? What are they made of? How can I create my own crude conductivity tester so I know a material can work in my circuit?

Science museums can be a great resource for teachers in particular to develop materials literacy. From recipes for bubbles to inexpensive ways to create easy circuit testers, there is a knowledge base that teachers can get from museums that is

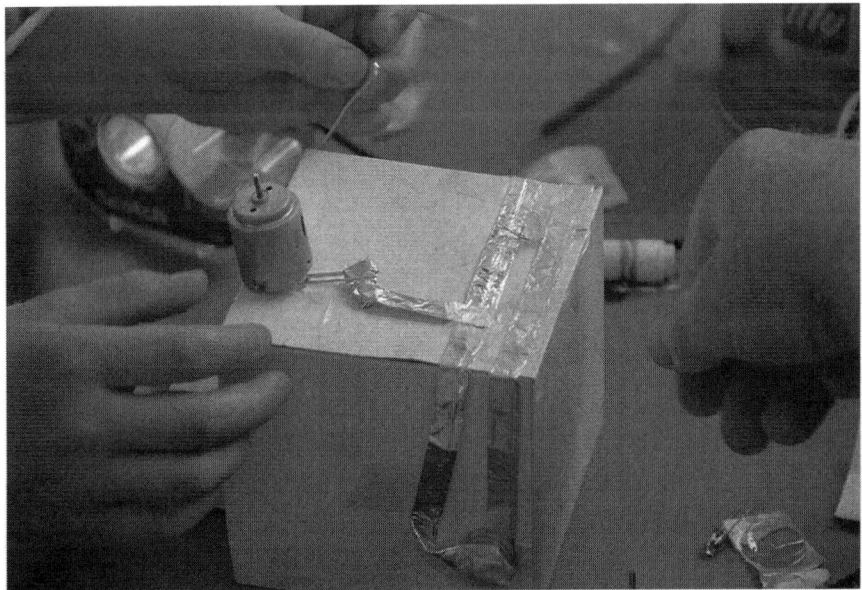

Fig. 3.2 Materials literacy—using familiar materials in new ways

extremely valuable yet rare to find elsewhere. Access to this expertise makes it possible for teachers to see that they "don't need kits" to teach science, a mindset they can pass on to their students. In all of our institutes and design workshops, exhibit developers and museum educators have been incredibly resourceful in helping teachers see new possibilities in materials they have in their kitchen and desk drawers.

Divergent Solutions

Contrary to inquiry-based science that aims for one elegant explanation to describe a wide range of phenomena, engineering design deals in situational tradeoffs where there are a number of right answers that depend on how you define the problem, the materials and tools available, and the context in which you intend to use your solution. Science inquiry activities reach toward a singular efficient explanation of a phenomenon, while the goal of engineering is to negotiate trade-offs to arrive at one of many possible solutions. One solution might be more efficient, one might be more exciting, and another still might be the easiest to use. Engineering design activities must invite divergent solutions in order to encourage learners to grapple with these trade-offs when solving problems and evaluating their designs.

Coming up with design activities that inspire divergent solutions involves a fine interplay of factors. We have found the need to carefully consider the materials provided, the problem definition, and the constraints given in order to open up an activity enough so that each solution is as unique as its designer. At the end of an activity, if all of the products of design look the same aside from some decorative elements, we have learned that is a tell tale sign that we really haven't developed a design experience that allows for divergent solutions. We haven't created a good invitation into the problem at hand.

We promote divergent solutions by providing opportunities for learners to define part of the problem they are solving and by giving them carefully curated materials that afford certain ideas to be investigated. This problem ideation or definition often increases investment, which in turn increases iteration, problem solving, and data collection. We learned this through many trials with both tried-and-true engineering activities and those of our own invention:

> *The prototyping activity on the exhibit floor, "Stranded," had already been through a long development process. It had started as "Tools for a Desert Island" where visitors were charged with building a device to catch food, and faux fish and animals had been available to encourage testing. But we weren't getting the deep investment we were used to in the prototyping space, and we saw little variation in the visitors' creations. They were making the same trivial traps and scoops again and again. Even when we varied materials, the solutions tended to converge around a single type of contraption. Today, we had decided to open the problem up. Way up. Today's version of the problem minimized the importance of catching food and even jettisoned the desert island setting, instead asking the students to define the setting and even the problem themselves. In the first run through, students' settings vary a great deal, from island to jungle, forest, the North Pole, and even an under-*

ground cave. But one of our materials in particular, a berry basket, is dominating children's creations, and all of their boats and shelters look alike. Encouraged by the variety of settings and problems, we pull the berry baskets off the tables to see what happens. This time, children build a wide variety of shelters, traps and tools, all with surprisingly detailed back stories that define the constraints of the student's chosen setting and the features of the problem they aim to solve. Now we know that we have a winner.

Sometimes it is a small tweak that can foster divergent solutions. Often it is in the problem framing itself. By opening up the problem and not prescribing what children had to design, children had ownership over the problem, which led them to think creatively about the kinds of simple machines that were most conducive to the problem they wanted to solve.

This is somewhat challenging for educators at first when they are trying to incorporate design activities into a standard curriculum where specific content has to be "covered": If learners are going to have such divergent responses to a problem, it's difficult to ensure that they converge on the specific content to be conveyed. A common inclination is to focus the problem with contrived constraints which make the problem feel like a riddle with a known answer rather than an engineering problem inviting a creative solution. In settings where children are used to finding the one right answer that the teacher is looking for, you have to work particularly hard to open the problem up and encourage an authentic engagement with engineering tradeoffs. One potential strategy is to give the students time during the problem ideation phase to identify for themselves the constraints that might come to bear on the problem. The teacher leading this process can introduce her own constraints during these class discussions, but they must be justifiable within the context of the problem.

While it might not be the most expedient path to content, this strategy incorporates key practices of engineering such as considering constraints, testing, and background research. For instance, while designing something that would help you survive on a desert island, you might need to investigate the kind climate you find yourself in, identifying the available flora and fauna, investigating simple water filtration methods, and the strength of materials to create structures that withstand extreme weather. At the heart of every good design problem is the opportunity to bump up against rich STEM content in the form of useful information, relevant concepts, and technical skills that help move you further along in enacting or improving your design.

Sharing and Reflection

A particular power of design activities is that they promote and enable sharing and reflection particularly well. Children have a tangible product that they are eager to describe and to demonstrate, and to relate their story of its creation. We actively encourage young people and educators to borrow and build off of others' ideas in the exhibition. This kind of appropriation is not cheating, it's an excellent way to learn how previous solutions work, what their limitations might be, and how they might be adapted or improved. In sharing and building on other people's ideas in an exhibit space, new solutions are morphed and born. The integration of a well-chosen good idea from elsewhere is not only a good avenue for learning, but it also defines the enterprise of engineering and innovation.

> *Children have been building jointed shadow puppets in the Design Lab prototyping space, and the wall is covered with their creations built from cardboard, chopsticks, brass fasteners and tape. Visitors have been constructing linkages to make their puppets move in evocative ways, both simple and complex. The explainers who have been facilitating the activity start to organize the kid-built puppets, exposing chains of influence and embellishment that have emerged as visitors' imaginations were inspired by the examples in the ever-evolving display. At first was a face profile with a hinged jaw. That inspired a few direct copies and then came an innovation: a big chomping fish head with pointed teeth. A few variations and embellishments later came a fish with a jaw and an articulate tail. Soon after the display was re-organized, a visitor starts working on an even more elaborate fish with several hinged sections on the tail, spending a long time perfecting the perfect wiggling swim.*

On the exhibit floor, we have been watching the way that visitors respond to the examples on display. We proactively curate these visitor artifacts to more effectively nurture visitor participation and inspire a deep engagement with the content we want them to explore. The display also becomes an incentive for visitors to show off and reflect on their solutions. We often ask visitors with a unique design to explain it to us in detail so that we can better help future visitors learn from their work (Fig. 3.3).

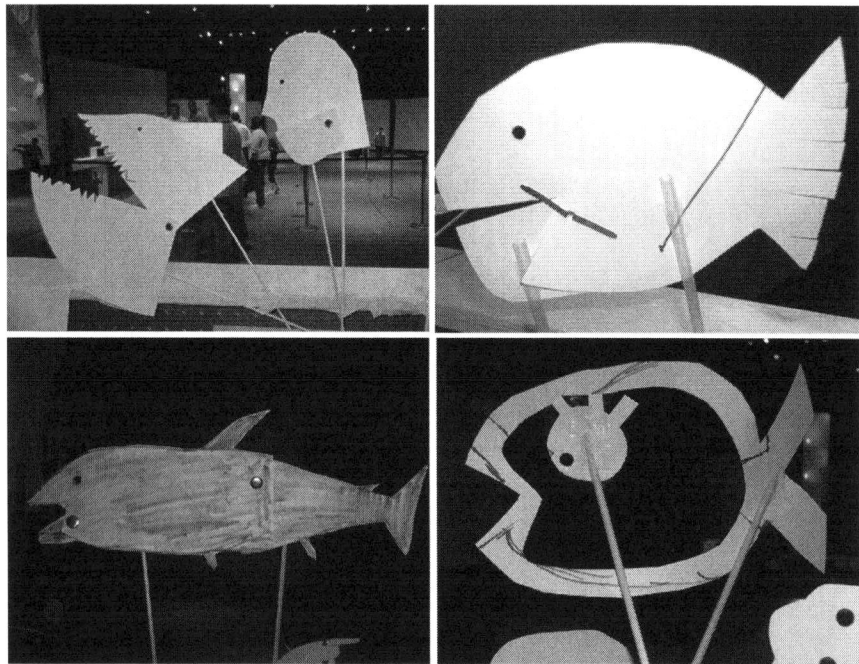

Fig. 3.3 Shadow puppets fostering divergent solutions

This power of example also holds true in the classroom. The teachers we worked with who piloted design activities in their classrooms spoke of how important it was for students to share their process and products of design through gallery walks, presentations, and even sharing their work online. Teachers used these discussions and demonstrations to draw out the ways that students had explored science and engineering processes and ideas while creating their design solutions. Students were eager to see how others solved problems their own groups were having, which deepened their perseverance and interests.

Bringing Design to the Classroom: What Museums Can Offer

Looking forward, we believe there is great potential for museums to be partners with formal education, providing new contexts for teachers to engage children in design and NGSS practices being advocated for. Through our instructional activities, field trip experiences and digital tools, we are offering strategies and supports for different ways for kids to be smart in STEM.

Museums are not bound by the same constraints that schools have when trying out something new. Though we aim to be useful to teachers who are bound by standards, we are not held to account for teaching those standards in the same way that teachers are. We have a degree of freedom to experiment and tinker with the ways that students can engage with the content. The combination of being free from standards and yet dependent on engaging our free-choice visitors defines the unique and pivotal role that museums play in the educational ecosystem, and the way that we can be helpful in melding the cross-cutting concepts with the engineering practices that NGSS calls for. Using the power of our setting for prototyping and experimentation, we are building bridges beyond the museum walls to make some of what informal learning environments do best: offering a place and resources for captivating and inspiring ways for children to be innovative and creative thinkers.

Power of Place: Museums as Tinkering Laboratory

Museums can serve as amazing prototyping spaces for teachers to iterate design projects that work for them in a low stakes environment with the help of our creative team of educators and developers. In this environment, teachers aren't on the hook for standards delivery, but they do need to engage visitors quickly or watch them walk away. For teachers, this can be terrifying and at the same time transformative. They are able to quickly learn where their project is going askew based on the reactions they get from visitors. There aren't any consequences—they aren't responsible for backtracking the next day for any mistakes they have made. They can quickly iterate and try again.

Alongside opportunities to test and iterate with visitors is the unique opportunity to work with exhibit developers and museum facilitators who are in the business of creating engaging experiences that captivate young people. The high school and college students serving as facilitators and design lab residents for our exhibit activities develop wide and deep knowledge of how to engage children in science and the kinds of questions and issues that children of different ages have. Museum educators and facilitators have a great deal to offer, as the following reflection from our Summer Institute prototyping sessions suggests:

> *On the last day of the week-long Summer Design Institute, sixteen teachers are busily engaging kids at tables prototyping their design projects on the floor. One teacher, Karen, is piloting a design activity involving boats that could sink or float using a range of materials. She explains the problem and the science in detail to the children that approach her. They politely listen for a bit, answering questions correctly when prompted, but run off to the next table before digging in to the problem. Carlos, one of our Design Lab facilitators, sees that the children aren't getting invested in the problem; they were just listening to a lesson. He gently suggests that Karen tweak her introduction to the problem. With the next group of children, he holds up the photos of a range of boats and asks the kids to share whatever they see that is special about each, and what makes them float. After they've shared some observations about these boats and thought for themselves about how boats float, they are then presented with the challenge. Kids cluster around the table and begin grasping for materials. As the next group rolls in, the teacher smiles and takes over.*

At NYSCI, we have honed our instincts to let us know quickly which problems are motivating and those that aren't. In the example above, a different introduction helped visitors invest in the problem sufficiently enough to get started. Other times, we might have to introduce compelling examples, incorporate an intriguing material, or even overhaul the problem itself.

Museums can also serve as powerful laboratories for creating and iterating engaging problems sets because of the sheer number of visitors across the ages that come to our setting. This opportunity to tweak activities and try them with all ages of children makes science museums a wonderful place to generate problem sets that captivate young people and at the same time foster learning of content.

Put Kids at the Center: Extending Design Problems through Digital Tools

Design Lab has also been committed to growing a strategy that we call "beyond the walls engagement." Irrespective of whether you ever visit our museum, we want to build on the methodology of design-based learning and everything we are learning through building the physical design lab environment, and create a set of digital tools that will bring the same kind of inspired learning to teachers and students in classroom contexts.

With the growing adoption of tablets and apps in K-12 schools, we have been developing new mobile tools that can play a supporting role in bringing STEM concepts and compelling design projects into schools. We were intrigued by the number of apps that could be used to investigate interesting scientific phenomena during the design process, either by gathering data during the iterative and testing phase of the design process or by revealing interesting phenomena at work in their projects (e.g., iSeismometer allows you to track vibrations on X, Y, and Z axes, which could reveal the magnitude and kind of vibrations their structures could withstand before failing). But many of these tools are decontextualized and unless you have a well-developed framework and a honed sense for what you are looking for, using these tools and interpreting the data can be difficult.

With support from the Bill and Melinda Gates Foundation, we have been creating Digital Design Lab, a suite of mobile apps, known as *noticing tools*, that enable children to use the world as a laboratory, making meaningful science and mathematics discoveries in the context of highly engaging problems kids would find worth solving. In collaboration with our media development partner, Local Projects, we are developing Digital Design Lab to be a larger digital ecosystem that includes orientation videos, instructional activities, and sharing and documentation functions that would allow children to share the products of their design process with others and to reflect on their work. To be useful to schools, these playful and compelling design projects are being directly aligned with Common Core Math Standards and the Next Generation Science Standards.

Case in point: SizeWise enables middle school students to use ratios and proportions while creating forced perspective photography shots where things appear wildly larger or smaller than they are in real life. Bundled with the app are a virtual objects library (e.g. images to pose with such as a giant soda can) and a set of computer-trackable objects, "stickpics," that can be printed, taped to a stick, and used as physical props. A suite of measurement tools including calipers for keeping track of real-life and onscreen heights, a ratio tool, and a distance meter are accessible in the app as you construct and set up your shots. Once students have taken a series of pictures, they can make their own comic strips and write out the ratios and math behind their photos so that others can recreate their shots. The strength of these tools is that students are at the center—they are the data under investigation.

Students and teachers alike have been enthusiastic about the possibilities of making mathematics and science content come alive through the use of tools that put students right at the center of the design process. They are yet another way of helping schools find exciting problem contexts and design projects outside the walls of the museum.

Hidden Strengths: Documenting New Ways to Be Smart

> *Engaging students in the practices of science and engineering ... is not sufficient for science literacy. It is also important for students to stand back and reflect on how these practices have contributed to their own development, and to the accumulation of scientific knowledge and engineering accomplishments over the ages. ... reflection is essential if students are to become aware of themselves as competent and confident learners and doers in the realms of science and engineering. NGSS (2012), Appendix F.*

As we look forward, we are thinking of powerful ways for the museum to offer schools a chance to reflect on the kinds of strengths and skills we see being exhibited by children every day in our museum setting and how they can take that excitement and insight back into the classroom. A major finding from our professional development work with teachers and museum prototyping was teachers' delight in uncovering previously hidden strengths in students who seemed unlikely to engage in science. The teachers enjoyed seeing these students find success, and were eager to build on that new enthusiasm and bring it to other areas of the curriculum. We realized that we could use the field trip activities we were creating toward a similar end. Our prototyping showed that the activities were engaging for a broad audience and could allow visiting teachers to see their students in a new light. We could deliver on the power of design to get teachers to take notice of students' abilities and skills in new ways.

In response to this feedback, we have begun work on a method of process documentation that is simple to execute and yet should aid teachers both in collecting evidence of student learning and in highlighting the ways that design activities can speak to traditionally unengaged students. Responses from teachers to our initial prototyping efforts have been very encouraging. We intend to emphasize process documentation in all aspects of the work moving forward.

There are a number of objectives for this work. First, in the museum, it will enable a teacher to focus on the ways in which her students are learning when she is not responsible for the activities that are happening. Teachers will have the time and impetus to visibly listen to their students, who will be occupied by their design projects and supervised by the museum explainers. In this way, a teacher can gain insights into her students (including their interests and preferred ways of STEM learning) that she may have otherwise missed.

Second, the documentation will enable teachers to revisit the museum experience with their students back in the classroom. The teachers and students will be able to easily label and annotate parts of the design process they photographed. The photographs and labels could then be used to refresh their students' memories of the museum visit, prompting the students to talk about their motivations and thought processes. This, in turn, could provide insights into the students and their learning preferences that teachers can use to develop future lesson plans and units. By enabling students to reflect on their own experiences, the process will also reinforce STEM concepts, including those that are advocated by the NGSS.

A third objective is to enable teachers to use the documentation to rekindle ideas and observations that they themselves gained from the process. These can also be translated into improved classroom experiences for students. For instance, during an initial prototyping session for the proposed project, a number of teachers discovered that through the design process, their students were working together more cooperatively and closely than they had previously. One teacher said she was "blown away" at how well a new girl in her class, a recent immigrant from Uzbekistan who spoke no English, was working with the other students. Another frequent remark from teachers was "I didn't know this student could do that." Teachers will be able to use information about an individual student's interests, ideations, and successes in order to increase that student's self-knowledge, confidence, and motivation. Additionally, they will be able to translate successful collaborative learning to the classroom. "Remember the way you worked together in the museum," a teacher might tell her students. "Let's work like that here."

A fourth objective of the proposed project is to provide teachers with a greater appreciation for design-based learning and specific ways in which they can translate this appreciation into lesson planning back in the classroom. Over and above the photographs and labels, which we hope will seed teachers' own design-based classroom activities, the project will provide access to a series of post-visit design activities that NYSCI is developing under another grant. By giving teachers a way to observe and document museum-based design activities, the museum hopes to make design "converts" of the teachers, providing them with the inspiration to incorporate the same strategies into their ongoing instructional repertoires.

Future Directions

NYSCI's Design Lab has been defining new ways for science centers to play a pivotal role in the educational ecosystem pushing for adoption of the standards in authentic ways. Through our in-person and beyond the walls efforts, we believe that helping schools embrace the core ingredients of design—problems worth solving, student-driven iteration, materials literacy, divergent solutions, and sharing can begin to bridge the content and practices gap in schools and broaden the ways in which kids get to be smart in STEM. Looking forward, research and development is needed to unpack how the core ingredients of design can be supported and sustained inside and outside of school. Questions for each of these areas include:

Problems Worth Solving How does learning compare when children are invited to work on problems they personally find worth solving versus more traditional inquiry-based problems? What kinds of strategies in problem definition promote authentic exploration of specific content? What kind of new examples are needed to round out the field-wide library of activities that invite a broader range of students to engage with science and engineering ideas?

Interest Driven Iteration How is it that educators in and out of the classroom can allow for more prolonged periods of iteration and redesign? What are the supporting strategies and structures that are needed to do this?

Materials Literacy What deliberate roles can science centers play in helping teachers and students develop materials literacy? How does materials literacy impact learners' abilities to seek and persist in solving new problems?

Divergent Solutions What kinds of problem frames invite divergent solutions from students? What kinds of supports and strategies can teachers use to uncover and leverage the content inherent in these potentially divergent design-based problems while serving their specific curriculum goals?

Sharing and Reflection How can digital tools support the kind of sharing and reflection that leads to prolonged engagement and iteration in informal and formal settings?

> *Design Lab and Digital Design Lab are made possible with generous support from the Bill and Melinda Gates Foundation, Phyllis and Ivan G. Seidenberg, Jim and Marilyn Simons, the Office of Naval Research, Verizon Foundation, and Xerox Foundation*

References

Caleb, L. (2000). Design technology: Learning how girls learn best. *Equity and Excellence in Education, 33*(1), 22–25.
Davidson, A. L., & Schofield, J. W. (2002). Female voices in virtual reality: Drawing young girls into an online world. In K. A. Renninger & W. Shumar (Eds.), *Building virtual communities: Learning and change in cyberspace* (pp. 34–59). New York: Cambridge University Press.
Duckworth, A. L., Peterson, C., Matthews, M. D., & Kelly, D. R. (2007). Grit: Perseverance and passion for long-term goals. *Journal of Personality and Social Psychology, 92*(6), 1087–1101.
Duckworth, A. L., Grant, H., Loew, B., Oettingen, G., & Gollwitzer, P. M. (2011). Self-regulation strategies improve self-discipline in adolescents: Benefits of mental contrasting and implementation intentions. *Educational Psychology: An International Journal of Experimental Educational Psychology, 31*(1), 17–26.
Dweck, C., Walton, G. M., & Cohen, G. L. (2011). *Academic tenacity: Mindsets and skills that promote log-term learning*. Paper presented at the gates foundation, Seattle, WA.
Holland, D., Lachicotte, W., Jr., Skinner, D., & Cain, C. (1998). *Identity and agency in cultural worlds*. Cambridge, MA: Harvard University Press.
Hull, G., & Greeno, J. G. (2006). Identity and agency in nonschool and school worlds. In N. C. Burbules & D. Silberman-Keller (Eds.), *Learning in places: The informal education reader* (pp. 77–97). New York: Peter Lang.
Johnson, S. (2010, September 25). The genius of the tinkerer. *Wall Street Journal*. Retrieved from http://online.wsj.com/article/SB10001424052748703989304575503730101860838.html
Margolis, J., & Fisher, A. (2002). *Unlocking the computer clubhouse*. Cambridge: MIT Press.
National Research Council. (2012). *A framework for K-12 science education: Practices, crosscutting concepts, and core ideas* (p. 218). Washington, DC: The National Academies Press.
NGSS Lead States. (2013). *Next generation science standards: For states, by state*s. Washington, DC: The National Academies Press. Available at http://www.nextgenscience.org
Partnership for 21st Century Skills. (2008). *A resource and policy guide*. Retrieved from http://www.p21.org/storage/documents/21st_century_skills_education_and_competitiveness_guide.pdf
Perry, D. L. (1994). Designing exhibits that motivate. In R. J. Hannapel (Ed.), *What research says about learning in science museums* (Vol. 2, pp. 25–29). Washington, DC: Association of Science-Technology Centers.
Rosser, S. V. (1990). *Female-friendly science: Applying women's studies methods and theories to attract students*. New York: Pergamon.

Schofield, J. (1995). *Computers and classroom culture*. New York: Cambridge University Press.
Turkle, S. (1984). *The second self: Computers and the human spirit*. New York: Simon and Schuster.
Turkle, S. (1988). Computational reticence: Why women fear the intimate machine. In C. Kramarae (Ed.), *Technology and women's voices* (pp. 41–61). New York: Routledge & Kegan Paul.
Turkle, S., & Papert, S. (1990). Epistemological pluralism: Styles and voices within the computer culture. *Signs: Journal of Women in Culture and Society, 16*(1), 128–157.
Turkle, S., & Papert, S. (1992). Epistemological pluralism and the revaluation of the concrete. In I. Harel & S. Papert (Eds.), *Constructionism* (pp. 161–191). Norwood: Ablex.

Chapter 4
Five Principles for Supporting Design Activity

Catherine N. Langman, Judith S. Zawojewski, and Stephanie R. Whitney

Even when teachers use highly supported and field-tested materials, the act of planning for implementation is a creative design process. As others in curriculum design research (*see* Chval, Wilson, Ziebarth, Heck & Weiss, 2012; *see also*, Ziebarth, Hart, Marcus, Ritsema, Schoen, & Walker, 2009) have described, although a teacher may be "delivering" a high quality lesson or activity to students, once that lesson is in the hands of the teacher, the nature of that delivery changes depending on the teacher, the needs of the students, the culture of the school, and other external factors. Assuming that the implementation of good curricular materials for teaching design varies across teachers, this chapter presents a set of principles that can guide planning in a way that supports and encourages students' engagement in authentic design activity.

Design and engineering have emerged as important areas of study and challenge for middle, secondary, and college classrooms (National Research Council of the National Academies, 2011). Engaging students in design and engineering is intended to develop students' capacity to adopt new thought processes and alternative ways of viewing the world, to establish methods of seeking and posing new problems, to willingly grapple with the complexities of emerging problems, and to apply and adapt tools for developing solutions to those problems. The goal is to engage students in important characteristics of the design process. For example, iterative cycles of expressing, testing and revising a design are a natural part of the design process. During these cycles, designers constantly engage in evaluating trade-offs, which are inevitable in complex settings. When making decisions about trade-offs, designers make assumptions, and keeping track of assumptions

---

C.N. Langman (✉) • J.S. Zawojewski
Illinois Institute of Technology, Chicago, IL, USA
e-mail: cnewman5@hawk.iit.edu

S.R. Whitney
DePaul University, Chicago, IL, USA

underlying a design is critical for providing rationales when the final design is presented and justified for the given context. Further, both the design process and whatever is designed are inevitably improved when designers actively seek to generate and understand external perspectives on each. These characteristics of design apply to what students should learn, as well as teachers who are planning the implementation of good design activities.

The purpose of this chapter is to propose and illustrate principles for planning the implementation of good design activities in a way that preserves opportunities for students to experience important characteristics of design. The chapter begins by introducing an activity that serves to illustrate the proposed principles. Then, the principles are presented and illustrated using two of the activities described in the first section. The chapter closes by showing how the authors have used the principles to implement a design activity drawn from a biomedical-engineering research project. Faced with implementing the activity in different contexts and with varied audiences, we share challenges faced, trade-offs made, and reasons for decisions. Our hope is to convey a set of *Implementation Design Principles* as flexible and elegant for use in planning for engaging students in design.

Part I: Getting to Know Model-Eliciting Activities

In this chapter, we use model-eliciting activities (MEAs) to illustrate the creative type of experience that students engage in while designing something—in this case, mathematical models. MEAs were selected as a site for proposing and illustrating implementation principles for three main reasons. First, the creation of mathematical models for non-routine science or engineering purposes is a powerful interdisciplinary design experience. Second, the design of the problem statement that drives the design experience of the MEA is already well established in the field, allowing the authors to focus explicitly on implementation issues. Third, asking students to design mathematical models captures important features of the Next Generation Science Standards and Common Core State Standards for Mathematics. These two curriculum reform initiatives contain overlapping and interlacing "engineering design standards" and "standards for mathematical practice," which strive to make the development of student scientific or mathematical thinking central to the course of study.

Background of MEAs

MEAs are designed and heavily field-tested to ensure that when small groups of students actively engage in the activities, they do indeed design mathematical models in response to the client-driven problem. MEAs have been productively used in elementary, middle, secondary, and collegiate classrooms (Magiera, 2013; Zawojewski, Diefes-Dux & Bowman, 2008; English & Watters, 2004; English, 2006), suggesting that designing mathematical models does not need to wait until students have learned some set of "basics"—rather, designing models is accessible to a broad range of students at a variety of levels. Diefes-Dux, Hjalmarson, Miller & Lesh (2008) describe how engineering education professors use these problems to teach design to their engineering students. They describe how these types of activities require students to work through iterative cycles of expressing or representing initial ideas for a mathematical model, testing the model, and revising it, and then returning to the beginning of the design cycle until the client's needs are met. Productive and well-communicated models are accompanied by assumptions and rationales that emerged during iterative design processes—just as in real-world modeling. Diefes-Dux et al., also describe how small teams of students working together on the problem and questions that arise during group presentations provide alternative perspectives that help students reconsider and improve intermediate models.

Description of MEAs

In MEAs, the problem statement, which specifically requests the creation of a mathematical model, is designed to elicit a range of reasonable mathematical models that vary in sophistication, depending on what knowledge and capabilities the modelers brings to bear. Therefore, most problem statements can be used across many different grade levels and backgrounds. When teachers decide to use a specific MEA with their students, what tends to change are the supporting materials and activities that facilitate students' engagement in design cycles (express-test-revise) and in making and articulating assumptions, evaluating trade-offs, and taking alternative perspectives.

The design of an MEA problem statement is driven by six long-established activity design principles (Lesh, Hoover, Hole, Kelly, & Post, 2000; Lesh & Doerr, 2003; Diefes-Dux, Hjalmarson, Miller, & Lesh, 2008) (see Table 1). The design principles ensure that the problem statement requires that students design mathematical models in response. While the principles ensure that the problems are open-ended design opportunities, the range of potential reasonable models is not wide open because the model must meet certain criteria—creating boundaries on what constitutes a "good" response to the problem statement. Our summary of the MEA design principles, adapted to engineering education contexts by Diefes-Dux, Hjalmarson, Miller & Lesh (2008), are in Table 4.1.

Together, these six design principles also ensure that teachers have a powerful educational tool, because when students engage productively in MEAs, the final model that students produce and the small-group interactions leading to its production serve as rich formative assessment sites for teachers. In particular, the designed model itself explicitly documents the mathematical procedure and representations developed by students, providing information about the mathematical elements, operations, and relationships the students found to be important. And, the collaborative model design process provides opportunities for the teacher to hear, see, and read what students are thinking in real time as they grapple with the problem situation, express their thinking to each other, and bring previously learned ideas to bear on the problem situation.

Table 4.1 MEA design principles

Design principle	Qualities of the MEA
The "Model Construction" principle	Students will be required to design a mathematical model in response to the problem
The "Reality" principle	Students will interpret the problem context as meaningful or realistic, and that the model they are asked to design is a compelling solution to the problem
The "Self-Assessment" principle	Criteria exist within the MEA for students to assess the effectiveness of the mathematical model they are in the process of designing
The "Model Documentation" principle	Students will be required to create documentation that explicitly reveals what the designed mathematical model is
The "Share-Ability and Re-Usability" principle	Students will be required to produce solutions that others can read, interpret and use on a new set of data for the same problem (including rationales and assumptions)
The "Effective Prototype" principle	Students will be required to design a mathematical model that will be useful for interpreting other situations in which a similar mathematical approach is relevant

Implementation of MEAs

Most MEA problem statements do not stand alone, due to the unusual contexts for the K-12 classroom. Students' initial experiences with designing a mathematical model in response to a problem also need additional support. To respond to these challenges, supporting materials and activities evolved over time and over teachers. Some types of supporting materials have become almost standard. For example, common supporting materials including an introduction to the problem context in the form of a newspaper article or video, accompanied by a series of questions to help students make sense of the context of the problem and identify critical prerequisite skills. Another example includes explicitly planned activities that require students to present their models and critique the models of others. This purposefully engages students in encountering other perspectives, while learning to communicate their models and underlying assumptions and rationales to each other.

The MEA Illustrations

Two MEAs are shared as illustrations and will be used to present the *Implementation Design Principles* in the second section. Both MEAs were extensively field-tested with a diversity of students to ensure the six MEA design principles from Table 1 were met. Both MEAs have been widely used for over 15 years by teachers across various educational levels (from middle school through college) and educational settings (e.g. urban, suburban, rural). Both MEAs also have an extensive history of research and publication (e.g., Magiera, 2013; Zawojewski, Diefes-Dux, & Bowman, 2008). We recommend you get to know each MEA by reading and attempting a first draft solution to each problem.

4 Five Principles for Supporting Design Activity

The "Choice of Aluminum Bats" MEA The *Choice of Aluminum Bats MEA* was originally designed for and implemented in middle school field test sites and subsequently used with Purdue University's first-year engineering course as part of a curriculum reform project funded by the National Science Foundation (Zawojewski, Diefes-Dux, Bowman, 2008). A shortened version of the MEA is shown in Fig. 4.1 below (and a version including supporting materials is attached as Appendix A).

The context for this MEA is the selection of aluminum softball bats that resist denting. Given that larger crystal sizes are associated with more bendable metals, the students are asked to design a mathematical way to determine typical crystal size of some samples. They are given microscopic pictures of crystals taken at different scales.

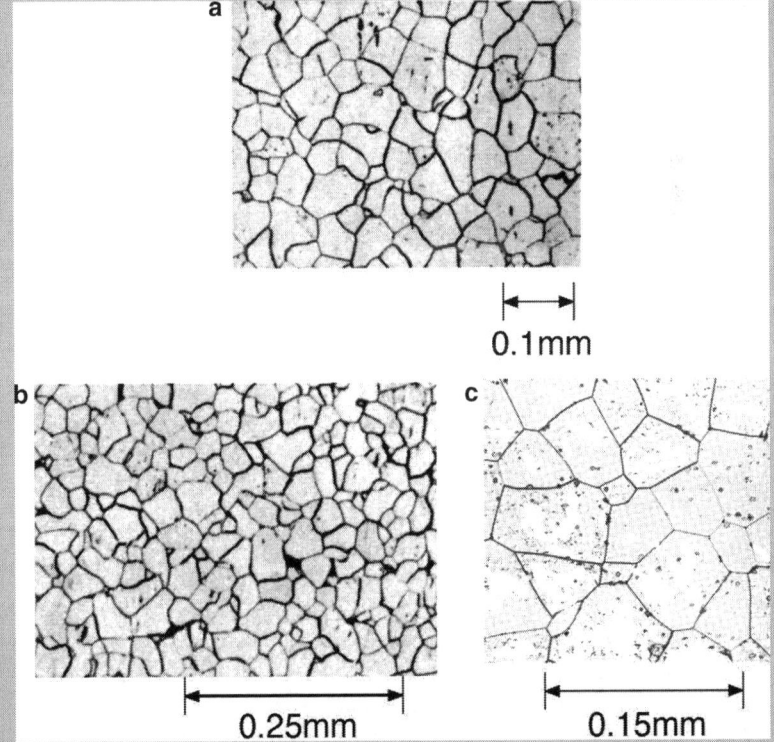

Fig. 4.1 The *"Choice of Aluminum Bats"* problem statement and images

A reasonable model for measuring crystal size, therefore, involves geometry (spatial reasoning and measurement), proportions (dealing with different scales) and sampling (deciding which crystals to measure). As part of the creative process, students need to identify and mathematize the variables that they selected to use in their design and respond in writing to the coach.

The "Paper Airplanes Contest" MEA The *Paper Airplane Contest MEA* was also originally designed for and implemented in middle school field test sites. It also has been used in Purdue University's first-year engineering class on problem solving and computer tools. A shortened version of the MEA is shown in Fig. 4.2 below (and a version that includes supporting materials is attached as Appendix B). The context for the MEA is the design of a fair scheme for awarding designations as "most accurate" and "best floater" to paper airplanes in a contest. Given are data from three trials for each of a number of paper airplanes: distance from start, distance from target, time in air, and angle from target. From this data, the modelers are to design and mathematize a procedure for determining "most accurate" (e.g., closest to the target) and "best floater" (e.g., goes slowly for a long time).

> **The *"Paper Airplanes Contest"* problem statement:** In past competitions, the judges have had problems deciding how to select a winner for each award (Most Accurate and Best Floater). They don't know what to consider from each path to determine who wins each award. Some sample data from a practice competition and a description of how measurements were made have been included. To make decisions about things like being the best floater, the judges want to be as objective as possible. This is because there usually are only small differences among the best paper airplanes—and it seems unfair if different judges use different information or different formulas to calculate scores. So, this year, when the planes are flown, the judges want to use the same rules to calculate each score.
>
> Write a brief 1- or 2-page letter to the judges of the paper airplane contest. Give them a rule or a formula which will allow them to use the kind of measurements that are given in table below to decide which airplane is: (a) the most accurate flyer and (b) the best floater. table below shows a sample of data that were collected from four planes last year. Three different pilots threw each of the four planes. This is because paper airplanes often fly differently when different pilots throw them. So, the judges want to "factor out" the effects due to pilots. They want the awards to be given to the best airplanes—regardless who flies them.
>
> Use the data in table below to show exactly how your rule or formula works—because the judges need to use your recommendation for planes that will be flown during the actual competition this year

4 Five Principles for Supporting Design Activity

| | | Information about Four Paper Airplanes Flown by Three Different Pilots ||||||||||||
| | | Pilot F |||| Pilot G |||| Pilot H ||||
Plane	Flight	Distance from Start	Time in Flight	Distance To Target	Angle from Target	Distance from Start	Time in Flight	Distance To Target	Angle from Target	Distance from Start	Time in Flight	Distance To Target	Angle from Target
A	1	22.4	1.7	15.2	16	30.6	1.6	14.5	23	39	1.8	7.5	-10
A	2	26.3	1.7	16.7	26	31.1	1.6	11.9	19	36.3	1.7	4.3	-6
A	3	31.6	1.7	7.1	10	26.7	2.2	8.9	-4	35.9	2.2	9	-14
B	1	32.1	1.9	7.6	-11	35.9	1.9	14.3	-23	43.7	2.0	9.5	6
B	2	42.2	2.0	9.2	-9	39	2.1	11.1	16	29	2.0	7.6	7
B	3	27.2	2.1	10.2	-11	25.6	2.0	11.7	12	36.9	1.9	12.4	19
C	1	19.2	1.8	16.6	-8	42.9	2.0	9.8	9	35.1	1.6	2.8	4
C	2	28.7	1.9	9.3	11	44.6	2.0	9.3	-1	37.2	2.2	2	-1
C	3	23.6	2.1	17.3	-25	35.7	2.2	3.2	-5	42	2.1	9.8	10
D	1	28.1	1.5	8.9	9	37.2	2.1	20.2	-32	41.7	2.2	10.1	11
D	2	31.6	1.6	14.8	-24	46.6	2.0	11.4	-2	48	1.9	14.1	-8
D	3	39.3	2.3	9.1	12	34.7	1.8	22.2	-36	44.7	1.7	11.5	-9

Fig. 4.2 The "*Paper Airplanes Contest*" problem statement and data table

A good solution to the *Paper Airplane Contest MEA* requires interpreting, sorting and weighting variables (choosing to what each of the four variables are important) and sampling (deciding whether to use all three data for each plane or to use the best performance event for each plane). Modelers may also use averages (for example, by averaging the three data points from each trial for each plane) and synthesize data (for example, combining their selected variables). The rules for making awards must be communicated to the judges to fulfill a sense of fairness.

Now that you, the reader, have produced a first draft solution to both MEAs, you can imagine that a variety of reasonable models can be designed. For example, in the *Choice of Aluminum Bats MEA*, two common models designed for determining crystal size have included calculating number of crystals per square area or calculating the average area per crystal. While each model of these two types is self-explanatory at a macro-level, at the micro-level a number of variations occur within each of these two approaches. For example, students need to describe and justify different ways to count the irregular crystal shapes and sizes, and students need to deal with varied scales. These demands prompt students to produce different methods in response. In the *Paper Airplane Contest MEA*, typical models for "most accurate" are pretty obvious (distance of plane from target). However, students also need to decide which data to use and how to sort and weight it. In the "best floater" designation, sometimes students use only the variable "time in air"—making the model quite simple—but there exist many different interpretations of what it means to be the "best floater" so students must define the construct "best floater" before creating a mathematical model.

Questions About Implementation Abound!

How can a teacher help students begin to comprehend the need to select and deselect variables (when there is simultaneously too much data and too little data)? How can a teacher help students to see that their initial designs are likely to be over-simplified and even somewhat incorrect (when they are used to getting math problems either right or wrong)? How can a teacher help students to expect and accept that express-test-revise cycles are part of a normal design process (when students think that going back and revising is akin to "getting it wrong")? How can a teacher help students understand that there are a variety of reasonable models, each based on trade-offs made and accompanying assumptions and rationales? These are among the issues faced by teachers designing implementation for individual MEAs. To begin to address these questions, we have pondered a wide variety of supporting material and activities that have been designed over the years to accompany and support MEAs. As a result, we have developed some *Implementation Design Principles* that are aligned with the goals of general design, and stated in general terms to apply to implementation of all types of effective design activities.

Part II: Introducing the *Implementation Design Principles*

In our review of the literature, materials and activities supported each MEA problem statement. These materials and activities intended to help engage students in design of mathematical models. While some supporting materials were usually designed by the MEA designer (e.g., newspaper articles, videos or other ways to introduce the context), most supporting materials have been created on an *ad hoc* basis by different teachers, instructors, or researchers working in different educational environments (*see*, for example, the University of Minnesota's database of MEAs at https://moodle2.umn.edu/mod/url/view.php?id=501011 and the Florida State Board of Education's teacher professional development initiative on MEAs at http://www.cpalms.org/cpalms/mea.aspx).

Based on the many types of support materials and activities, we developed a set of *Implementation Design Principles* for supporting students' engaged in design. We found these principles highly useful, not only in describing the materials and activities that have been developed for MEAs, but also for helping us design implementation materials. Our experience is that the core characteristic of each of the principles applies to a broad array of design activities, including bridge-building design projects and a program to help high school students design exercise and nutrition plans to teach to third graders. The five *Implementation Design Principles* are presented, and described, in Table 4.2. The table also includes some helpful verification questions and statements of usefulness for each principle.

Table 4.2 Five implementation design principles

The "Familiarity" principle	This principle ensures that the designed implementation supports help a particular group of students relate to, and care about, the context of the problem *Verification Questions:* Do the designed supports: • help bring the context of the problem to life for these particular students? • provide references to the real lives of these targeted students? • help generate student interest in the problem? *Usefulness:* The implementation of these supports helps bring the context of the problem to the doorstep of the students
The "Prerequisites" principle	This principle ensures that designed implementation supports target critical vocabulary, concepts and context information appropriate for the particular students and educational context *Verification Questions:* Do the designed supports: • help students (and teachers) identify the prerequisite knowledge needed to understand the problem statement? • provide sufficient vocabulary? • reveal to students the relationships between key ideas? • give students opportunities to practice this new knowledge or these prerequisite skills? *Usefulness:* The implementation of these supports helps students (as well as their teachers) to identify strengths and weaknesses in their prerequisite knowledge
The "Accessing Complexity" principle	This principle ensures students can productively engage in designing a model (or other object) for a complex, intellectually challenging situation that might otherwise be outside students zone of proximal development *Verification Questions:* Do the designed supports: • contain information about the meaning of the variables that are likely to be needed in the students' model? • help students identify which variables to use when there are simultaneously too many and too few variables, and they have varying levels of relevance to the model being designed? • help students access the complexity of the problem at a level appropriate for their capabilities? *Usefulness:* The implementation of these supports helps students identify important variables, recognize the relationships between variables that are likely to be captured in the model (or other object), and select variables that are useful to them when considering their own skills and capabilities

(continued)

Table 4.2 (continued)

The "End-in-View" principle	This principle ensures that the designed implementation supports help the students to keep present in their thinking what their final product might be *Verification Question:* Do the designed supports: • help students think about what their final product might be? • help students keep track of their end product as they engage in the design process? *Usefulness:* The implementation of these supports help students keep in mind what the characteristics of their final product will be. (For MEAs, it is a mathematical model that needs to be clearly communicated to a client for the client's specific need.)
The "Alternative Perspective" principle	This principle ensures that the designed implementation supports provide opportunities to put the students in the position of taking an alternative perspective on their model (or other designed object) *Verification Questions:* Do the designed supports: • provide opportunities for students to hear and interpret other points of view? • reward students for changing their mind when they hear what they consider a new and better idea? • reward students who actively seek other perspectives? *Usefulness:* The implementation of these supports help students enter the express-test-revise cycle initially through small group interactions, then large all-class discussion, and with maturity gain the ability to actively generate different points of view independently during the design process

The "Familiarity" Principle

When classroom activities are designed to appeal to the interests of students, academic engagement in the activity can be enhanced. While the compelling nature of the context and problem will vary with individual students, this principle ensures that the supporting materials and activities are designed to help make connections between the students' experiences and the context.

Consider the *Paper Airplane Contest MEA*. While only a few students may have made and flown paper airplanes, the context of paper airplanes is reasonably familiar. Some teachers who have designed implementation have collaborated with science teachers to have students make and fly paper airplanes prior to implementing the MEA in order to increase students' familiarity of the context, whereas other teachers have made judgments that the context is familiar enough as described in the newspaper article (see Appendix B). Notice that the designer-created newspaper article works hard to introduce the students to this particular paper airplane contest—making explicit connections between what students are likely to know (i.e., flying paper airplanes) and what they will be asked to do (mathematize features of different types of airplane performance in this contest). Similarly, for the *Choice of*

Aluminum Bats MEA, the context of softball and aluminum bats is familiar to most students. In this case, the newspaper article ("Batter, Batter Swing"—see Appendix A) brings the unfamiliar notion of the dentability of aluminum bats to the forefront, and then works hard to engage students in a context that will help bring meaning to the idea that there can be a mathematical relationship between crystal size and the performance of metal. In some cases, such as the newspaper article that accompanies the *Paper Airplane Contest MEA* (see Appendix B), the setting can be modified so that it is local to the students who are doing the MEA. The point of the "Familiarity" Principle is to allow students to begin in a context that has personal relevance to them as a point of entry to the context of the problem.

The "Prerequisites" Principle

When classroom activity involves true integration of content, as is often the case in design activities, high levels of engagement depend on students' understanding of prerequisite material that may or may not be in their prior schooling. Some of the most popular MEAs deal with science or engineering content that is unfamiliar to the students, and the accompanying implementation support materials and activities are designed to make the needed prior scientific and engineering knowledge available to provide students with access to the context and problem statement. Further, when students are immersed in the richness and complexity of such design activities, students may not activate their relevant "math class" capabilities. Thus, the purpose of this principle is also to activate prior knowledge that is likely to be useful in the modeling process.

Consider the *Choice of Aluminum Bats MEA*, which draws on concepts from materials science uncommon in school science. The facts that all metals are made up of crystals and that, while the crystals "tend" to be of a certain size, they are not uniform in shape or size are unknown to most students. To address this bit of prerequisite science knowledge, the designers of the *Choice of Aluminum Bats MEA* included a newspaper article to introduce the context, followed by what might be called a "science supplement" in the form of a report on Coach Hart's conversation of Professor Louisa Rodriguez (see Appendix A accompanying "Batter Batter Swing" article, followed by a narrative—in italics—reporting on Coach Hart's conversation with the professor). At the end of that report, the reader is formally introduced to a photograph of a traffic pole that shows the crystals that most of us have seen every day and a close up of the crystals (see Fig. 4.3 below). Further, the close-up superimposes an outline of three crystals to enhance students' conceptual images. Both images provide scale marking, which position the "science supplement" piece to serve as a focal point for classroom discussion about the fundamental concepts of metallic crystals.

As a follow up to the combined "Batter, Batter Swing" newspaper article and the science supplement (narrative of Coach Hart's and Prof. Rodriguez's conversation), the designer of the MEA wrote a variety of "Readiness Questions" (common to most MEA pre-reading material, and also in Appendix A, following the article and

Fig. 4.3 Traffic light pole crystals and the "Prerequisites" principle

science supplement). Two of the questions included were intended to activate specific pre-requisite mathematical knowledge. In particular, Question #3 asks, "How is the size of an aluminum crystal related to the bat's resistance to denting?" The purpose of this question is to bring to the surface students' recognition and articulation of the relationship between two quantitative notions: crystal size and bendability. The first is a straightforward quantitative feature (crystal size), whereas the second—bendability—is one a student may need to think about, to realize that perhaps bendability can be quantified. However, this question only asks for a general statement of the relationship between the attributes: that the larger the crystal, the more bendable the metal; or the smaller the crystal, the less bendable the metal. Question #6, "Given the scale marker below the picture of the traffic light pole, how wide is the pole?" addresses prerequisite knowledge about using scales to interpret images, and is more standard to school mathematics curriculum.

The "Accessing Complexity" Principle

The challenges of designing a model for a real context involves the need to notice, attend to, and mathematize critical features of the problem situation. The recent curriculum reform Common Core State Standards states that in mathematical modeling, ". . . real-world situations are not organized and labeled for analysis; formulating tractable models, representing such models, and analyzing them is appropriately a creative process. Like every such process, this depends on acquired expertise as well as creativity" (CCSS.Math.Modeling, retrieved from http://www.corestandards.org/Math/Content/HSM). Thus, the goal of this implementation principle is not to decrease the cognitive demand of the problem, but to ensure that students encounter information that makes it clear that certain variables and features are especially important to their design of a mathematical model (or the need to explicitly make a case not use a variable in one's model). In other words, the goal is to afford students with opportunities to intellectually engage in challenging situations that might otherwise be outside the students' zone of proximal development.

The targeting of critical variables is explicit in the *Paper Airplane Contest MEA* Readiness Question #2, "What types of measurements do you believe should be taken for each throw to fairly judge the contest?" (see Appendix B following the newspaper article). Given that Readiness Questions are discussed in small groups or as a class prior to engagement in the MEA problem statement, in this question students are alerted to a need for identifying critical variables that contribute to the mathematical complexity of the problem. Another example is in the discussion of Readiness Question #3, "How would you decide which airplane is the best floater?" The intent of this question is to begin to grapple with a construct that does not have a standard meaning. This question can be used in a discussion in which students are asked to think about real world contexts in which people use the word "float." For example, people seldom say that birds "float" but they will use it to describe how a feather travels in the air. The question provides an opportunity to consider a variety of variables that may be used to determine a "best floater," providing students with access to the complexity of the model to be designed. Some teachers have asked modelers to compare jet aircraft to blimps as "floaters," and students usually agree that the blimp is a more compelling image of "best floater"—because it moves slowly for long periods of time. Designing a model that mathematizes "moving slowly for a long time" is much more complex and much more compelling. On the other hand, a small group of students may make the case that everyone is using paper airplanes (i.e., no mechanically propelled air planes are in the contest), and thus time in air captures floating perfectly well.

The "End-in-View" Principle

The term "end-in-view" is adopted from the work of English and Lesh (2003). They, in turn, adopted the term from the work of John Dewey (Archambault, 1964) who addressed the importance of evaluating the means for accomplishing a task:

> It is simply impossible to have an end-in-view or to anticipate the consequences of any proposed line of action save upon the basis of some consideration of the means by which it can be brought into existence. Propositions in which things (acts and materials) are appraised as means enter necessarily into desires and interests that determine end-values. Hence the importance of inquiries that result in the appraisal of things as means (pp. 91–92)

According to English and Lesh (2003), "Although solvers do not know the exact nature of the product that is required of an ends-in-view problem, they know when they have developed one. This is because the given criteria or design specifications serve not only as a guide for product development but also as a means of product assessment" (p. 300). This is also true in MEAs.

MEAs are deliciously complex, and that very richness can result in students losing sight of what they need to produce in the end—a mathematical model that responds to the problem statement. Prior school experiences often prime students to expect to answer a question, rather than to design a model. For example, in one of the author's experience with the *Choice of Aluminum Bats MEA*, students often think that they are to simply tell you which sample to use and stop at that point—thinking they have answered the question. Similarly, in the *Paper Airplane Contest MEA*, students often think they are supposed to answer the question: Who won the contest? Even though the MEA problem statement requires a model be designed as an answer to the problem, this type of "answer" is so foreign to students that they tend to misinterpret it. Therefore, supplementary implementation materials and activities are often created to help the students keep the *end in view*.

To illustrate, one team of implementation designers (Diefes-Dux & Imbrie, 2008) of the *Choice of Aluminum Bats MEA* wrote a set questions to explicitly call students' attention to the end-in-view. After reading the newspaper article and science supplement, answering "Readiness Questions", and then reading the problem statement together, students answered a set of three "Team Readiness" questions: "Who are you working for? What do you need to create for them? How will you provide them this information?" These questions are certainly implied and embedded in the problem statement, so one might think that the Team Readiness questions are redundant and not needed. But for this population, based on the professors' prior experience, the implementation designers decided that these questions were needed to help the students stay in front of the goals of their work.

Another example of an end-in-view implementation strategy is based on Schoenfeld's (1992) work, where he taught college students in a mathematical problem-solving course. Schoenfeld found a way to support students without trying to give hints or clues about what he thought would make a good solution to the students. By asking, "What are you doing? Why are you doing it? Is it helping?" Schoenfeld found that, over multiple problem-solving experiences, students began to ask these questions of themselves and began to learn how to struggle, productively, in solving problems. One of the authors has used this mechanism in MEA implementation at many levels and with many audiences. Similar to the Purdue professors' "Team Readiness Questions", this questioning strategy was found to help students keep the end in view—while avoiding the urge to replace students' initial ideas for designing a model with the instructor's ideas.

The "Alternative Perspectives" Principle

The use of small groups in modeling activity has inherent in it the "Alternative Perspectives" Principle. While one could assign MEAs to individuals, there is much greater opportunity to design a better model by engaging in the negotiation process that can happen naturally in collaborative work—essentially engaging in numerous cycles of expressing-testing-and-revising initial ideas, and eventually honing a common model. Team members verbally express their individual ideas, testing these ideas out on each other. This provides opportunities to give and receive critical feedback. Further, when critiquing someone else's publicly expressed model, the critique-er has an opportunity to gain new perspectives about other types of models that may in turn inform improvement in his or her own model. At the very least, the critique-er has an opportunity to develop a deeper understanding about why one's own model does addresses "this" and not "that."

Other implementation activities that exhibit this principle include selecting models to be presented and defended to the class as a whole, and putting small groups together to make presentations to each other. The challenge in this latter type of activity is eliciting the "revised" version after entertaining alternative perspectives—as students are unaccustomed to revising work in mathematics and science classes.

Blind peer review—as done in many language arts classes—is another way to motivate the revise phase of the design cycles while meeting the "Alternative Perspectives" Principle. For example, at Purdue, an online, blind peer-review process requires students to engage in giving and receiving alternative perspectives on models in various stages of design. While details are described in various publications (e.g., Zawojewski, Hjalmarson, & Diefes-Dux, 2013; Diefes-Dux, Zawojewski, Hjalmarson, & Cardella, 2012), the critical feature of this implementation activity is that each student is engaged in giving feedback on another group's model via a blind peer review, receiving feedback from peers to use in revision of their own model, receiving feedback from their instructor on the revised model, and revising it once again prior to formal evaluation. The Purdue experience has institutionalized the "Alternative Perspective" Principle.

Integrating the Implementation Design Principles

While the proposed implementation principles for teaching design are illustrated above with separate instances, in reality, the work of planning for implementation can be, and should be, more integrated. For example, the "Readiness Questions" that typically follow the newspaper article frequently address a number of principles simultaneously. In the case of the *Paper Airplane Contest MEA*, one cluster of questions, when asked as a sequence, provides more powerful support for entering a design activity than each question would provide on its own. As described in the

"Accessing Complexity" Principle above, Readiness Question 2 ("What types of measurements do you believe should be taken for each throw to fairly judge the contest?") and Readiness Question 3 ("How would you decide which plane is the best floater?") are intended to help students bring to the fore the important variables and characteristics that will be needed in their model. Question 4 follows immediately, similarly asking students to consider "How would you decide which plane is the most accurate?" Together, these three questions engage students in considering both the variables (i.e., different types of measurement) and requires a metacognitive reflection that doesn't ask for a design to be produced, but instead asks students to begin formulating ideas about what their modeling process will be and what their model might look like. Students typically respond with primitive versions of the models that they will later produce in response to the problem statement. In a sense, these questions as a group also function to address the "End-in-View" Principle—prior to reading the problem statement. In concert, these three questions together provide a powerful implementation mechanism for students to establish an image of what their design activity will involve.

Thus, in this section, we have described the *Implementation Design Principles* using illustrations. In the next section, the actual process of designing support materials and activities is shown to be more a complex and intellectually stimulating design process.

Part III: Applying the *Implementation Design Principles*

This section illustrates how to design materials and activities to support students' engagement in the MEA problem statement, using the *Implementation Design Principles*. We describe our experience with a MEA that was created from the work of an interdisciplinary team on an angiogenesis simulation project. Natural iterative cycles of expressing, testing, and revising implementation support materials and activities are illustrated in the description of our analysis of what was needed by students in a particular context compared to what supports were already available from earlier implementations. Based on our actual implementation, we make recommendations for how to revise the supports for future audiences in a similar educational setting.

During these implementation design cycles, we were constantly engaged in evaluating trade-offs, such as grappling with the time factors versus lengthy engagement with different components of the implementation, as well as making decisions about what content to emphasize given that there are a variety of reasonable options. By working as a team, we continually encountered alternative perspectives, and when sharing the plans with the students' instructor, we gained insights from an external perspective. In other words, we were engaging in the design process of producing implementation supports that would enhance students' development of their design capabilities.

4 Five Principles for Supporting Design Activity

The model-eliciting activity we use to illustrate this section focuses on simulating blood vessel growth, which was analogous to the core activity of the research project. Figure 4.4 presents the MEA problem statement that we used to engage students in design and to elicit their mathematical models. For students who have not previously learned about the process of blood vessel growth, supplementary implementation materials and activities are important to help the students enter into this design activity.

The *"Blood Vessel Growth MEA"* problem statement:

To: Engineering Team
From: Dr. Cinar, Director of the Center for Tissue Engineering
RE: Method for estimating healthiness of a blood vessel network in porous scaffolds

To advance our research in tissue engineering, we are trying to determine a procedure for measuring the amount of blood vessel growth and healthiness of blood vessel networks in porous bioscaffolds.

We are asking you to help us by creating a mathematical procedure for scoring these samples based on the amount of blood vessel growth and the overall healthiness of the blood vessel network. The procedure will be used to score future samples, when we run lab experiments using other pore sizes, and different types of material for scaffolds.

To assist in your work, we are providing you with sample images of blood vessel growth in bioscaffolds. These images are the fourth week of blood vessel growth for scaffolds with pore sizes 270 microns, 160 microns, 135 microns, and 45 microns. The images represent 800 × 800 micrometer regions of porous bioscaffolds. This size will be standard in all future experiments, as will the placement of the host blood vessels at the top and the bottom of the region, and the VEGF source in the center of the region.

Deliverable: A memo that includes:

- A written description of your mathematical procedure or series of steps that will be used to determine the amount of new blood vessel growth and score the overall healthiness of the blood vessel network for all future samples produced in our lab.
- A demonstration of you procedure by applying it to one of the samples provided to you. Please attach the sample you use to the memo.

Thank you in advance for your attention to this matter.

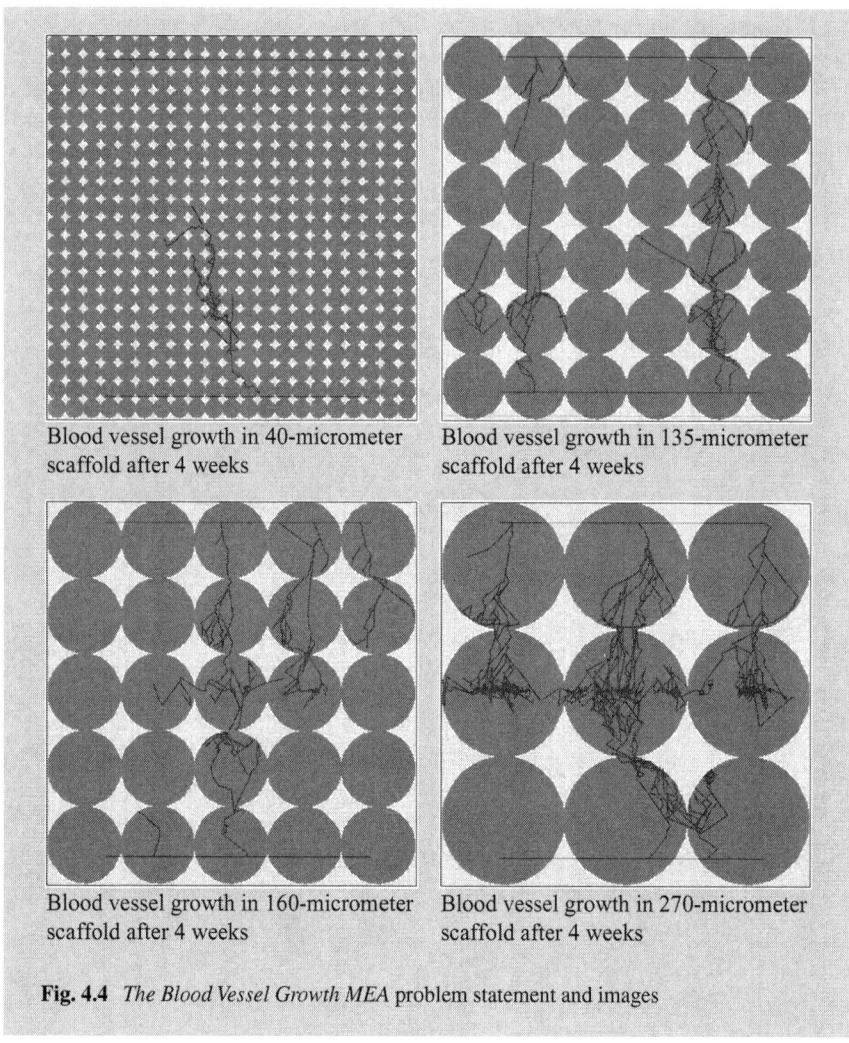

Fig. 4.4 *The Blood Vessel Growth MEA* problem statement and images

Designing Implementation Supports to Meet the Five Principles

The *Blood Vessel Growth MEA* is based on active research in tissue engineering (NSF Award Number IIS-1125412, "CDI-Type II: Optimization of Engineered Tissue Growth by Active Learning"). Part of the challenge of implementing this MEA lies in the fact that the context of the problem, which is grounded in a complex research setting, needs to be communicated in a way that is relatable to students, while capturing essential features of blood vessel formation and biodegradable porous scaffolds, which are complex systems in their own right. To address these needs, we chose to focus on a famous image and story from the field to help relate the concept of seeding a porous scaffold to something that is within the realm of public knowledge, in the newspaper article titled "Growing Ears!" (See Appendix C).

To flesh out the content of the "Growing Ears!" article and to focus in on the "Prerequisites" Principle, a science supplement was written to explains more about technical aspects of tissue engineering and to provide details that relate to specific aspects of blood vessel growth—the basis of the MEA. Given that blood vessel growth is very complex, we had to make decisions about what will be of critical importance for students to engage in a meaningful design process, while staying within most students' reach. While students would be working in groups—that usually amplifies individual's capability to work with complexity—the entry into understanding blood vessel growth was daunting.

To begin the process of designing implementation supports, two supports were pursued. One was the identification of educational videos about angiogenesis (recommended by the project scientists and engineers) to address science prerequisites. The other was to design a game that simulates angiogenesis growth, which would further solidify their understanding of the science and also highlight variables they might focus on in their design of a mathematical model. Like the *Choice of Aluminum Bats* MEA, this is content that is not typically in students' prior school science experience. To reinforce students attention to critical variables needed for the MEA problem statement, a variety of "Readiness Questions" were written, which we knew would likely change depending on the audience and educational setting for any implementation.

The videos selected from the popular video media site YouTube.com introduced the concept of angiogenesis, or blood vessel growth from existing vessels. The first video reveals how a host blood vessel has special cells on it that responds to chemical signals from cells in need—of oxygen, for example. These special cells sprout into newly growing blood vessels that grow toward the direction of the source of the chemical, branching along the way, and finally making connections through the cell. The second video shows actual footage of the sprouting process, blood vessel fusion, and blood beginning to circulate when two blood vessels connect.

Field tests indicate that students are drawn into these videos, but that the information is compact, comes very quickly, and has a very high level of complexity. The game that simulates blood vessel growth—which provides more access to the complexity by simplifying the variables in the process—is welcomed by and engaging for students. The concept of the game was initially drawn from the scientists and engineers on the project. In it, students pair up to simulate the process of blood vessel growth through a set of rules provided to them (See Appendix C following the article and science supplement.). The game asks students to roll a die a set number of times and make moves according to a rule sheet. Sample student results are in Fig. 4.5 below, which shows how these students took turns drawing in paths of blood vessel growth in the direction of the chemical signal, when seeking points for greatest length, most branches and most connections.

The MEA problem statement asks students to design a procedure that will measure and score the amount of blood vessel growth and healthiness of blood vessel networks. The game helps students unpack the complexity of the desired model into the three critical variables of blood vessel networks: (1) length (longest blood vessel), (2) density (most blood vessels), and (3) anastomosis (connections between distinctly

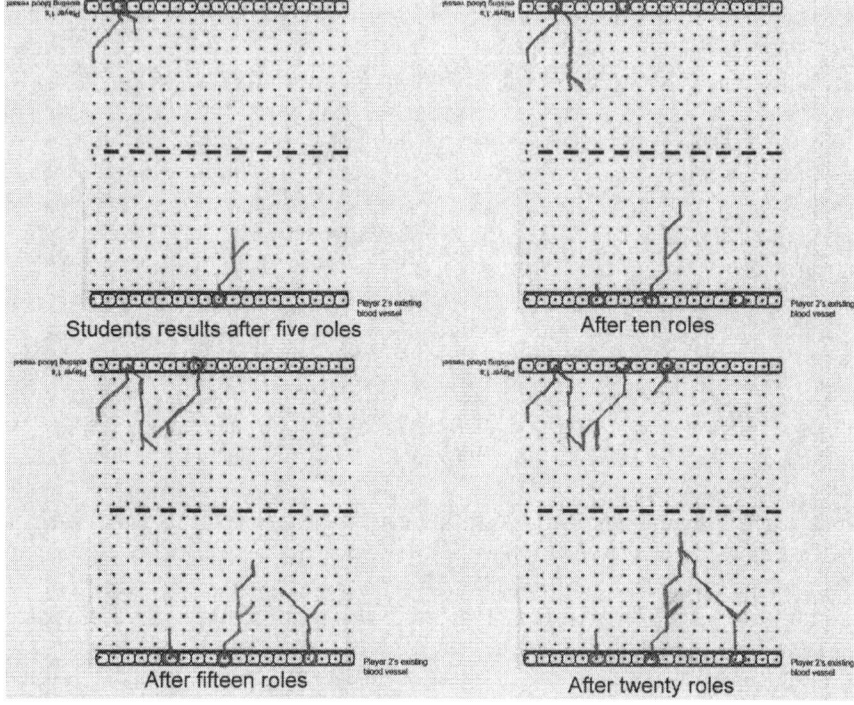

Fig. 4.5 Sample student results from full circuit game

growing blood vessels). The game also helps students relate the concept of new blood vessels growing from existing ones, the patterns that blood vessels form and the role of randomness as a way of representing things that might be in the way of the path of blood vessel growth or the blood vessels' response to multiple sources of chemical stimulant. The game is won by accumulating the highest points, which are assigned for making the most connections between distinctly-growing blood vessels, growing the longest blood vessels, and growing the most blood vessels. In this way, the game also provides an "end-in-view"—an early example of quantifying blood vessel growth, albeit under different circumstances than the problem statement. As a result, students gain a glimpse into the type of model that might be generated in response to the problem statement before actually receiving the MEA problem statement.

With each of the field tests, including an all-girls private school in a major city, a STEM-oriented summer program for high school juniors and seniors, a conference for mathematics teachers, a class of community college students in a pharmacy technician program, and a class of community college students in a business program, both of these mechanisms were useful and well-received, making them each a likely candidate as a permanent part of any implementation package. The videos provide opportunities to meet the "Familiarity" Principle due to participants' instant interest in the science, and the game provides opportunities to meet the "Prerequisites" Principle in providing engagement in relevant science (concept of angiogenesis) and mathematics (quantifying variables), as well as "End-In-View" Principle through their engagement in a simulation (i.e., using a relevant mathematical model).

Finally, after students receive the problem statement, students work independently to list which features from the images that they think could be used to indicate a healthy blood vessel network then compare lists. This mini-task follows the "Alternative Perspectives" Principle and the "End-in-View" Principle—it asks students not only to consider what their final product might be, but it also prompts them to begin small group interactions to kick-start the design process.

A Tool for Designing MEA Implementation

The questions we use for our own implementation design process prior to, during, and after any field tests are: *Do students have access to the MEA problem statement?* and *How do we know?* Earlier researchers (Chamberlin, 2004) have used an analogous emphasis in professional development with teachers who are using MEAs with their students. Chamberlin emphasizes teachers' externalization and documentation of their interpretations of their students' thinking for the purpose of making future implementation decisions. She finds that when teachers reflect upon and revise their interpretations of their students thinking, they are more likely to engage the express-test-revise design cycles in planning for students' engagement in MEA problem statements. Applying Chamberlin's emphasis to our work, we produced an "Implementation Design Sheet" (template attached as Appendix D) to produce a trail of documentation about what changes are made to MEA implementation supports, relative to the prior available implementation materials. The Implementation Design Sheet is organized around the implementation design principles so that we can systematically analyze whether we are encouraging important aspects of design with students.

Figure 4.6 is a summary of our analysis of needed implementation supports for taking the *Blood Vessel Growth MEA* into a new educational context with a different audience. It also includes our forward-looking thoughts about further revisions to make to the implementation for a similar audience and setting. We had previously implemented the *Blood Vessel Growth MEA* with high school anatomy students and with high school mathematics teachers during a professional development session. Working closely with the regular instructor for the new field test site, we were preparing to bring the MEA into a collegiate setting, where the students are described as "non-traditional," and take courses as cohorts—thus, students are familiar with each other and have often worked collaboratively with peers. The instructor explained that one course was a problem-solving course for business majors that had used three MEAs previously, and the other was a course for pharmacy tech students who had never used MEAs.

We shared the MEA and the existing implementation materials with the regular instructor and, based on her feedback and our own professional judgment, began the process of analyzing the existing implementation materials and activities using the Implementation Design Sheet. Given that this tool is organized around the *Implementation Design Principles*, we were able to consider the existing implementation supports with respect to the audience and educational setting of the upcoming field test.

Class Profile: About 40 students in an early college course at a university that offers associates, bachelors, and masters degree programs to traditional and non-traditional students. This particular group has done three model-eliciting activities earlier in their coursework.

Constraints specific to this group of students: Each class would meet for only 1 h and 50 min. To manage the time constraints, students are assigned to watch the video and read the newspaper article and science supplement as homework

Implementation Principle	Identified needs for intended audience.	Implementation support.	Implementation revisions for future similar audience in the same context.
"Familiarity" Principle	We expect the science supplement to draw the students into the biomedical engineering context and science content.	We revised the science supplement to simplify the language and descriptions used and to describe the engineering team that works on the project.	We will revise the science supplement even more to reveal a diverse team of engineers who wrote it.
"Prerequisites" Principle	We need to introduce the term and concept of "micrometer" as a unit of measure.	We included a verbal introduction to include definition of term and real world reference.	We will include a prepared image of the real world referent in the verbal introduction. Have student practice converting between microns and meters in the Readiness Questions.
"Accessing Complexity" Principle	We need to better highlight the "connections" variable.	We changed the name of the game to the "Full Circuit" Game.	We will consider how to also highlight the other two variables.
"End-In-View" Principle	We need to help these students think ahead about how they might go about designing a mathematical model.	We included Readiness Questions to follow the game that will prime them to think about the features of the model they are about to design.	We will refine the Readiness Questions to include only those that directly accomplish the goal to prime thinking about designing the model.
"Alternative Perspectives" Principle	We need more structured approaches to ensuring students encounter alternative and external perspectives.	We asked individuals to write a list of features they believe are important to their model after reading the newspaper article. When beginning group work, the students shared their lists	We will move the individual listing of important features to include in their model to right after introducing the problem statement.

Fig. 4.6 Implementation Design Sheet for Blood Vessel Growth MEA

4 Five Principles for Supporting Design Activity

The "Familiarity" Principle Given the expected population of students, especially the business focus of one of the classes, we did not expect the science supplement to bring the needed science ideas to the doorstep of most students. We were confident that the "Growing Ears!" newspaper article would entice students into the general context due to the famous photo of the mouse with an ear growing on its back. We also thought that the science explanations in the "Growing Ears!" article would help students understand what that photo was really about. On the other hand, the science supplement—while satisfying other principles (e.g., "Prerequisites" Principle, "Accessing Complexity" Principle, etc.)—needed revision to better meet the "Familiarity" Principle. It was then revised to include more specific information to make it appear to be a report from tissue engineers that would connect the information more tightly to people engaged in research and to the contexts and presentations in the newspaper article. After implementation, we decided to continue work on the supplement by more specifically describing the research team in all of its diversity—making it easier for students to relate to the researchers, hopefully finding that personal connections will make the technical aspects more motivating to the students.

The "Prerequisites" Principle Our analysis of the available implementation supports and the students' background suggested that the term and concept "micrometer" as a unit of measure was something most of these students would not have encountered previously. While they may have heard the word, we did not have confidence that they could deal with the scale factor effectively. For the implementation with the non-traditional college students, then, we decided to include in our verbal introduction to the MEA problem statement the definition of micrometer (as a unit of length and its relationship to microns), and pointed out the relationship between a micrometer and a real life example. We referred to the diameter of a human hair as about 25.4 µm, while a capillary—the smallest type of blood vessel in the human body—is only 1 µm in diameter. While this treatment of prerequisite knowledge helped, after observing and listening to students, the decision was made that in the implementation next semester, the goal would be to broaden accessibility of this idea by incorporating an image of a human hair, in order to create a vivid scale reference—similar to the aluminum pole in the *Aluminum Bats MEA*. In addition, we would plan to have students practice measuring and converting between meters and microns in the Readiness Questions.

The "Accessing Complexity" Principle An important aspect of blood vessel growth is that the vessels connect within a reasonable amount of time (before dying off) to create a circulatory system in which blood will travel. The game, which was designed to help students to simulate three variables of healthy blood vessel growth (length, number of branches, number of connections), was previously called the "Angiogenesis Game". To better highlight the critical variable of connections, we wondered about changing the name of the game: Would it better scaffold students' understanding of the science? For the field test we changed the name of the game to the "Full Circuit Game." When implemented, the "Full Circuit Game" had the intended effect of focusing the students on the connections between the blood vessels—this showed up later in the activity, when many groups prioritized connections over all other feature of a blood vessel network. The name of the game is still under consideration, as the name seems to affect how students attend to the complexity of the game.

The "End-In-View" Principle We needed to help students, especially those who had not done any MEAs previously, think ahead about how they might go about designing a mathematical model. In the previous implementation support package, students were asked the questions, "Who is the client?" and "What does the client want you to produce?" We discovered that these questions were helpful in getting students to understand that they needed to write a memo to describe a series of steps, or a procedure. On the other hand, the questions did not prompt students to think about how they might go about this, so Readiness Question #3 was written: Name three qualities of blood vessel growth that could be used to indicate a healthy network of blood vessels. The goal was to prime students to think about the features of the model they would be designing in the MEA problem statement, providing a way to highlight the end-in-view.

The "Alternative Perspectives" Principle Although students are expected to work in collaborative groups, naturally putting individuals in the position of encountering other points of view, we decided that students needed a structured approach to ensure encountering further external perspectives. After reading the newspaper article, individuals were asked to write a list of features they believe were important to their model. Then they shared their lists with others in their small group. We found that this helped provide every member of the group an opportunity to bring their thoughts to the process. But, we decided that this individual listing and sharing would be more powerful if placed immediately after reading the MEA problem statement as a class. This way, the ideas about designing a model would be very recent as students embark on a design process.

Part IV: Planning Implementation of Design *Is* Design

General design themes permeate the experience of teachers planning for the implementation of design activities. Adapting, implementing and revising implementation supports for the various groups of students and for various educational settings requires working through express-test-revise cycles, identifying and stating assumptions, evaluating trade-offs, and seeking other perspectives when producing effective implementation support packages.

Teachers, as designers, engage in iterative cycles of expressing, testing and revising components (support materials and activities), and the overall organization of the implementation package. The design cycles are embedded in the *Implementation Design Sheet* when looking across columns. The first column is where the designer identifies needs that are not met in currently available supports. The second column requires the teacher to explicitly describe the proposed support. After designing the new support material or activity, the designer tests it in the new implementation, reflects on the results, and makes a recommendation for future implementations in the third column—launching a new express-test-revise cycle.

Teachers, as designers, are engaged in identifying and stating assumptions, evaluating trade-offs and making decisions. "Throughout the design cycle, the designers should return to an examination of the problematic situation in order to identify whether the product is meeting an objective, if the new products need to be developed, if the problematic situation was changed by the design process, and to document how the needs of the situation has been addressed" (Hjalmarson & Lesh, 2008, p. 102). The *Implementation Design Sheet* facilitates these aspects of design. The first column requires the designer to state assumptions about the needs that should be addressed. In order to decide what new or revised component will be reported in the second column, trade-offs must be considered and then decisions made. Entering recommendations into the third column requires that the designer evaluate the results of the implementation and propose what will be tested in the next implementation. Careful tracking of designer reflections after each implementation, alongside careful documentation of the nature of the audience and educational situation, can be useful data for more effective and efficient planning of implementation for different audiences and educational contexts.

Part of the challenge of acting as a designer for implementing MEAs is in knowing a particular group of students without underestimating them. Students bring surprising and interesting ideas with them to the classroom and are often capable of more than one might expect. In many of our implementation sessions, we initially accompany the MEA and co-implement the MEA with the regular instructor. When teachers have not taught design activities to students before, they find it difficult to believe their students will be capable of designing mathematical models for complex situations. But, they also express trepidation about their own lack of experience in supporting students' designing solutions, compared to teaching them various concepts and skills for mastery. The purpose of the *Implementation Design Principles* is to help teachers make decisions about how to implement design activities effectively with their students, by structuring the supporting materials with respect to what students need in order to gain traction on the problem statements, rather than giving them ways to solve the problem.

No implementation component or package of materials and activities can fully anticipate all the problems, issues and needs that come up when students are engaged in design. The *Implementation Design Principles* can also be used in real time for teachers to respond to and react to students' questions and anxieties during the designing episode. They can help the teacher select what to attend to and what to let go. Just as the Next Generation Science Standards specifically seek to help students learn to "design a solution to a complex real-world problem, based on scientific knowledge, students generated sources of evidence, prioritized criteria, and tradeoff considerations" (NGSS, HS-ETS 1–2), so, too, are teachers required to nurture design based on knowledge of design, responding to evidence of students thinking, and prioritizing the types of scaffolds to be provided without taking away the design experience for students.

Finally, the teacher, as designer, actively engages in perspective taking. One type of perspective-taking, for the teacher, is observing and listening to students, identifying the ideas at the heart of the students' reasoning, perceiving and interpreting the interactions between students in a group and within the whole class, and devising strategies to attend to these different views. Another type of perspective taking occurs when teachers work together in planning and implementing design activities, as a personal professional development strategy. To teach design teachers, themselves, engage in cycles of design. To have students successfully engage in the modeling practices outlined in Next-Generation Science Standards and Common Core State Standards, teachers also act as designers and acknowledge implementation of design activities evolves and changes with the teacher and the context in which they are teaching.

Acknowledgments This paper was supported by a grant from the National Science Foundation (NSF Award Number IIS-1125412, titled "CDI-Type II: Optimization of Engineered Tissue Growth by Active Learning"). The authors wish to thank the National Science Foundation for their generous support of the advancement of discovery and understanding while promoting teaching, training, and learning.

The authors wish to thank all of the members of the engineering team at the Illinois Institute of Technology working on the grant, including faculty members Dr. Ali Cinar, Ph.D., Dr. Eric Brey, Ph.D., and Dr. Mustafa Bilgic, Ph.D., and graduate students Hamid Mehdizadeh, Elif Bayrak, Caner Komerlu, Banu Akar, Brianna Roux, and Sami Somo. The support, feedback, and encouragement from every one of these people made this work possible.

Appendix A: The Choice of Aluminum Bats MEA

Original Source: Keith A. Bowman, 2002.
Copyright by Keith A. Bowman, Permission to reproduce for classroom use granted.

Background Information

Batter, Batter....SWING!!

Stillwater, MN – The Lady Ponies are ready to charge! Coach Hart verified today that a new summer league softball team will be forming and joining the league.

"We have begun signing up players, and we still have two positions open – third base and center field. So, if you know of anyone that might be interested in playing these positions or even other positions, please have them contact me," said Hart. "We are also beginning to make decisions about our uniforms and the pieces of equipment that we need to purchase."

The Lady Ponies will wear uniforms of red and black after their team colors. The Heritage Embroidery on Market Street is designing the uniforms, and the uniforms will be available for purchase by next Friday. Players will be responsible for purchasing their own uniforms, cleats, and mitts.

Since deciding on the team's colors and uniforms, Coach Hart has been investigating the purchase of the necessary equipment for practice and games. Plenty of softballs have been purchased and batting helmets are being priced. Gart Brothers Sports has helmets available for $34.99 and Outpost Sports has them available for $32.95. "I'll probably purchase the helmets from Gart Brothers because they are better quality than the helmets at Outpost," said Coach Hart. "Besides, I can pick up the helmets when I also purchase the catcher's mitt and the catcher's mask from Garts."

The only remaining equipment for the coach to purchase will be the softball bats. The coach is considering three styles of aluminum bats, each of which costs about the same amount. "Since bats are so expensive and last year the bats dented too easily, I want to purchase bats that are more resistant to denting," commented Coach Hart.

The first game for the Lady Ponies will occur on June 6 at home. They will be playing the Oakdale Lady Stingers at Varsity Field. "I'm looking forward to helping the women get ready for our first game. I've heard the Oakdale Stingers have some good players, so we'll need to be ready to go!" explained Coach Hart.

> Coach Hart knew that Eva, who plays first base for the Lady Ponies, has an older sister who works as a materials engineer. Her name is Louisa Rodriguez, Ph.D. When he contacted Dr. Rodriguez, she explained that the size of the crystals in the aluminum is often a good indicator of the relative resistance to denting or strength of the material. She said that aluminum consisting of smaller crystals was stronger than aluminum consisting of larger crystals. Dr. Rodriguez volunteered to provide microscopic photographs of the crystal size called 'micrographs' because they were the standard way to compare the size of crystals. Materials engineers can chemically treat polished pieces of aluminum to make the boundaries between the crystals more visible. Using a camera attached to a microscope, a picture of the boundaries between the crystals can be estimated.

Coach Hart was fascinated and asked if it is ever possible to see metal crystals without a microscope. Dr. Rodriguez suggested that Coach Hart check out the new metal poles supporting the traffic lights on a nearby corner. These steel poles are coated with a thin layer of zinc metal that helps prevent rust formation. The zinc metal forms very large crystals that can be seen by the naked eye. The pictures below show the metal pole and a close-up picture of the crystals on the surface of the pole. The letters a, b, and c indicate three crystals that have had a line drawn along the boundaries between the crystals. The arrow on the drawing is the scale marker for this picture.

Fig. 1 Traffic light pole

Fig. 2 Close-up of crystals

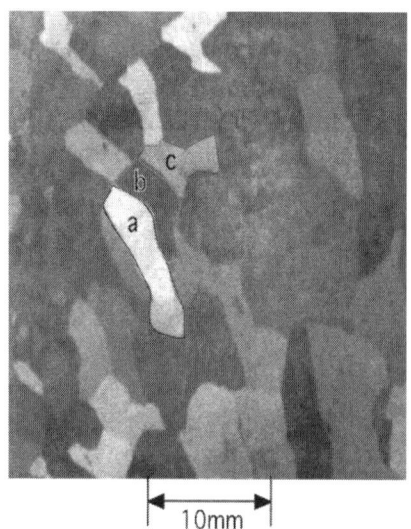

(A) Individual work – questions:

Readiness Questions

1. Why is Coach Hart purchasing the batting helmets from Gart Brothers when they are cheaper at Outpost Sports?
2. How is Coach Hart going to decide which bat to purchase?
3. How is the size of an aluminum crystal related to the bat's resistance to denting?
4. How can material engineers view crystals when they are too small to be seen by the naked eye?
5. Can some crystals be seen with the naked eye? Where?
6. Given the scale marker below the picture of the traffic light pole, how wide is the pole?

(B) Team work – questions:

First:

In your team, read the "problem statement".

Second:

In your team, answer these questions:

1. Who are your working for?
2. What do you need to create for them?
3. How will you provide them this information?

Third:

Work together in your team on the problem presented in the "problem statement".

The Choice of Aluminum Bat

Your Mission Using the three microscopic pictures of the samples of aluminum below, determine the typical size of crystal in each sample for Coach Hart. Also, write a letter to Coach Hart explaining how you found the typical crystal size so that he may share your process with other softball players and coaches that plan to purchase aluminum bats.

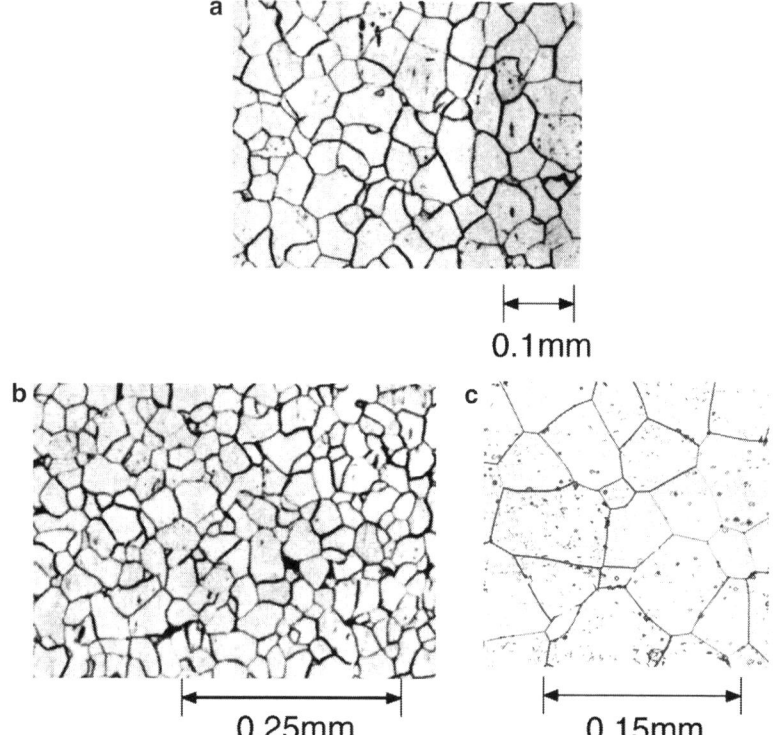

4 Five Principles for Supporting Design Activity

Appendix B: The Paper Airplane Contest MEA

Original Source: Richard A. Lesh

This activity development was supported by the Twenty First Century Conceptual Tools (TCCT) Center, Purdue University, West Lafayette, IN, under the direction of Richard Lesh. Copyright by Richard A. Lesh. Permission to reproduce for classroom use granted.

This activity was subsequently modified through the University of Minnesota.

Students to Fly Away with Paper Airplane Contest in the Twin Cities

St. Paul, MN – If you stop by Amy Frank's eighth grade science classes this week, you are likely to find a very busy group of kids. Ms. Frank's students will follow in the footsteps of the Wright Brothers, engineers, and pilots as they design and fly paper airplanes in the Twin Cities Annual Paper Airplane Rodeo held in the Metrodome.

Frank's students will be designing, creating, and flying paper airplanes throughout the week. The students are learning how engineers work as they plan, create, test, and redesign their paper airplanes. They won't be using aluminum parts or jet engines for these planes. All they will need are pieces of paper – and a whole lot of imagination.

Students will need to design planes that are able to fly long distances as well as stay in the air for a long period of time. Each contestant will design a plane to try to win prizes in one of two categories: Best Floater and Most Accurate. Said Frank, "The contest is designed to require the students to be very thoughtful about making their planes, so students who want to enter the paper airplane contest must follow a few rules." The rules are as follows: each plane must be made using a single sheet of $8.5'' \times 11''$ paper. No cuts can be made in the paper, and no tape, staples, glue, or paper clips can be used to hold the plane together or to change the plane's weight or balance. Also, each entry must qualify as being able to fly. For example, last year, a spitball and a dart were disqualified because they didn't really fly – even though it was possible to throw them so that they stayed in the air for a long time. Parachutes and helicopters also were disqualified because they didn't go anywhere. For each throw, the judges will measure the time spent in the air, the distance the plane lands from the starting point, the distance the plane lands from the target, and the angle the plane lands from the target.

Because all paper airplanes are minutely different, it is difficult to make decisions about which plane is the best. In order to make the competition as fair as possible, the judges are implementing two new processes for the contest. First, to minimize thrower advantages, the contest will have three neutral pilots to throw all planes in the contest. Second, the judges are designing a new scoring system to fairly judge the two winners.

"Some students are really getting into this contest – I've heard a couple who said they're bringing in-flight refreshments, crash helmets, and parachutes," said Frank. "It will be lots of fun and very interesting."

Questions to Get You Started
1. What are the categories for which the airplanes will be judged?

2. What types of measurements do you believe should be taken for each throw to fairly judge the contest?

3. How would you decide which airplane is the best floater?

4. How would you decide which airplane is the most accurate?

5. What are the judges doing differently this year than in years past? Why are they doing it?

Problem

In past competitions, the judges have had problems deciding how to select a winner for each award (Most Accurate and Best Floater). They don't know what to consider from each path to determine who wins each award. Some sample data from a practice competition and a description of how measurements were made have been included. To make decisions about things like being the best floater, the judges want to be as objective as possible. This is because there usually are only small differences among the best paper airplanes – and it seems unfair if different judges use different information or different formulas to calculate scores. So, this year, when the planes are flown, the judges want to use the same rules to calculate each score.

Write a brief 1- or 2-page letter to the judges of the paper airplane contest. Give them a rule or a formula which will allow them to use the kind of measurements that are given in Table 1 to decide which airplane is: (a) the most accurate flyer and (b) the best floater. Table 1 shows a sample of data that were collected from four planes last year. Three different pilots threw each of the four planes. This is because paper airplanes often fly differently when different pilots throw them. So, the judges want to "factor out" the effects due to pilots. They want the awards to be given to the best airplanes – regardless who flies them.

Use the data in Table 1 to show exactly how your rule or formula works – because the judges need to use your recommendation for planes that will be flown during the actual competition this year.

Note The paper airplanes were thrown in a large 40-ft by 40-ft area in the arena. Each paper plane was thrown by a pilot who was standing at the point that is marked with the letter S in the lower left-hand side of each graph in Fig. 1. So, this starting point is located at the point (0,0) on the graph. Similarly, the target is near the center of each graph, and it is marked with the letter X. So, the target is located at the point (25,25) on the graph (Fig. 2).

In Table 1, the angles are measured in degrees. Positive angles are measured in a counter-clockwise direction – starting from a line drawn from the lower left-hand corner of the graphs to the upper right-hand corner of the graphs (or starting from the point S and passing through the point X). Negative angles are measured in a clockwise direction starting from this same line.

(continued)

4 Five Principles for Supporting Design Activity

Table 1 Information about four paper airplanes flown by three different pilots

Plane	Flight	Pilot F				Pilot G				Pilot H			
		Distance from start	Time in flight	Distance to target	Angle from target	Distance from start	Time in flight	Distance to target	Angle from target	Distance from start	Time in flight	Distance to target	Angle from target
A	1	22.4	1.7	15.2	16	30.6	1.6	14.5	23	39	1.8	7.5	−10
	2	26.3	1.7	16.7	26	31.1	1.6	11.9	19	36.3	1.7	4.3	−6
	3	31.6	1.7	7.1	10	26.7	2.2	8.9	−4	35.9	2.2	9	−14
B	1	32.1	1.9	7.6	−11	35.9	1.9	14.3	−23	43.7	2.0	9.5	6
	2	42.2	2.0	9.2	−9	39	2.1	11.1	16	29	2.0	7.6	7
	3	27.2	2.1	10.2	−11	25.6	2.0	11.7	12	36.9	1.9	12.4	19
C	1	19.2	1.8	16.6	−8	42.9	2.0	9.8	9	35.1	1.6	2.8	4
	2	28.7	1.9	9.3	11	44.6	2.0	9.3	−1	37.2	2.2	2	−1
	3	23.6	2.1	17.3	−25	35.7	2.2	3.2	−5	42	2.1	9.8	10
D	1	28.1	1.5	8.9	9	37.2	2.1	20.2	−32	41.7	2.2	10.1	11
	2	31.6	1.6	14.8	−24	46.6	2.0	11.4	−2	48	1.9	14.1	−8
	3	39.3	2.3	9.1	12	34.7	1.8	22.2	−36	44.7	1.7	11.5	−9

(continued)

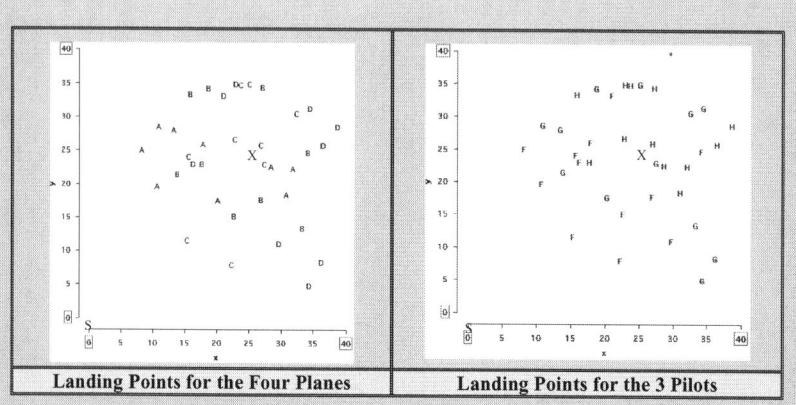

Fig. 1 Landing points for four paper airplanes thrown by three pilots

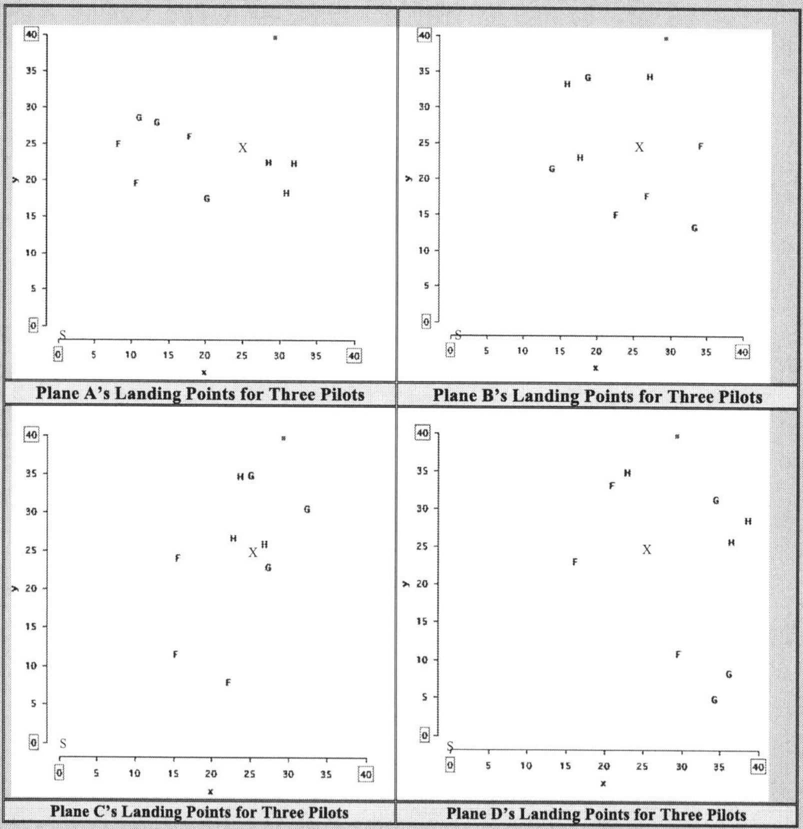

Fig. 2 Separate graph for four paper airplanes

Appendix C: The Blood Vessel Growth MEA

Activity development supported by NSF Award Number IIS-1125412, Illinois Institute of Technology, Chicago, IL, under the direction of Dr. Ali Cinar. Copyright held by Catherine Langman, Judith Zawojewski, Ali Cinar, and Hamidreza Mehdizadeh. Permission to photocopy granted for classroom use and research.

Growing Ears!

Does this sound like science fiction? It's not. In 1997, newspapers across the country introduced Americans to the groundbreaking work of Dr. Charles Vacanti and his brothers, Drs. Jay and Marty Vacanti. Charles had seeded cartilage cells on a biodegradable mold in the shape of an ear. The cells grew into cartilage (the tissue that holds the skeleton together) to cover the shape, and Vacanti implanted the whole structure under the skin of a mouse. The result—an (artificial, non-hearing) human ear growing on a mouse!

How did this happen? In the 1980s, scientists had already found ways to grow skin in a lab. Dr. Vacanti and his brothers asked, why not grow larger, more complex organs in the lab? Drs. Vacanti worked with a chemical engineer from MIT named Dr. Langer and together they hit on the idea of using biodegradable polymers—chemical structures that, when placed in the body, slowly degrade in the presence of water into harmless substances. The scientists realized that they could mold a polymer into a three-dimensional shape, seed it with living cells that would then grow into tissue, and implant the new tissue including the scaffold in a living animal. Over time, they reasoned, the polymer should dissolve like medical sutures, and the implanted tissue would attract blood vessels and grow. It worked, but there was, and still is, a limit to how big—and how complex—a hunk of tissue they can grow.

One challenge is that the tissue needs an ongoing, very close source of oxygen and nutrients to survive, as well as a system to take away waste products. Blood vessels deliver oxygen to cells and take away waste, so it is important for a healthy blood vessel network to rapidly form in the new tissue. This is why tissue engineers study angiogenesis—the formation of new blood vessels from existing blood vessels. Angiogenesis occurs when cells need oxygen and send out a chemical signal, which stimulates the nearest blood vessel to grow toward the distressed cell. Scientists are addressing other practical problems in tissue engineering. Which material makes the best scaffold? How fast will the scaffold degrade in the body once it is implanted? How should the scaffold be constructed to support blood vessel growth?

Lab-grown tissue has seen some success in medical applications. As early as 1998, Charles and Marty Vacanti used lab-grown bone to replace the thumb of a man who had lost his in an accident. The idea of whole organs grown in labs for transplant, using a patient's own cells, is now in the realm of possibility. Even closer on the horizon is the ability to repair tissue damaged by diseases like diabetes and atherosclerosis. Diabetes, in particular, is approaching an epidemic among the American population. This disease afflicts thousands of people, with complications like wounds that do not heal. The hope is that tissue grown externally on a scaffold, and derived from the individual's own living cells, can be implanted into the wounds to help the wounds heal.

Source: Foreman, J. (2003, December 30). Scientists at work—Joseph, Charles, Martin and Francis Vacanti; From old cars to cartilage, brothers like to tinker. *The New York Times*. Retrieved from http://www.nytimes.com/2003/12/30/health/scientists-work-joseph-charles-martin-francis-vacanti-old-cars-cartilage.html?pagewanted=all&src=pm (October 15, 2013).

Growing Ears! The Science Supplement

Recent research in growing healthy tissue has been motivated by the need to repair damaged and diseased tissue in human bodies, such as wounds that will not heal for many diabetics. A hoped-for treatment is to harvest a person's healthy cells and use those cells to seed a scaffold that will grow healthy tissue. The new tissue and its scaffold would be implanted into the wound area, enhancing the wound's ability to heal.

One major challenge of creating tissue is helping the new tissue get oxygen and remove waste products. In healthy tissue, this work is done by blood vessels, which transport oxygen-rich blood to the tissue and carry waste products away from the tissue. Therefore, it is important to study how new blood vessels form and connect with each other. Angiogenesis is the scientific term for the growth of new blood vessels from existing blood vessels.

In angiogenesis, an existing blood vessel is lined with endothelial cells, each of which can be stimulated to sprout a new blood vessel when it detects a chemical distress signal from a cell. When the cell does not have a source of oxygen nearby, the cell secretes a chemical called vascular endothelial growth factor (or VEGF). The VEGF stimulates endothelial cells to start growing.

Another important part of angiogenesis happens when two blood vessels cross pathways and fuse. The connection between those two blood vessels is called anastomosis. Looking under a microscope, scientists report that when tissue is healthy, they can see that the blood vessels grow throughout the tissue and have connections to each other.

Scientists and engineers are trying to find ways to use computer simulations to predict how blood vessels will grow. This is an image from a computer simulation of blood vessel growth.

4 Five Principles for Supporting Design Activity

The image represents a cross-section of a porous scaffold—the white regions between the circles are the polymer material of the scaffold and the gray circles are the cross-sections of the pores, in which blood vessels can grow. Two existing blood vessels are located at the top and at the bottom of the scaffold. The dotted line in the center represents a source of VEGF, which radiates chemical signals in all directions. The scientists use this computer simulation to estimate the amount of new blood vessel growth in different scaffolds week by week. Using this approach, they are able to compare the effect of different types of scaffolds and different pore sizes on healthy blood vessel growth.

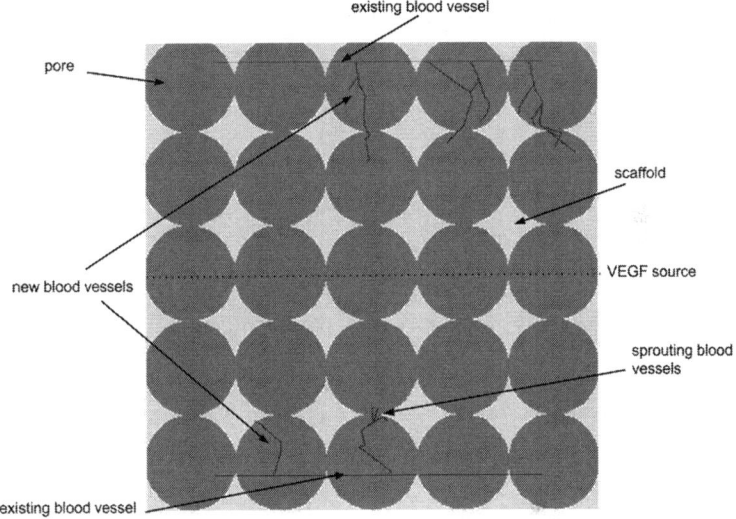

800 x 800 micrometer porous scaffold simulation results

Today, research in tissue engineering involves teams of bioengineers and scientists who set up experiments that involve implanting animals with scaffolds of different pore sizes or polymer material, gather data concerning the changes in blood vessel growth from week to week, and analyze the data to determine which condition produces the best quality of the blood vessel growth. Computer scientists also contribute to the work by creating simulations of blood vessel growth. They model how blood vessels grow in porous scaffolds. The data from the simulations help inform decisions for later laboratory experiments.

Source: Artel, A., Mehdizadeh, H., Chiu, Y. C., Brey, E. M., & Cinar, A. (2011). An agent-based model for the investigation of neovascularization within porous scaffolds. *TISSUE ENGINEERING: Part A, 17*(17 and 18), 2133–2141.

Questions on *Growing Ears!* Newspaper Article and Science Supplement

Part I: Blood Vessel Growth

1. Explain the role of each of the following terms during angiogenesis:
   (a) endothelial cells
   (b) vascular endothelial growth factor (abbreviated VEGF)
2. How would you describe or show healthy blood vessel growth to someone who has not seen this video?

Part II: Connections Between Blood Vessels

3. What happens when two distinctly-growing blood vessels connect to each other?
4. Why is it important for distinctly-growing blood vessels to connect?

The Full Circuit Game

This game is designed to help you simulate the growth of new blood vessels from existing blood vessels.

Objective The purpose of this two-player game is to form as many connections between distinct blood vessels as possible, to create a longer blood vessel than your opponent, and to create as many new blood vessels as possible in a given amount of time.

Setup Each player gets one die and one colored pencil or marker in a different color from the other player. Each game is played on one shared game board. Each player starts with his or her own existing blood vessel. Each existing blood vessel is lined with starting cells. The center of the board has invisible molecules of a chemical that stimulates blood vessel growth, which diffuses over the whole game board at the same rate.

Play the Game

1. To start, each player selects and circles a starting cell on his or her main blood vessel.
2. To play, each player rolls his or her own die. For each roll of the die, the player makes a move according to the chart below. Each player rolls one die a total of twenty times.
3. When a player's turn results in two blood vessels connecting, circle the point of intersection using the player's colored pencil.

Roll of 1 or 2	**Extend** one segment of short distance (leg of right triangle) (e.g., ↑)
Roll of 3 or 4	**Extend** one segment of long distance (one hypotenuse of right triangle)
Roll of 5	**Branch** using two segments of either kind (one leg and one hypotenuse of a right triangle or two hypotenuses)
Roll of 6	**Mark** a new starter endothelial cell on the existing blood vessel. Do not extend or branch in any direction on this turn

Legal moves	During any roll of the die, the player CAN move:
	   from their selected starting cell on their main blood vessel to a neighboring point that is forward or diagonal to the right or left;
	   from the tip of a new blood vessel to a neighboring point that is forward or diagonal to the right or left;
	   from the tip of a new blood vessel to connect to the opposing player's blood vessel at a dot on the game board or on the opposing player's main blood vessel, and the player who connected the blood vessels together draws a circle around the connection

Illegal moves	During any roll of the die, the player CANNOT move:
	   backwards (towards your own main blood vessel);
	   sideways (parallel to your main blood vessel);
	   circling blood vessel connections that do not occur on a game board dot

The game ends when each player has rolled their die 20 times.

How to Win the Game Assign 10 points to the player that has:

- the greatest number of circled connections between distinctly growing blood vessels
- the blood vessel with the greatest number of contiguously connected dots
- the greatest number of blood vessels that begin at different starting cells

The player with the most points wins the game. (Scoring sheet attached.)

Samples of types of moves

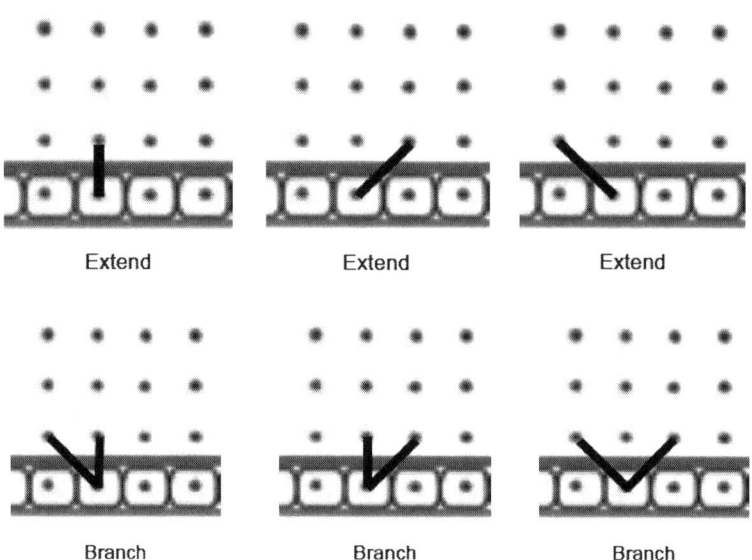

Scoring sheet

Category	Name Player 1:			Name Player 2:		
	Game 1	Game 2	Game 3	Game 1	Game 2	Game 3
10 points: player with the greatest number of circled connections						
10 points: player with the blood vessel that has the greatest number of contiguously connected dots						
10 points: player with the greatest number of blood vessels that begin at different starting cells						
GAME TOTAL						

Winner of Game 1:_____ with _____ points
Winner of Game 2:_____ with _____ points
Winner of Game 3:_____ with _____ points

4 Five Principles for Supporting Design Activity

After You've Played the Full Circuit Game at Least Three Times...

1. (a) What percent of the time can you expect to lose a turn? How do you know?
   (b) What percent of the time can you expect the move to be an extend move? How do you know?
2. (a) How is time represented in the game?
   (b) In the body, sometimes blood vessels have to grow around bone and other obstacles in the tissue. How is this represented in the game?
   (c) In the body, cells give off a chemical distress signal called vascular endothelial growth factor (VEGF, for short) and blood vessels respond to the chemical signal by growing towards the source of the signal. How is this represented in the game?
3. Name three qualities of blood vessel growth that could be used to indicate a healthy network of blood vessels.

Full circuit game board

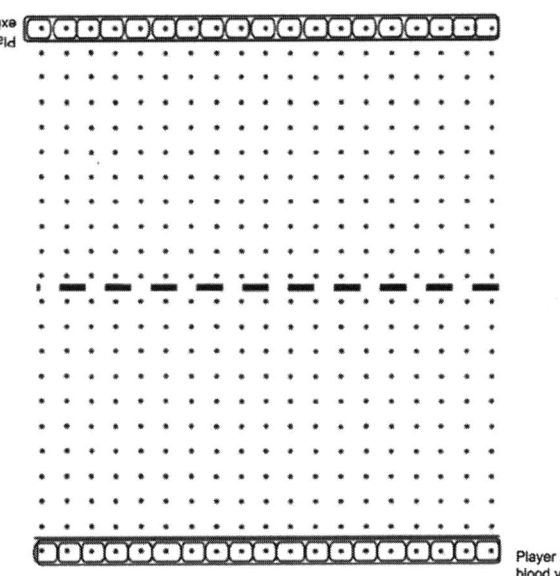

Model Creation Activity

Individually

1. Individually, read the attached memo.

INTEROFFICE MEMORANDUM

To: Engineering Team
From: Dr. Cinar, Director of the Center for Tissue Engineering
Date: November 24, 2013
RE: Method for estimating healthiness of a blood vessel network in porous scaffolds

To advance our research in tissue engineering, we are trying to determine a procedure for measuring the amount of blood vessel growth and healthiness of blood vessel networks in porous bioscaffolds.

We are asking you to help us by creating a mathematical procedure for *scoring* these samples based on the amount of blood vessel growth and the overall healthiness of the blood vessel network. The procedure will be used to score future samples, when we run lab experiments using other pore sizes, and different types of material for scaffolds.

To assist in your work, we are providing you with sample images of blood vessel growth in bioscaffolds from a computer simulation. These images are the fourth week of blood vessel growth for scaffolds with pore sizes 270 microns, 160 microns, 135 microns, and 45 microns. The images represent 800 x 800 micrometer regions of porous bioscaffolds. This size will be standard in all future experiments, as will the placement of the host blood vessels at the top and the bottom of the region, and the VEGF source in the center of the region.

> Deliverable: A memo that includes:
> - A written description of your mathematical procedure or series of steps that will be used to determine the amount of new blood vessel growth and score the overall healthiness of the blood vessel network for all future samples produced in our lab.
> - A demonstration of you procedure by applying it to one of the samples provided to you. Please attach the sample you use to the memo.

Thank you in advance for your attention to this matter.

Getting Started

As a team, answer the following questions:

1. Who is asking you for help?
2. What do they want you to produce?
3. Why does the client want a procedure or a series of steps, rather than a determination of which sample has the healthiest blood vessel growth?

After answering the above questions as a team and *before beginning* on your collaborative work, *spend 3–5 min in silence during which each team member independently lists features from the images that could be used to indicate a healthy blood vessel network.* Then, as a team, decide: which features you will include in your scoring procedure, how you will quantify each feature, and how you will synthesize those quantities into a single score.

Sample images of blood vessel growth in porous bioscaffolds

Blood vessel growth in 40-micrometer scaffold after 4 weeks

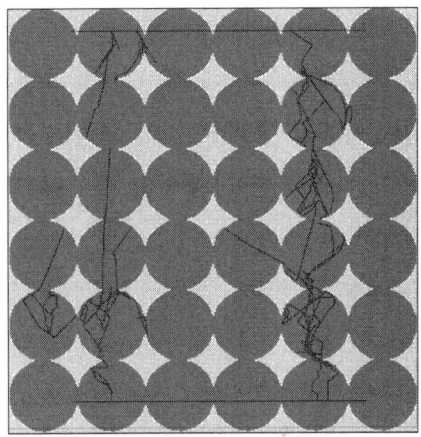

Blood vessel growth in 135-micrometer scaffold after 4 weeks

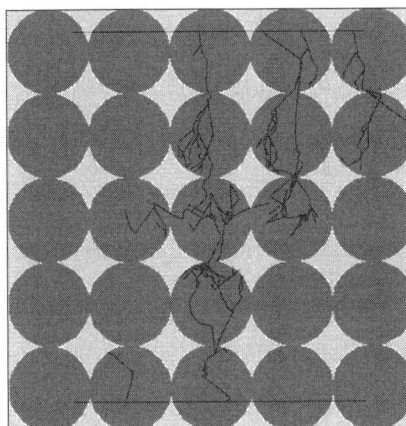

Blood vessel growth in 160-micrometer scaffold after 4 weeks

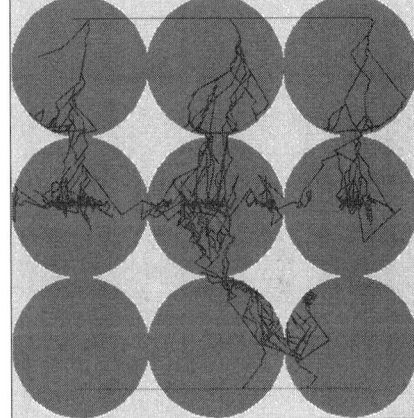

Blood vessel growth in 270-micrometer scaffold after 4 weeks

Appendix D

Template of Implementation Design Sheet

Class Profile:
Constraints specific to this group of students:

Implementation Principle	Identified needs for intended audience	Implementation support.	Implementation revisions for future similar audience in the same context
"Familiarity" Principle			
"Prerequisites" Principle			
"Accessing Complexity" Principle			
"End-In-View" Principle			
"Alternative Perspectives" Principle			
Before Implementation….	After Implementation….		
1. What do you think will go well? 2. What challenges do you think you will face?	1. What went well? 2. What challenges did you face? 3. Based on your observations and data collected during the implementation, what changes could you make to the MEA for the next revision of the MEA for a similar population and educational setting?		

Note The National Science Foundation supported the research reported and described in this chapter. The National Science Foundation (NSF) funds research and education in most fields of science and engineering. Grantees are wholly responsible for conducting their project activities and preparing the results for publication. Thus, the Foundation does not assume responsibility for such findings and their interpretation. Any opinions, findings, and conclusions or recommendations expressed in this material are those of the authors and do not necessarily reflect the views of the National Science Foundation

References

Chamberlin, M. (2004). Design principles for teacher investigations of student work. *Mathematics Teacher Education and Development, 6*, 52–62.

Chval, K. B., Wilson, L. D., Ziebarth, S. W., Heck, D. J., & Weiss, I. R. (2012). Introduction. In D. Heck, K. Chval, & I. Weiss (Eds.), *Approaches to studying the enacted mathematics curriculum* (A volume in research in mathematics education, pp. 1–18). Charlotte: Information Age.

Diefes-Dux, H. A., Hjalmarson, M. A., Miller, T. K., & Lesh, R. (2008). Model-eliciting activities for engineering education. In J. S. Zawojewski, H. A. Diefes-Dux, & K. J. Bowman (Eds.), *Models and modeling in engineering education: Designing experiences for all students* (pp. 17–35). Rotterdam: Sense.

Diefes-Dux, H. A., & Imbrie, P. K. (2008). Modeling activities in a first-year engineering course. In J. S. Zawojewski, H. A. Diefes-Dux, & K. J. Bowman (Eds.), *Models and modeling in engineering education: Designing experiences for all students* (pp. 55–92). Rotterdam: Sense.

Diefes-Dux, H. A., Zawojewski, J. S., Hjalmarson, M. A., & Cardella, M. E. (2012). A framework for analyzing feedback in a formative assessment system for mathematical modeling problems. *Journal of Engineering Education, 101*(2), 375–406.

English, L. D. (2006). Mathematical modeling in the primary school. *Educational Studies in Mathematics, 63*(3), 303–323.

English, L., & Lesh, R. (2003). Ends-in-view problems. In R. Lesh & H. Doerr (Eds.), *Beyond constructivism: A models and modeling perspective on problem solving, learning and instruction in mathematics and science education* (pp. 297–316). Mahwah: Lawrence Erlbaum Associates.

English, L., & Watters, J. (2004). Mathematical modeling in the early school years. *Mathematics Education Research Journal, 16*, 59–80.

Hjalmarson, M. A., & Lesh, R. A. (2008). Engineering and design research: Intersections for education research and design. In A. E. Kelly, R. Lesh, & J. Baek (Eds.), *Handbook of design research methods in education: Innovations in science, technology, engineering, and mathematics learning and teaching* (pp. 96–110). New York: Routledge.

Lesh, R., & Doerr, H. M. (2003). Foundations of a models and modeling perspective on mathematics: Teaching, learning, and problem solving. In R. A. Lesh & H. M. Doerr (Eds.), *Beyond constructivism: Model and modeling perspective on mathematics, problem solving, learning and teaching* (pp. 3–34). Mahwah: Lawrence Erlbaum Associates.

Lesh, R., Hoover, M., Hole, B., Kelly, A., & Post, T. (2000). Principles for developing thought-revealing activities for students and teachers. In A. E. Kelly & R. A. Lesh (Eds.), *Handbook of research design in mathematics and science education* (pp. 591–646). Mahwah: Lawrence Erlbaum Associates.

Magiera, M. T. (2013, February). Model-eliciting activities: A home run. *Mathematics Teaching in the Middle School, 18*(6), 348–354.

National Research Council of the National Academies. (2011). *A framework for K-12 science education: Practices, crosscutting concepts, and core ideas*. Washington, DC: The National Academies Press.

Schoenfeld, A. H. (1992). Learning to think mathematically. Problem solving, metacognition, and sense making in mathematics. In D. Grouws (Ed.), *Handbook of research on mathematics teaching and learning* (pp. 334–370). New York: McMillan.

Zawojewski, J. S., Hjalmarson, M. S., & Diefes-Dux, H. A. (2013). Student team solutions to an open-ended mathematical modeling problem: Gaining insights for educational improvement. *Journal of Engineering Education, 102*(1), 179–216.

Zawojewski, J. S., Diefes-Dux, H. A., & Bowman, K. J. (Eds.). (2008). *Models and modeling in engineering education: Designing experiences for all students*. Rotterdam: Sense Publishers.

Ziebarth, S. W., Hart, E. W., Marcus, R., Ritsema, B., Schoen, H. L., & Walker, R. (2009). High school teachers as negotiators between curriculum intentions and enactment: The dynamics of mathematics curriculum development. In J. T. Remillard, B. A. Herbel-Eisenmann, & G. M. Lloyd (Eds.), *Mathematics teachers at work: Connecting curriculum materials and instruction* (pp. 171–189). New York: Taylor and Francis.

Chapter 5
Studio STEM: A Model to Enhance Integrative STEM Literacy Through Engineering Design

Michael A. Evans, Christine Schnittka, Brett D. Jones, and Carol B. Brandt

Studio STEM: A Model to Enhance Integrative STEM Literacy through Engineering Design

Interest in science, technology, engineering and mathematics (STEM) during the middle school years is a predictor of future involvement in those fields (Maltese and Tai 2010). Science-rich out-of-school programs have the potential to sustain interest during this formative period. Informal learning programs offer youth opportunities to engage in meaningful hands-on, minds-on science, often resulting in conceptual change and more positive attitudes toward science (Schnittka and Bell 2011; Gerber et al. 2001). Learning activities in out-of-school programs are often designed to be collaborative in nature, lending to both social and cognitive development while

---

This material is based upon work supported by the National Science Foundation (NSF) under Grant No. DRL 1029756 and 1239959, the Institute for Society, Culture and Environment (ISCE), and the Institute for Creativity, Arts, and Technology (ICAT) at Virginia Tech. Any opinions, findings, conclusions or recommendations expressed in this material are those of the author(s) and do not necessarily reflect the views of NSF, ISCE, or ICAT. Studio STEM (http://studiostem.org) includes the authors and a talented team of graduate and undergraduate research assistants.

M.A. Evans (✉)
Department of Teacher Education and Learning Sciences, North Carolina State University, 1890 Main Campus Drive, 27606 Raleigh, NC, USA
e-mail: michael.a.evans@ncsu.edu

C. Schnittka
Department of Curriculum and Teaching, Auburn University, Auburn, AL, USA

B.D. Jones
School of Education, Virginia Tech, Blacksburg, VA, USA

C.B. Brandt
Department of Teaching and Learning, Temple University, Philadelphia, PA, USA

© Springer International Publishing Switzerland 2016
L.A. Annetta, J. Minogue (eds.), *Connecting Science and Engineering Education Practices in Meaningful Ways*, Contemporary Trends and Issues in Science Education 44, DOI 10.1007/978-3-319-16399-4_5

providing a safe space to engage with peers and adults other than their teachers (Durlak et al. 2010). Although out-of-school learning experiences can promote interest in STEM, there remains much to be learned about how youth engage in out-of-school programs and the role that more knowledgeable others and technology play in the process (Evans 2009). A greater contribution could be made if out-school programs focused on engineering design, an approach increasingly promoted yet insufficiently investigated.

The value of introducing STEM education programs in middle school curricula for youth has become an increasingly important issue (Katehi et al. 2009). There is an ongoing national US agenda to reform science and mathematics education and to increase youth interest in STEM. Leaders in areas of government, business, and educational policy have expressed a need for this reform (National Academy of Engineering and National Research Council 2012). Multiple problems related to STEM education have been identified, including: US students score lower in standardized mathematics and science tests compared to students in many other countries, an insufficient number of students pursuing STEM careers, and a lack of diversity within STEM fields (Moore and Richards 2012).

Attracting youth to STEM fields is necessary long before they apply to college because many youth formulate ideas about possible careers by adolescence (Riegle-Crumb et al. 2011). By high school, many students' opinions about science have been formed and remain somewhat fixed (Archer et al. 2010). Sadler et al. (2012) found that an important predictor of STEM career interest at the end of high school was youth's interest in STEM at the beginning of high school. In another study, many advanced science students in high school reported that their interest in science developed in middle school (Maltese and Tai 2010). Further, enrolling in science and mathematics courses in high school has been shown to predict the pursuit of a science or mathematics college major in college (Trusty 2002). These findings highlight the importance of getting students interested in STEM early in their education and have served as a driving force for targeting this age group for Studio STEM.

This chapter describes Studio STEM, a engineering design-based out-of-school program with an interdisciplinary curriculum that utilizes a technology-rich context. The goal of Studio STEM is to assist youth in learning about energy conservation while motivating girls and boys to one day pursue careers in STEM. First, we describe the role of integrating STEM into out-of-school curricula and the ways in which research on adolescent youth in the middle grades (ages 11–15) has influenced the development of Studio STEM. Next, we describe the Studio STEM model and the theoretical underpinnings that guided our program development. Having completed two years of our program funded by the National Science Foundation (DRL 1029756), we share preliminary findings of successful implementation of Studio STEM in rural communities in southwestern Virginia. Finally, we discuss the ways that Studio STEM has been translated across other contexts and the implications for the Studio STEM model as a way to re-conceptualize STEM education inside *and* out of schools.

The Challenges of Adopting Integrative STEM Curricula for Middle School-Aged Youth

By adolescence, many youth have begun to formulate ideas about future career possibilities, likely making educational choices that correspond with these ideas about their futures (Reigl-Crumb et al. 2010). Decisions about career futures are based in part on the values that students place on the topics (Does it seem interesting, important, or rewarding?) (Eccles 2005; Osborne and Jones 2011) and the degree to which students believe that they can be successful in activities related to that topic (Eccles 2005). These values and expectancies can be influenced by teacher feedback and encouragement (Chouinard et al. 2007), interactions with peers (Fraser and Kahle 2007), and experiences outside of the school setting. Studio STEM was developed around the notion of providing designed opportunities for positively influencing values and expectancies to foster identification with science and engineering (Schnittka et al. 2012).

During this period in which career aspirations are formed, interest in science, engineering, and mathematics often wanes (Kanter and Konstantopoulos 2010). In fact, Maltese and Tai's (2010) work suggests that students who report a strong interest in science by grade eight are significantly more likely to go on to a science career than those students who do not report similar strong interests. As interest decreases, so does enrollment in high school science and mathematics classes. Those course decisions in high school often limit access to STEM majors in college (Tai et al. 2006). Success in middle school and high school mathematics may, in particular, act as a filter that limits access to other STEM fields (Evans and Biedler 2012; Shapka et al. 2006).

For youth from rural, low-income communities, positive experiences with science outside the classroom are often limited. As an example, parents might not be able to offer advice about career options, and youth might draw their understanding of science careers from television shows, social media, or textbooks. Consequently, out-of-school programs have the potential to narrow this gap by offering middle school students opportunities to engage in STEM curriculum in ways that extend, or provide different types of experiences than, classroom curricula. These program choices show promise in creating experiences in which youth can make personal connections to scientific language, ideas, and methods (Barton and Tan 2010; Rahm 2008). Studies of out-of-school science and engineering programs suggest that hands-on, inquiry-driven experiences potentially increase enthusiasm about science (Rahm et al. 2005), expand youth's understanding of career options (Markowitz 2004), and help youth to understand the role that science plays in their everyday lives (Barton et al. 2008). These experiences are enhanced and extended when more knowledgeable peers serve as mentors, and youth have unfettered access to social network forums and mobile technologies to deepen meaningful, academically oriented discourse (Evans et al. 2014a, d).

Cognitive, Social, and Affective Justifications for Integrative STEM

Integrative STEM is by nature fundamentally associated with context-bound, relevant problems connected to the everyday life of youth. When developing STEM programming, we take into consideration problems to which youth can relate. By focusing on open-ended real life problems, Studio STEM examines energy sustainability, a pressing issue in the coal country of southern Appalachia. Similarly, Diefes-Dux et al. (2004) have argued that for students to benefit from STEM design-based instruction they need more experience in working with real-life problems through Model Eliciting Activities (MEAs). This approach in engineering provides students with real-world, context driven problems and supports the development of higher-thinking skills. MEAs use open-ended problem solving that foster conceptual development through creative design, model testing, and re-design, which consequently extends the learner's thinking. Moore et al. (2013) also describe how modeling through MEAs is a social practice that requires students to externalize their thinking and to adequately communicate their emerging ideas about their design. Thus, we argue that integrative STEM should take into account the affective domains – the ways that youth relate to a problem that is meaningful in their lives, as well as how youth work together to externalize and communicate their emerging ideas and conceptual knowledge with others (Deater-Deckard et al. 2013).

Integrative STEM education is based on the idea that real-world issues require multiple perspectives, skills, and knowledge to be productively addressed (Wang et al. 2011). Integrative STEM can have positive effects on youth achievement, especially at the K-12 level. The largest effects are seen when all four components of STEM are integrated, though the relative weight of those components could vary depending on context and intent (Becker and Park 2011). Although there is still some debate about what defines true STEM integration, Morrison (2006) emphasizes a combination of problem solving, innovation, invention, and logical thinking. A dominant theme in the literature is that integrative STEM involves problem solving and inquiry (Wang et al. 2011), two key aspects of all curricula developed for the Studio STEM project.

Using a social constructivist approach, Studio STEM utilizes in-service teachers (site leaders), and engineering and science undergraduates (who act as facilitators), to work with Studio STEM youth using a curriculum with real-life problems in energy sustainability. Our program emphasizes: (a) a content-rich curriculum that links students to their environment; (b) support and scaffolded discussions with mentors; and (c) an online network that supports the creation and maintenance of relationships. The informal character of this program allows students the freedom to explore and self-identify with topics.

Studio STEM is designed to introduce rural, at-risk youth from low socioeconomic level communities to topics in science and engineering through engineering design-based activities facilitated by undergraduate mentors from related disciplines (Evans et al. 2014a, d; Schnittka et al. 2012). Youth are introduced to background information about an energy issue and its effect on an animal or ecosystem. Information is presented through information and communication technologies (ICTs) with video clips, audio, and images that can be presented in lecture or, in more recent iterations, via a webquest format that allows for self-directed inquiry on a need-to-know basis. Youth are encouraged to contemplate the impact that humans and human-made technologies might have on the planet and ecological subsystems comprised of humans and other living creatures. This approach is designed to relate the academic material more strongly to youth on a personal level, which has demonstrated to influence their engagement with the project (Evans et al. 2014a). Science concepts are presented in the form of hands-on experiments and demonstrations. Youth are challenged to design and construct an artifact of some sort, depending on the curriculum. For example, in the case of the *Save the Penguins* curriculum a dwelling is constructed from materials that include wood, cotton, and Mylar. Groups are given a limited "budget" that participants may use to purchase such materials to construct these artifacts. Through an iterative design process, attentive youth correct errors to improve earlier prototypes. Design is the iterative selection and arrangement of elements to form a whole by which individuals create artifacts, systems, and tools intended to solve a range of problems.

Teaching STEM content using engineering design is a potentially powerful instructional method appropriate for out-of-school, informal settings. When youth identify a problem, consider options and constraints, and then plan, model, and test multiple iterations, they are engaged in higher-order thinking skills. Design-based learning engages youth as critical thinkers and problem solvers and aides in productively and purposely using science and technology as means to greater ends (Honey and Kanter 2013). Added to this curricular mix is the scaffolding provided by site leaders (teachers and experts recruited from the base school or local community) and facilitators (STEM undergraduates from a nearby university). The role of site leaders is to serve as "conduits" for the content and pedagogy developed by STEM educators, educational psychologists, and learning scientists who lead Studio STEM. The role of facilitators is to probe and guide youth without lecturing or merely providing answers (Evans et al. 2013). A social networking forum (SNF), Edmodo, provides a platform where individuals can communicate with teams on-site or elsewhere to ask self-generated questions, share design prototypes, and serve as emergent *experts* of topics or tools associated with a particular curriculum. The design of Studio STEM has benefitted from prior investigations into knowledge building communities and intentional learning environments (Evans et al. 2014d).

In Studio STEM, meaningful activities, social practices, discussion, and collaborative meaning making are inextricably linked and are fundamental to the learning of science and engineering, on-site and online. As site leaders, facilitators, and youth engage in attempts to identify and resolve design problems within the space of the studio, they develop social norms, participate in discussions, and use technological tools while making sense of the design problems that are presented to them (Evans et al. 2013). The goal of Studio STEM is to encourage a community of learners in science and engineering who use technological tools and social media in the design process. Technological tools are also important as youth learn to effectively communicate their emerging design ideas and conceptual understanding as they begin to self-identify with science and engineering. We contend that these kinds of out-of-school experiences assist youth in seeing themselves as capable of doing science and engineering, and thus, more likely to pursue STEM careers as they progress in formal schooling (Schnittka et al. 2012).

Scientific, Technological, and Engineering Literacy in Studio STEM

Students' ideas about energy begin at a young age and are transformed through experience and education. The term *energy* is used informally in everyday language so often, that the scientific meaning is often obscured. Youth may think about having enough energy to get through a school day or think that energy is a fluid that flows from one place to another to make things work, like juice or electricity or gasoline. They may think energy sources are unlimited, and not even think about what happens for their lights to work. Without a basic understanding of energy, a more complex understanding of energy transformations, energy security, and energy sustainability is untenable. Studio STEM includes explicit interventions that are designed to target misconceptions that youth might have about the science of energy, which can help them become more literate in science, technology, and engineering along the way. Though one aim of Studio STEM is to encourage more youth into the STEM workforce pipeline, another aim is to improve STEM literacy in general because it is a more broadly achievable goal for many youth. Youth from poor, rural communities that do not have a tradition for movement to postsecondary education or professional degrees and occupations cannot be expected to change within the scope of this project. Nevertheless, there is a higher probability that these youth, their parents, and the surrounding communities will be open to becoming more literate about STEM that could have immediate impact in two-year college settings and satisfy local employment needs.

Energy Sustainability and Concepts of Energy Conservation Energy literacy encompasses understanding what energy is and where energy comes from. Energy literacy is vital because it leads to informed decisions about energy use at home, consumer choices, and to national and international energy policies. "Current national and global issues such as the fossil fuel supply and climate change highlight the need for energy education" (ED 2012, p. 4). Energy literacy takes three forms, and involves cognitive constructs (knowledge about the science and technology), affective constructs (attitudes), and behavioral constructs; all three help citizens make informed decisions about energy use (Dewaters and Powers 2011).

Recently STEM educators have issued the call for more curricula and teaching that emphasizes a critical "place consciousness" in which youth's attachments to location are examined in terms of economic, environmental, and cultural sustainability (Aikenhead et al. 2006; Gruenewald 2003). In what ways can youth participate and imagine themselves as being connected to issues surrounding energy sustainability that seem remote and distant from their own experiences? Moreover, how can out-of-school programs encourage youth to link local practices to a global perspective of environmental sustainability? Yet, the dilemma of how to connect rural youth who have rarely ventured far from their local context to consider global environmental concerns has typically gone unaddressed in educational research. Although science educators advocate an approach that emphasizes placed-base education (Sobel 2004), through Studio STEM we offer one demonstrable approach to expand the awareness of rural youth to understand the global environmental issues far beyond their immediate experience, leveraging social media and mobile technologies as one example toward of this goal.

"Save the Animals" Theme The *Save the Animals* curriculum used in Studio STEM was designed to encourage youth to recognize how their energy behaviors at home might affect animals all over the world. Most youth do not realize that electricity is primarily produced by burning coal and that transportation primarily relies on fossil fuels, a matter of deep importance in rural Appalachia where Studio STEM is currently offered. The fossil fuel energy used in power plants and transportation has been linked to increased levels of carbon dioxide in the atmosphere, which, in turn, is having widespread effects on life on Earth (Gross 2005; Jenouvrier et al. 2009). When engineers design better building materials to conserve energy, and when builders use these materials, it has the potential for positive impacts on the environment. When engineers consider alternative sources of energy for transportation or electrification, the environment benefits. With a finite supply of fossil fuel energy, energy security represents the ability we have as a society to be more self-reliant on energy sources that are clean and readily available, such as sunlight, wind, and things that naturally fall; such as rain and water. This is the problem presented to youth: given requisite knowledge of science and engineering, how we can think about alternative sources of energy at home, and conserve energy to reduce the impact of CO_2 emissions on the environment? The theme of *saving animals* was chosen after the first curriculum module was used in Studio STEM, Save the Penguins. Afterwards, students reported that they wanted to save more animals, so subsequent curriculum modules were modeled on that theme, including saving snails and slimy creatures, seabirds, and the black-footed ferret. We have found that affection for animals, and empathy for caring for them, brought out an aspect of human emotion that motivated the youth to learn the concepts and complete required designs.

The Studio STEM Model

Theoretical and Research-Based Foundations

Studio STEM is grounded on the premise that learning is the result of social production and communicative acts. Learning requires youth to engage in dialogue and involves being assimilated into a new discourse community that includes new conceptual objects, signs, terms, technology, and phrases for which the learner has no, or little previous experience. As youth and their instructors undertake STEM inquiry in the design studio, their discussions introduce youth to the implicit and explicit rules of science practice and engineering design that are accepted by the wider STEM community. These social practices – for instance, conducting a fair test – involve fostering new mental habits and ways of thinking that are connected to the learner's sense of self, motivation, and identity as a participant in the learning community. Similarly, science discourse communities are found beyond the classroom walls: afterschool science clubs, science centers and museums, or interactions at home conducting a hobby are contexts where learners become assimilated into science discourse communities (Brandt et al. 2011). Consequently, Studio STEM draws upon theory and research from science education, technology, and educational psychology that offer socially situated, positive and motivating activities for learning.

Scientific Inquiry and Conceptual Change The curriculum designed for Studio STEM is founded not only on the principles of engineering design, but also on the principles of scientific inquiry and conceptual change in science. Scientific inquiry involves answering a scientific question through data analysis (Bell et al. 2005). Throughout each curriculum module, youth are engaged in inquiry activities: they measure voltage to determine which solar panel to use, they mass cubes to see which motor pulls the strongest, they measure temperature to see which insulator blocks heat transfer the best, and they measure time to see how their gear train slows down the descent of a water bottle. The data they collect is analyzed to answer scientific questions that inform the engineering design. Called "predictive analysis" by Merrill et al. (2009) because the scientific results of inquiry questions predict the success of a design, it is often the first component to be left out of the design process in K12 curriculum (Gattie and Wicklein 2007; Katehi et al. 2009).

Although scientific inquiry has been linked to gains in science understandings (Anderson 2002), the ways in which inquiry is implemented are crucial to its effectiveness. The ultimate goal, other than having students learn and practice process skills, is to promote deep science learning through conceptual change. Conceptual change is the process by which students' naïve or preconceived notions about how the world operates are identified, targeted, and re-formed. Conceptual change theory has been an active area of discussion in the science education literature for decades (Driver et al. 1985; Osborne and Freyberg 1985; Driver et al. 1994; Duit and Treagust 2003). Before a person's naïve conceptions are modified to be more in line with current scientific thought, the person must consciously become dissatisfied with their current ability to explain or act. Once this awareness of dissatisfaction is present, the person is ready to accept an alternative, scientifically rigorous explanation for natural phenomena. The new explanation must make sense, and fit within the network of scientific ideas already accepted by the person (Strike and Posner 1982). One successful method for identifying and targeting youth's naïve conceptions is to present discrepant events: events that were predicted one way, but turned out another way. For example, youth may believe that aluminum foil wrapped around a cold can of soda helps keep the can cold or believe that aluminum foil wrapped around a hot baked potato keeps it hot. When a "more knowledgeable other" presents data that conflict with preconceived ideas, the cognitive dissonance can lead to a desire to understand and a willingness to discard former ideas (Hewson and Hewson 1984; Piaget 1980). In the curriculum used in Studio STEM, inquiry activities and discrepant events are embedded to provide the conditions necessary for conceptual change. However, the model for Studio STEM provides the other vital piece thought necessary for conceptual change- motivation (Dreyfus et al. 1990; Lee and Anderson 1993; Pintrich et al. 1993). The atmosphere of the studio, the support of the facilitators and site leaders, and the social collaboration with peers provide the motivation to accept new scientific concepts. These new concepts are then used to truly use predictive analysis and design more robust artifacts (Schnittka and Bell 2011).

Technological Literacy and New Media Youth are increasingly accessing the Internet (Madden et al. 2013) and social networking forums (SNFs) in their personal lives, making it an attractive area of research for the purposes of education and specifically, integrative STEM. The incorporation of SNFs into the Studio STEM curriculum previously examined, is one way in which we have attempted to integrate the technology part of the STEM equation more effectively. The platform, Edmodo, which serves much like an age-appropriate version of Facebook for middle-school youth, allows participants a forum to explore the social and cognitive space of the curriculum and studio, seeking assistance, sharing ideas and iterations, and offering solutions with peers in the service of collective effort. Most recent iterations of Studio STEM have incorporated mobile technology by giving students access to iPads, providing quicker access to SNFs and other online resources, and providing the ability to photograph and video record designs and processes.

The reasons that youth access SNFs are diverse. However, Ito et al. (2010) have described three different methods of engagement related to SNFs and other forms of social media and digital technologies. Collectively, these genres of participation are referred to as the "hanging out, messing around, geeking out," or HOMAGO, model. *Hanging out* refers to engagement with technology for the purposes of social interaction and casual exchange of information. *Messing around* refers to engagement for the purposes of experimentation and investigation of topics that youth find interesting. Finally, *geeking out* refers to engagement for the purposes of discussing topics of interest in greater depth. It is at this point that youth may contribute as cyber "experts" in their topic area (Ito et al. 2010). Previous work investigating

Studio STEM utilizes the HOMAGO model in coding and characterizing the discourse of middle school youth through a social networking website (Evans et al. 2014d). Joseph et al. (2010) also utilized the HOMAGO model in examining the inclusion of SNFs in a library based learning program. The opportunity to interact with SNFs encouraged students to move from the *hanging out* form of engagement to the *messing around* form of engagement requiring deeper commitment to learning the material.

While the HOMAGO framework is still relevant and applicable to the examination of how youth interact through and with SNFs, the "connected learning" framework is perhaps a more recently evolved and appropriate framework for current research for out-of-school STEM learning (Ito et al. 2013). Connected learning describes the collaborative nature of learning in digital environments. Social interaction paired with interest can result in the increased opportunity for youth to engage in supported STEM learning. Digital media are promoted as a way to connect the learning environments of school, home, and the community in order to create more meaningful insight and connections (Ito et al. 2013). This process is facilitated by the inherent interest that youth appear to have for exploring and engaging with SNFs. The collaborative nature of the connected learning framework is consistent with the problem-based learning focus of Studio STEM curricula in which knowledge is shared by a group and applied towards reaching a defined goal or solving a defined problem. Environments designed for the purposes of problem-based learning lend well to the integration of technologies including SNFs and tablet computers. Collaboration among youth can be important for learning in a physical learning environment such as the design studio as well as virtual environment that allows for enhancement and expansion of these experiences.

Through interaction with SNFs, youth may also establish a sense of identity, which is important overall to engaging youth with STEM and promoting STEM literacy. Parker et al. conducted a study in which middle school youth were encouraged to critically analyze the messages found in advertisements for food. During the course of the analyses, youth were found to express attitudes consistently and frequently indicating stability and identification with certain healthy eating concepts. The identification of youth by username and avatar were also contributors to the establishment of identity. Specifically, the number of comments was logged for each user resulting in a sort of status hierarchy for those youth who interacted frequently with the system. Our research efforts have produced similar results that encourage continued use and refinement of the social media and digital tool features of Studio STEM (Evans et al. 2014d).

Motivation in and Identification with STEM Two key purposes of the Studio STEM model are: (a) to motivate students to participate in STEM activities and, (b) to provide foundational experiences that can lead to longer-term identification with STEM. To motivate students to participate in STEM activities, both the curriculum and teaching approach are consistent with current motivation research and theory. To explain how studio STEM activities are motivating, it is useful to compare the studio STEM design principles with the MUSIC^{SM} Model of Motivation (Jones 2009, 2015) because the MUSIC model summarizes five key research-based principles that instructors can use to increase student motivation. The MUSIC model states that students are more motivated when they perceive that: (1) they are *eMpowered*, (2) the content is *Useful*, (3) they can be *Successful*, (4) they are *Interested*, and (5) they feel *Cared* for by others in the learning environment (MUSIC is an acronym based on these five principles; see Jones 2009, 2015 for further explanation).

The *empowerment* component of the MUSIC model refers to the amount of perceived control and decision making that students have over their learning. Students are more motivated when they feel empowered and have control over their learning environment. The curricula and teaching approaches used in studio STEM are consistent with the Next Generation Science Standards (NGSS Lead States 2013) that emphasize student-centered learning environments where active inquiry is a primary vehicle for learning. When students are active learners, they are empowered because they are making choices and decisions related to their learning. In the studio STEM model, instructors and facilitators serve as guides to support youth in their decision-making processes. As they engage in solving the problems, learning is self-directed to a significant degree and students learn skills and facts as they progress through the process of solving the problems (Boud and Feletti 1997). The informal nature of the studio STEM model can also contribute to students' feelings of empowerment. In the informal learning environment of studio STEM, students are not in a formal schooling environment where they are provided with grades and subjected to high-stakes tests that can lead students to feel external pressures and reduced autonomy (Jones et al. 2003). In contrast, they are able to have more choices and feel less constrained by external pressures.

The *usefulness* component of the MUSIC model involves the extent to which students believe that the coursework (e.g., assignments, activities, readings) is useful to their short- or long-term goals. The studio STEM curriculum presents problems that are relevant (i.e., valuable, important, and useful) in today's world; and thus, students should perceive the curriculum to be useful to their own goals. Through the studio STEM curriculum, students learn science and engineering concepts that may be useful to their current schoolwork and/or their future career plans. Moreover, the use of real-life problems allows students who have been historically underrepresented in STEM to apply learning to their lived experience. Basu and Barton (2007) have argued that students from low-SES communities develop a sustained interest in science when learning experiences are connected with their own futures and when students can envision their role in solving real-life problems. These authors and others (e.g., Fusco and Barton 2001; Seiler et al. 2001) note that a sense of one's ability to act on real-life problems and their perceptions being useful in the problem-solving process were centrally connected to the ways that students began to see a future in STEM careers. A report on informal science learning by the National Academy of Science (Bell et al. 2009) concluded that learners thrive in informal settings where their needs and experiences are valued and where adult mentors and facilitators play a critical role in supporting science learning. In a sense, Studio STEM provides a "practice field" (Barab and Duffy 2000; Senge 1994) where learners can engage in activities that simulate the ones they would find in the real world.

The *success* component of the MUSIC model is based on the idea that students need to believe that they can succeed if they put forth the appropriate effort. Studio STEM is designed to support student success in a variety of ways. The curriculum was designed specifically for middle school students by including activities that could be reasonably completed by this population. This is important because students feel successful when they complete challenging activities. To help ensure that students feel successful, the Studio STEM model allows students to work together and with facilitators who can guide their experiences and help them navigate challenges as they solve problems. Further, the engineering design model used in Studio STEM allows students to try things, test them, redesign them, and try them again. Thus, this process serves as a safe place for students to explore and try new ideas. Being unsuccessful is okay because it is part of the design process. As an example, in the *Save the Penguins* curriculum, students solve the problem of how to keep penguins from becoming warm by designing a home for them that reduces heat transfer. After the youth complete their initial design of the house and test it under the heat lamps, the youth discuss which design features worked well and poorly and they are provided with feedback from peers, facilitators, and the instructor to use in the redesign of their home. After the redesign, they test it again, share their results, reflect on what they learned, and document their findings (in text and image) in online blog.

The *interest* component of the MUSIC model includes situational interest, which refers to the immediate, short-term enjoyment of or interest in instructional activities. To interest students, Studio STEM uses a curriculum that involves solving problems related to saving animals. Results from research studies (e.g., Baram-Tsabari and Yarden 2009; Schnittka et al. 2012) indicate that many middle school students are interested in topics related to animals and environmental issues affecting animals and humans. An important component of interest is emotional engagement (Hidi and Renninger 2006) and the Studio STEM model is designed to stimulate emotional engagement by eliciting concern from students about wanting to save and protect the animals. Further, the curriculum is designed to elicit emotions such as excitement and empathy through the types of activities presented. Part of this excitement can be attributed to the novelty of the activities used in Studio STEM. Many of these activities and technologies are ones that students will not have encountered previously. For example, students use information and communication technologies (ICTs) similar to popular social media sites.

The *caring* component of the MUSIC model includes the degree to which students believe that the instructor cares about their well-being and whether they succeed in the coursework. Student interaction is highlighted in the Studio STEM model through the groups in which students work and the design studio that is used as a model for introduction. Ideally, these pedagogical elements allow for positive student interactions where students can help one another in a manner that allows youth to feel cared for by not only the instructor and facilitators, but also by their peers. In addition, the ICTs allow youth to communicate with their peers and more knowledgeable others on a regular basis. Such communications should also lead students to feel cared for and supported in their learning. Finally, the facilitators are trained to work with the youth in a manner that fosters academic caring.

In addition to motivating students to participate in STEM activities, Studio STEM is designed to provide foundational experiences that can lead to identification with STEM fields. Identification with a domain, such as science or engineering, has been defined as the degree to which an individual values the domain as an important part of the self (Osborne and Jones 2011). Being identified with a domain is important because higher domain identification has been linked to outcomes such as higher GPAs (Osborne and Walker 2006), amount of deep cognitive processing of course material and self-regulation (Osborne and Rausch 2001), grade point average and academic honors (Osborne 1997), and behavioral referrals and absenteeism (Osborne and Rausch 2001; Osborne and Walker 2006). Conversely, a lack of academic identification has been shown to be related to a variety of negative outcomes, such as problem behavior (Gold and Mann 1984), lower GPAs (Osborne 1997), school absenteeism (Reid 1981), and dropping out (Elliot and Voss 1974; Osborne 1997). The process through which students become identified in STEM fields involves many factors and includes those that are part of the formal educational curriculum and those that are outside of formal schooling contexts. For example, Osborne and Jones (2011) discussed several factors that can influence a student's identification with an academic domain, including group membership (race, gender, social class); family, peers, and community environment; school climate; and formal and informal educational experiences.

We contend that Studio STEM can affect youths' identification with STEM fields in a manner similar to that proposed by Osborne and Jones (2011) and presented in Fig. 5.1 as a simplified version of that model. Researchers have documented that the five components of the MUSIC Model of Motivation not only motivate and engage students in activities, but also can lead to increased identification with a domain (see Jones et al. 2014, for evidence in engineering; see Jones et al. 2015a, for evidence in science; see Osborne and Jones 2011, for a general discussion). That is, youth can begin to identify with a domain when they (a) believe that they are *empowered* to act with some autonomy within it, (b) believe that the domain is *useful* to their goals, (c) believe that they can *succeed* in it, (d) are *interested* in it, and (e) believe that they are *cared* for in a supportive environment (Jones 2009). Figure 5.1 shows that the pedagogical approach used in Studio STEM can affect students' identification with STEM fields and their goals and beliefs (about their abilities, interests, and utility value in STEM fields). These factors can then affect students' choice of activities and future class selection (e.g., physics, calculus), level of engagement in STEM activities, and academic outcomes, such as their level of success in STEM activities. In Fig. 5.1, the arrow from academic outcomes back to the other factors indicates that these outcomes also affect students' domain identification, goals, beliefs, choices, and effort. As an example, students who are successful in science activities will likely believe that they have a higher level of science ability than students who are unsuccessful. In sum, this figure provides a conceptual model of how the Studio STEM curriculum and pedagogy can affect youths' STEM identification. Because of the many factors that can affect a youths' domain

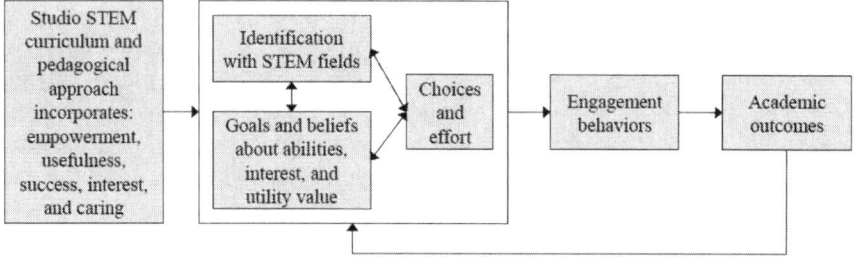

Fig. 5.1 Model of how Studio STEM can affect students' identification with STEM fields (Modified from Osborne and Jones 2011)

identification, it is unrealistic to expect all of the youth who participate in studio STEM to be highly identified with a STEM field. However, we believe that even the students who do not become highly identified with the STEM field can take away several positive outcomes from participating in Studio STEM, such as: enrolling in future STEM-related courses, considering the possibility of a STEM career, feeling more confident in their STEM-related abilities, having an increased value and appreciation for STEM-related activities and fields, becoming a more informed citizen who is involved in and cares about STEM-related issues that affect their community and world.

Facilitation and Discussion In a previously published account (Motto et al. 2011), an informal science educator, who was a doctoral student, acted as site leader for an iteration of Studio STEM. She guided the weekly sessions, serving primarily to ensure that milestones in the curriculum were reached according to the prescribed timeline. Small group activities were facilitated by undergraduate science and engineering students from a nearby large public research university. Each undergraduate mentor worked with two or three youth participants. Specific interactions that took place within these small groups were the focus of that study.

One goal of the Studio STEM program was to provide youth with support and mentoring from undergraduate students from the local university. In this rural, low-income community, youth had inadequate exposure to higher education; thus, the university students acted as positive role models, perhaps, making college seem a tangible possibility. According to Rhodes (2004), "Faced with fewer curricular demands than teachers, afterschool staff are often afforded unique opportunities to engage in the sorts of informal conversations and enjoyable activities that can give rise to close bonds with youth" (p. 146). Intentionally, this feature was designed into Studio STEM.

Influenced by Rogoff's (1990) concept of guided participation, Studio STEM provides opportunities for youth to be guided by older, more experienced, and more knowledgeable others, to positively influence their academic and social skills. The undergraduate students provided one-on-one and small group guidance, kept youth focused on the goals of the project, and supported conceptual understanding. Youth also benefited from exposure to and interactions with mentors who acted as role models from STEM-related fields (Schnittka et al. 2012).

The resource constraints imposed on recruitment for Studio STEM, nevertheless, meant that mentors might have had minimal experience as teachers or coaches, receiving only requisite training in small group facilitation as a condition of participating in the program. As such, the approaches used to guide and support youth can vary, which is of interest analytically and practically to the investigators. As anticipated, the elicitation strategies used by each mentor influenced the ways that youth made meaning and engaged in the design process. Thus, investigators determined that a micro-level analysis of those elicitation strategies could assist in understanding the role of mentors in youth engagement and serve to inform mentor training in subsequent iterations of the project. This in turn, may provide insights for similar afterschool STEM learning programs.

Curricular Principles and Examples A pilot version of Studio STEM, using *Save the Penguins* was reported as an 8-week unit designed to guide youth to understand environmental issues through an exploration of heat transfer, thermodynamics, and engineering design in an afterschool studio setting (Schnittka et al. 2012). Youth worked with undergraduate mentors from a nearby large research university to construct, test, re-assess, and re-construct miniature "dwellings" designed to insulate penguin-shaped ice cubes from a radiant heat source. Through collaborative, problem-based learning, enhanced by personal blogs and team wikis, students participated in weekly activities that allowed them to identify with STEM topics and disciplines. In doing so, they developed an understanding of the ways in which energy consumption at home may impact the global climate, and created strategies for improving energy efficiency in their own homes. By exploring materials and processes related to energy transfer, environmental issues, and impact on other living organisms, the curriculum encouraged students to:

- Make connections between the natural and designed worlds;
- Interact with students and professionals in science and engineering fields;
- Understand the influence actions can have on the local and global climate; and
- Understand the role that science, information and communication technologies, and engineering play in the improvement of local and global conditions.

During the first sessions of the club, youth were introduced to current living conditions of Antarctic wildlife, and how their own lives are connected to the global environment. In the second and third sessions, youth observed discrepant events involving the transfer of heat through various media (plastic, metal, fabric), and made connections between the physical properties of those materials and energy conservation in their homes. Youth were then tasked with designing a "dwelling" that could prevent a penguin-shaped ice cube from melting under extreme heat. In the next several sessions, youth worked in small groups with undergraduate mentors to research online their topics, design and test multiple iterations of penguin houses, and reflect on the results. The final two sessions were devoted to reporting their findings through storyboards, self-directed videos, and multimedia presentations.

Throughout the 8-week unit, participants were encouraged to document and exchange their experiences via wiki entries and blog posts hosted on the project-dedicated learning management system. The online platform served as a data repository for participant work and data source for investigators.

Assessment and Evaluation Protocols We have employed both formative and summative evaluations to assess the success of Studio STEM. The formative evaluation has focused on the implementation of Studio STEM and has included a mixed methods approach (Cresswell and Plano-Clark 2007), combining surveys of students, site leaders, and facilitators; observations of program activities; student focus groups; and interviews with facilitators. Specific formative evaluation questions have focused on students and site leaders, including: (1) To what degree does the professional development of facilitators and mentors prepare them for teaching the curriculum? (2) Are learning objectives clear and guiding for facilitators? (3) Which activities do facilitators find easy or challenging to implement, and what adaptations do they make? (4) What supports and resources are available to implement the program effectively and are there additional needs? (5) Does ICT become a fluid component of the design and inquiry processes for students? (6) Do the activities address students' prior conceptions constructively? The formative evaluation describes the implementation of Studio STEM, generates hypotheses about the mechanisms by which Studio STEM generates effects, and outlines the contextual constraints within which the program operates and must address. In addition, the evaluation describes the role of site leaders in Studio STEM, and evaluates the preparation they require to effectively support students' participation in program activities.

The summative evaluation has addressed the short-term outcomes that are within the scope of the program's period of implementation:

1. *To what extend did Studio STEM participants develop an understanding of key concepts in energy and the environmental impact of energy production and use?*

One of the most important aspects of the Studio STEM curriculum was the ways in which the conceptual content was folded into the design activities. Students either had to think about the science ideas and use them as they developed and improved their design solutions, or science ideas became evident in their engagement with the designs. Students grappled with the content in their work on the designs. They had to think about the science content to solve problems. They grappled with conflicting ideas and misconceptions. While we currently don't have content learning outcomes to report on, we can nevertheless infer that students had vivid experiences with the concepts, had their misconceptions challenged, and presumably have real experiences to fall back on as they encounter similar conceptual content in other settings to solidify their learning. For example, in the Penguins unit, students explored different types of materials to insulate their penguin dwellings and had to take into consideration conceptual ideas around heat transfer.

2. *To what extend did Studio STEM participants develop skills, such as engineering design, experimental design, applied math, and technological fluency with digital tools?*

The qualitative data strongly show, however, that Studio STEM provided convincing opportunities for students to learn engineering skills. Students spent considerable time problem solving, iterating on their design, and having discussions with each other about the process. Site leaders and facilitators consistently reported that students improved their skills with the design process. Students also described what they were doing as engineering, according to staff. Staff said that students used information about content and what they were learning through the process to improve their designs as well. Interview data also showed that many students viewed the idea of failing as a productive part of the process, and that they were willing to work beyond mistakes and failures in spite of often feeling frustrated. In other words, the qualitative data suggest that students learned persistence – or at the very least had ample opportunities to experience the positive results of persisting – which is an important part of the engineering design process as well as learning in general.

3. *To what extend did Studio STEM participants, including girls, gain greater interest in and identification with STEM-related subjects or careers?*

Our findings indicate that the program did positively impact students' choice of STEM careers. McNemar tests conducted indicated a significantly larger number of students chose STEM careers in the post-test than they did in the pre-tests ($\chi^2 = 7.117$, $p < .05$). And these changes were more prominent among girls ($\chi^2 = 6.4$, $p < .05$) than boys ($\chi^2 = 1.284$, n.s.). In conclusion, thus, the Studio STEM program positively impacted girls, who are historically underrepresented in STEM fields, to show more interest and identify with STEM-related careers. Some of these jobs they listed include the following: ecologist, accountant, veterinarian, video game designer, computer programmer and scientist. Clearly, these were girls that previously did not identify with STEM before, opting for becoming a writer, ranch owner, fashion designer and police officer. Furthermore, boys were not more likely to choose STEM careers over girls either at pre-test ($n = 37$ for girls and $n = 53$ for boys at pretest, $\chi^2 = 2.159$, n.s.) or at post-test ($n = 45$ for girls and $n = 56$ for boys at pretest $\chi^2 = .111$, n.s.), thus indicating that there was no gender difference among students choosing STEM careers at pre or post-test.

4. *To what extend did site leaders and facilitator become more knowledgeable of the concepts targeted by Studio STEM and more confident of their ability to support student learning of these concepts?*

When considered across all five implementations of the Studio STEM curriculum, the feedback and interview data overwhelmingly show that site leaders and facilitators across all three sites reported high levels of satisfaction with and confidence in implementing the curriculum as a result from participating in the professional development trainings. At the same time, in terms of content knowledge and adequate preparation for facilitating the designing the activities, the PD did not always suffice to provide site leaders and facilitators what they needed. These issues will be discussed in more detail below. The role of the curriculum coordinator was very important for staff as they frequently turned to her for advise on implementation and instructional problem solving, which has some implications for possible efforts to scale a curriculum like this to more sites and students. We also found that site leaders' and facilitators' backgrounds and different types of expertise made an important difference.

Assessing students' motivation in an informal, afterschool, inquiry environment can be challenging. To address this challenge, we used a version of the MUSIC Model of Academic Motivation Inventory (Jones and Skaggs in press) that was designed to assess youths' motivation-related beliefs for the five key components of the MUSIC Model of Motivation (Jones 2009) using the following constructs: *autonomy* for the empowerment component, *utility value* for the usefulness component, expectancy for success component, *situational interest* for the interest component, and *caring* for the caring component (the inventory and other assessments are available at www.theMUSICmodel.com). Jones and Wilkins (2013b) examined the use of this inventory with over 300 fifth-, sixth-, and seventh-grade youth and found that the inventory produced valid scores. The inventory consists of 18-items that are rated on a 6-point Likert-type scale and includes three or four items for each MUSIC component. We have administered the inventory at the end of each Studio STEM curriculum unit and asked the youth to report their perceptions with respect to the Studio STEM activities in that unit. An example item for the interest component of the MUSIC model is: I enjoyed completing the Studio STEM activities (rated from *strongly disagree* to *strongly agree*). By querying youth about their perceptions across five important motivation-related constructs for specific curriculum units, we have documented youths' motivation in a manner that provides useful data without taking a lot of time away from the program curriculum. The data is easily analyzed by computing averages for each MUSIC component.

To complement the quantitative data obtained from the inventory, we collected qualitative data by interviewing students on a range of motivation-related beliefs. Some semi-structured interview questions were designed specifically to address students' beliefs related to each of the MUSIC model components about certain curriculum units and others were more open-ended to assess students' beliefs about science and engineering more generally.

Projects Similar to Studio STEM Studio STEM is but one project out of many recent concerted efforts to leverage afterschool and other out-of-school settings to enhance STEM learning and literacy. Afterschool learning environments offer many advantages over formal classrooms for engaging students in STEM material. For example, afterschool instructors may not be compelled to cover topics mandated by state or national educational standards. The time allotted and pacing is generally more flexible, allowing youth to explore and develop new ideas at an individualized pace. Also, afterschool environments are non-evaluative, meaning that students are able to experiment with STEM ideas without the pressure of grades or following a regimented procedure (Bevan et al. 2010). The collaborative nature of informal environments broadens participation by allowing youth to share ideas and prior knowledge, as opposed to instruction delivered primarily through lectures and individual assignments (Bell et al. 2009).

To shift students' thinking from fact memorization to thinking like scientists and engineers, youth can benefit from participating in programs that allow them to explore, ask questions, solve problems, and think critically (Asghar et al. 2012). This is important because engineers apply content knowledge and cognitive skills to an ill-structured problem through the process of designing, analyzing, and troubleshooting (Brophy et al. 2008). In fact, numerous STEM programs already explicitly or implicitly use this engineering design process to create the opportunity for self-guided inquiry and application of science knowledge to a real-world problem (Bevan et al. 2010; Bouvier and Connors 2011; Brophy et al. 2008). Because it is consistent with the engineering design process, problem-based learning (PBL) is an instructional strategy that fits well into the goals of STEM education. PBL involves experiential learning through the investigation, explanation, and resolution of meaningful problems (Barrows 1998; Torp and Sage 1998). The design, test, and rebuild process that engineers use parallels PBL. In PBL, youth work collaboratively in groups, while a teacher acts as a facilitator to guide learning. Youth learn what is needed to solve the problem, analyze the problem, and subsequently consider possible solutions. Then they identify what it is they do not know, gather new knowledge, and apply this new knowledge to reform the hypothesis (Hmelo-Silver 2004). Studio STEM uses the method of PBL in an afterschool, informal learning environment by having youth work to solve an open-ended engineering problem by applying science and mathematics content knowledge.

Evidence of Successful Implementation in Informal Settings

Gains in Understanding in STEM Concepts

Guided by a mixed methodological framework (Creswell 2013) we have attempted to identify the different ways in which the SNFs, collaborative teams, the curricular activities, the design challenge, and teachers' and facilitators' words and actions helped youth with problem-solving and conceptual understanding of science. Additionally, pre- and post-tests on the science content were administered at the beginning and end of the curriculum units. The instruments used were designed to target common alternative conceptions that youth have about physical science concepts. Analysis of pre- and post-test results indicates significant gains at most sites on most curriculum modules. The Penguins and Sea Birds instruments were 12-item multiple-choice instruments that had gone through a series of evaluations to demonstrate reliability and validity. The Snails instrument was an open-ended writing prompt with a 10-point rubric. For the Save the Penguins curriculum, we saw significant gains at two of the three sites. The small sample size at the South Middle School site was problematic because the results are not necessarily representative of the students who participated in the project at that school. See Table 5.1.

Table 5.1 Pre- and post-test science scores at three sites for three modules

		n	Pre-test	Post-test
Save the penguins	East Middle School	20	4.75	7.3***
	South Middle School	3	4.33	4.66
	North Middle School	25	4.64	6.88***
Save the sea birds	East Middle School Teacher J	13	2.69	5.00***
	East Middle School Teacher M	14	3.50	6.07***
	North Middle School	11	5.00	7.36***
Save the snails	East Middle School	14	0.71	1.64*
	North Middle School	15	1.00	5.00***

Note: $* = p < .05$, $*** = p \leq .001$

While raw scores on the post-tests were low compared to what one would typically see on a teacher-made unit test administered in a school setting, the results were aligned with what has actually been observed in classroom settings (Schnittka and Bell 2011). Also, these tests were designed with an upper limit not typically achieved by college students or even veteran teachers. For example, Schnittka (2009) reported that mechanical engineering seniors taking the Heat Transfer Evaluation typically scored between 10 and 11 out of 12 points. Schnittka et al. (2014) reported that middle school science teachers taking the Force Motion Evaluation also typically scored between 10 and 11 out of 12 points. Schnittka (2012) also reported that mechanical engineering students typically scored between 5 and 6 on the Coal Assessment. The assessments can be found at the following links: (1) Heat transfer (http://www.auburn.edu/~cgs0013/ETK/Heat_Transfer_Evaluation.doc); (2) Force and Motions (http://www.auburn.edu/~cgs0013/ETK/Force_Motion_Evaluation.doc) and; (3) Coal (http://www.auburn.edu/%7Ecgs0013/ETK/coal_assessment.pdf).

Integration of Technology and New Media

One way in which instructors attempt to incorporate STEM into the school systems is through the development of out-of-school programs and informal learning settings. The Studio STEM curriculum, *Save the Penguins* developed by Schnittka (2009) is an example of a curriculum that is geared towards teaching middle school aged students about the concepts of heat transfer and engineering. Students participating in *Save the Penguins* are also given information on the different types of projects that engineers work on in the real world. Informal learning settings cannot only involve spending time after the typical school day, but also interaction with STEM concepts at home through the use of technology.

Social media are extremely popular among youth for personal use. As such, there is great potential for the use of social media in connecting formal classroom learning with informal learning and inquiry (Chen and Bryer 2012). Since social media are already attractive to students, and since students are already engaged with social media for personal use, being able to integrate STEM learning has the potential to be very effective. Social media provide a way for students to self-regulate their learning and their learning environment. Through their interactions with peers, the instructor, and the technology, they are able to customize their learning experience allowing them a sense of control that they may not have in the formal classroom setting (Kitsantas and Dabbagh 2011). The difficulty appears to lie in that the instructor lacks the necessary skills and confidence in their ability to integrate technologies such as social media into their curriculum successfully (Campbell and Ellingson 2010). This issue is true of many technologies that are being incorporated into classrooms today. However, the potential for increasing student participation and interest in STEM topics may outweigh the extra support systems and training that may be required for instructors to implement these strategies.

A strong advantage of social media in instruction is that there is a preserved record of discourse between students, other students, and instructors. This makes it an attractive area for qualitative research. Discourse analysis involves the study of how people communicate, and how that communication leads to action (Potter 2003). The communication analyzed can be either dialogue in person, or back and forth through text-based methods such as instant messenger or Facebook. Discursive psychology is a field that utilizes discourse analysis in order to examine language and how people ascribe meaning to that language. Language is situational (appears within a context), action oriented (utilized to achieve an objective), and constructive, as if is made up of much smaller components (Roth 2008). Discourse between students and instructors about STEM concepts could therefore provide clues about the way that a subject is perceived, and how much the student understands the material. A student's identity, for example, whether he perceives himself as being capable or incapable of learning difficult scientific concepts, can affect the way that he will perform in the classroom. This can be researched through the use of discursive psychology methods (Hsu and Roth 2010).

The Studio STEM curriculum developed by Schnittka (2009) incorporates interaction of students through social media (Edmodo). This allows researchers the chance to examine changes in student discourse related to STEM, and specifically to heat transfer and engineering concepts. The objective of a recent study was to see whether participation in Studio STEM increased student understanding of science concepts through the use of discursive psychology (Evans et al. 2014d). This was then related to a model describing engagement of students with technology. The HOMAGO model developed by Ito et al. (2010) describes three distinct levels of youth engagement: *hanging out*, *messing around*, and *geeking out*. The *hanging out* portion of the model describes interactions with technology that are geared towards developing social relationships with peers. *Messing around* is the term used to describe interactions with technology for the purpose of informally seeking information of interest to the individual. Finally, *geeking out* describes interactions with technology that are specifically directed towards increasing individual expertise and knowledge of a particular subject area of interest. Since the model specifically applies to youth and technology, HOMAGO is a good resource for analyzing text-based discourse through Edmodo.

Students were expected to engage in discourse across all three major categories, but to gravitate more frequently towards the *geeking out* side of the spectrum with the progression of Studio STEM, and increasing exposure to experimentation methods and heat transfer concepts. This shift would be facilitated by the input and encouragement of site leaders and other facilitators including undergraduate and graduate students involved with the project.

Youth Motivation in and Identification with STEM

Evidence from Studio STEM indicates that youth are motivated to engage in the project and that some youth have developed or maintained identification for one or more STEM fields. We discuss some of the motivation-related outcomes in this section and in the order of the MUSIC Model of Motivation: empowerment, usefulness, success, interest, and caring. Although we present the results separately for each MUSIC component, these components are related and increases in one component may lead to increases in another component (Jones and Wilkins 2013a). Also, the percentages provided in this section refer to responses on open-ended interview questions; therefore, the actual percentages may have been higher if students had been asked directly about the topic with a closed-ended question.

Overall, students have reported that they feel empowered during Studio STEM (the mean rating for empowerment was 5.2 on a scale from 1 [*strongly disagree*] to 6 [*strongly agree*]; Jones et al. 2015b). Students reported that they had choices during many of the activities, ranging from specific tasks within activities (e.g., choice of how to build a motor to pull up the basket with cubes; 56 % of students) to more general choices (e.g., choice of how to build the solar cars; 89 % of students; Jones et al. 2015b). Students have also reported that they have a better understanding of the usefulness of science and engineering as a result of participating in studio STEM (Schnittka et al. 2012). Some of the students specifically noted that the project was useful for figuring out what to study in college (18 % of students), for succeeding in college (27 % of students), or for becoming a scientist or engineer (45 % of students; Jones et al. 2015b). Students have also reported that the Studio STEM activities are useful for their present lives. For example, some students said that it was useful to learn about motors, electricity, or solar cells or cars (73 % of students; Jones et al. 2015b).

Support from the facilitators and instructors help the youth to feel successful (Schnittka et al. 2012). One of the ways that instructors provided this support was by questioning youth about their designs and activities (Evans et al. 2014d). Further, youth noted that the specific tasks with clear goals provided them with feedback as to whether they were successful or not. When they did not meet their goals, they were often motivated to improve on their designs and achieve success in their redesign (Schnittka et al. 2012).

Students also reported enjoying their participation in Studio STEM (Schnittka et al. 2012) and found specific aspects of the activities interesting (the mean rating for interest was 5.4 on a scale from 1 [*strongly disagree*] to 6 [*strongly agree*]; Jones et al. 2015b). For example, students were interested in building the solar car (67 % of students) and working with motors to lift up a basket (45 % of students; Jones et al. 2015b). However, some of the pedagogical approaches were found to be less interesting, such as when the instructor presented information (73 % of students reported some aspect of the presentations uninteresting; Jones et al. 2015b). These findings indicate that students' interest can vary from activity to activity and that instructors must be cognizant of these differences and how they are affecting students.

With respect to the caring component of the MUSIC model, students generally reported that they felt supported by the instructors and facilitators and enjoyed the attention they received from them (the mean rating for caring was 5.7 on a scale from 1 [*strongly disagree*] to 6 [*strongly agree*]; Jones et al. 2015b). Overall, students felt less cared for by their peers (Schnittka et al. 2012). Theory predicts that students should be more motivated when they feel cared for, so it might be the case that the caring students feel from their instructors and facilitators is sufficient and that they do not need to feel a lot of support from their peers. Further research is needed to understand the importance of peer caring in motivating students in studio STEM.

As predicted by the MUSIC model, students who reported high levels of the five MUSIC model components were also motivated to engage in the studio STEM activities. For example, students self-reported that they put forth a high level of effort during Studio STEM (the mean rating for effort was 5.5 on a scale from 1 [*strongly disagree*] to 6 [*strongly agree*]; Jones et al. 2015b) and observations of students' behaviors have substantiated this finding (Evans et al. 2014d). We do not claim that all of the students are always motivated and engaged. Some students reported that they were bored at some points and other students were less engaged when one of the students in their group dominated the decision-making processes (Evans et al. 2014d). It may be the case, however, that the lack of motivation for some students it is not related to the design of Studio STEM, but rather to how the Studio STEM model was implemented. For example, it might be possible for facilitators to help ensure that one student does not dominate a group, which in turn may lead to higher levels of engagement by all students.

Finally, there is some evidence that students who participate in Studio STEM are more likely to become identified with STEM fields. For instance, as a result of participating in Studio STEM, students believed that learning science and engineering was more important and interesting than before participating in Studio STEM (Schnittka et al. 2012). In addition, because students believed that their science and engineering abilities increased as a result of participating in studio STEM (Schnittka et al. 2012), they should be more likely to engage in related science and engineering activities in the future (Bandura 1986; Wigfield and Eccles 2000). In fact, students reported that they were more interested in taking a course in science and engineering, even if it wasn't required was more than before participating in Studio STEM

(Schnittka et al. 2012). These types of findings are encouraging, but it would be useful in the future for researchers to follow youth longitudinally over a longer period of time to assesses how Studio STEM can contribute to youths' identification with STEM fields.

Productive Questioning in Small Group Discussion

In our work, we are concerned with describing and characterizing learning through interactions between youth participants and their undergraduate mentors. We have been interested not only in student learning outcomes, but also how the approaches used by undergraduate mentors influenced the ways in which youth made meaning of the project and engaged in the design process. To understand the role that mentor talk plays in guiding student learning, we have employed a discursive psychology (DP) approach. DP shifts the focus of psychological analysis away from cognitive processes toward social interactions situated in everyday activities (Wiggins and Hepburn 2005). DP is valuable in this instance as it assists to understand how the elicitation strategies of mentors influence the ways that youth discursively construct science understandings within a pre-engineering design process.

Using Engineering Design-Based Approaches toward STEM Literacy through Professional Development

Through the biannual professional development, site leaders new to Studio STEM are introduced to the programming model and curricula. In addition, professional development provides veteran site leaders an opportunity to deepen their understanding of the core science concepts behind each curricular unit. Instructors and undergraduate facilitators meet for a one-day workshop every September and January to review the curricula, practice the design problems themselves, and to discuss potential obstacles they foresee in the implementation of the unit. These conversations with site leaders have not only been important in expanding their knowledge, but have also led to modifications of the curriculum that improved the delivery of Studio STEM. Most of the instructors had never taught an integrative STEM lesson that used an open-ended problem driven approach. Yet, they quickly adopted the engineering design process and were excited about the possibilities to improve their teaching in the school classroom as well. One site leader, a sixth grade math teacher was especially appreciative of the professional development that allowed her to expand her science content knowledge. She and other site leaders noted that in the school classroom where time is tightly managed, teachers had little time to conduct investigations that were open-ended.

Conclusion

To understand youth engagement in an out-of-school program, we must recognize that a variety of social and environmental considerations contribute to the sustainment of engagement in every moment of interaction. Studio STEM proposes an integrative approach to engaging youth by implementing inquiry-driven experiences supported by facilitators and information and communication technologies. The *save the animals* theme of all Studio STEM curricula challenge youths' misconceptions regarding the science of energy while providing opportunities to engage in the engineering design process to enhance technological literacy. Targeting middle school audiences leverages evidence that youth are most receptive to STEM concepts and careers, increasing the probability that they will continue to pursue STEM-related courses and careers. Of course, not all youth will enter the workforce as scientists and engineers.

Nevertheless, Studio STEM is positioned to broaden the scientific and technological literacy of students not only in rural communities, but in urban and suburban ones as well. Several years of Save the Animals implementation in cities, towns and rural areas across country have produced encouraging results as we continue to test and refine the current curricula, assessments, and training (Schnittka et al. 2014; Schnittka and Ewald 2013; Griffin et al. 2015).

References

Aikenhead, G., Calabrese, A. B., & Chinn, P. W. (2006). Forum: Toward a politics of place-based science education. *Cultural Studies of Science Education, 1*(2), 403–416.

Anderson, R. D. (2002). Reforming science teaching: What research says about inquiry. *Journal of Science Teacher Education, 13*(1), 1–12.

Archer, L., DeWitt, J., Osborne, J., Dillon, J., Willis, B., & Wong, B. (2010). "Doing" science versus "being" a scientist: Examining 10/11-year-old schoolchildren's constructions of science through the lens of identity. *Science Education, 94*(4), 617–639.

Asghar, A., Ellington, R., Rice, E., Johnson, F., & Prime, G. M. (2012). Supporting STEM education in secondary science contexts. *Interdisciplinary Journal of Problem-Based Learning, 6*(2), 4.

Bandura, A. (1986). *Social foundations of thought and action: A social cognitive theory.* Englewood Cliffs: Prentice-Hall.

Barab, S. A., & Duffy, T. M. (2000). From practice fields to communities of practice. In D. H. Johassen & S. M. Land (Eds.), *Theoretical foundations of learning environments* (pp. 25–55). Mahwah: Lawrence Erlbaum.

Baram-Tsabari, A., & Yarden, A. (2009). Identifying meta-clusters of students' interest in science and their change with age. *Journal of Research in Science Teaching, 46*(9), 999–1022.

Barrows, H. S. (1998). The essentials of problem-based learning. *Journal of Dental Education, 62*(9), 630–633.

Barton, A. C., & Tan, E. (2010). We be burnin'! Agency, identity, and science learning. *The Journal of the Learning Sciences, 19*(2), 187–229.

Barton, A. C., Tan, E., & Rivet, A. (2008). Creating hybrid spaces for engaging school science among urban middle school girls. *American Educational Research Journal, 45*(1), 68–103.

Basu, S. J., & Barton, A. C. (2007). Developing a sustained interest in science among urban minority youth. *Journal of Research in Science Teaching, 44*(3), 466–489.

Becker, K. H., & Park, K. (2011). Need a title here. *Journal of STEM Education: Innovations and Research, 12*(5–6), 23–37.

Bell, R. L., Smetana, L., & Binns, I. (2005). Simplifying inquiry instruction. *The Science Teacher, 72*(7), 30–33.

Bell, P., Lewenstein, B., Shouse, A. W., & Feder, M. A. (Eds.). (2009). *Learning science in informal environments: People, places, and pursuits.* Washington, DC: National Academies Press.

Bevan, B., Michalchik, V., Bhanot, R., Rauch, N., Remold, J., Semper, R., & Shields, P. (2010). *Out-of-school time STEM: Building experience, building bridges.* San Francisco: Exploratorium, Retrieved April, 29, 2013.

Boud, D., & Feletti, G. (1997). Changing problem-based learning. In D. Boud & G. Feletti (Eds.), *The challenge of problem-based learning* (2nd ed., pp. 1–14). London: Kogan Page.

Bouvier, S., & Connors, K. (2011). *Increasing student interest in science, technology, engineering, and math (STEM): Massachusetts STEM pipeline fund programs using promising practices.* Report Prepared for the Massachusetts Department of Higher Education, 74.

Brandt, C., Motto, A., Schnittka, C.G., Evans, M., & Jones, B. (2011). Socio-cognitive scaffolding in the studio: Informal STEM learning and identity. *Proceedings of the National Association for Research in Science Teaching*, Orlando.

Brophy, S., Klein, S., Portsmore, M., & Rogers, C. (2008). Advancing engineering education in P-12 classrooms. *Journal of Engineering Education, 97*(3), 369–387.

Campbell, K., & Ellingson, D. A. (2010). Cooperative learning at a distance: An experiment with wikis. *American Journal of Business Education (AJBE), 3*(4), 83–90.

Chen, B., & Bryer, T. (2012). Investigating instructional strategies for using social media in formal and informal learning. *The International Review of Research in Open and Distributed Learning, 13*(1), 87–104.

Chouinard, M. M., Harris, P. L., & Maratsos, M. P. (2007). Children's questions: A mechanism for cognitive development. *Monographs of the Society for Research in Child Development, 72*(1), vii–ix.

Creswell, J. W. (2013). *Research design: Qualitative, quantitative, and mixed methods approaches.* Los Angeles: Sage.

Cresswell, J., & Plano-Clark, V. (2007). *Designing and conducting mixed methods research.* Thousand Oaks: Sage.

Deater-Deckard, K., Chang, M., & Evans, M. A. (2013). Engagement states and learning from educational games. *New Directions in Child and Adolescent Development, 139*, 21–30. doi:10.1002/cad.20028.

Department of Energy [ED]. (2012). *Energy literacy: Essential principles and fundamental concepts for energy education.* Retrieved from http://www1.eere.energy.gov/education/pdfs/energy_literacy_1_0_high_res.pdf

DeWaters, J. E., & Powers, S. E. (2011). Energy literacy of secondary students in New York State (USA): A measure of knowledge, affect, and behavior. *Energy Policy, 39*(3), 1699–1710.

Diefes-Dux, H. A., Moore, T., Zawojewski, J., Imbrie, P. K., & Follman, D. (2004). A framework for posing openended engineering problems: Model-eliciting activities. In *Frontiers in Education, 2004. FIE 2004. 34th Annual* (pp. F1A-3). New York: Institute of Electrical and Electronics Engineers.

Dreyfus, A., Jungwirth, E., & Eliovitch, R. (1990). Applying the "cognitive conflict" strategy for conceptual change: Some implications, difficulties and problems. *Science Education, 74*, 555–569.

Driver, R., Guesne, E., & Tiberghien, A. (Eds.). (1985). *Children's ideas in science.* Philadelphia: Open University Press.

Driver, R., Squires, A., Rushworth, P., & Wood-Robinson, V. (1994). *Making sense of secondary science: Research into children's ideas.* London: Routledge.

Duit, R., & Treagust, D. F. (2003). Conceptual change: A powerful framework for improving science teaching and learning. *International Journal of Science Education, 25*(6), 671–688.

Durlak, J. A., Weissberg, R. P., & Pachan, M. (2010). A meta-analysis of after-school programs that seek to promote personal and social skills in children and adolescents. *American Journal of Community Psychology, 45*(3–4), 294–309.

Eccles, J. S. (2005). Subjective task values and the Eccles et al. model of achievement-related choices. In A. J. Elliot & C. S. Dweck (Eds.), *Handbook of competence and motivation* (pp. 105–121). New York: The Guilford Press.

Elliot, D. S., & Voss, H. L. (1974). *Delinquency and dropout*. Lexington: D. C. Heath and Company.

Evans, M. A. (2009). Promoting mediated collaborative inquiry in primary and secondary science settings: Sociotechnical prescriptions for and challenges to curricular reform. In R. Subramaniam (Ed.), *Handbook of research on new media literacy at the K-12 level: Issues and challenges* (Vol. I, pp. 128–143). Hershey: Information Science Reference.

Evans, M. A., & Biedler, J. (2012). Playing, designing, and developing video games for informal science learning: *Mission: Evolution* as a working example. *International Journal of Learning and Media, 3*(4). doi:10.1162/IJLM_a_00083

Evans, M. A. Won, S., Drape, T., & Smalls, D. (2013). *STEM Club Hang Out: Social media use in an informal learning space*. Paper presented at the American Educational Research Association Conference, San Francisco, 27 Apr–1 May 2013.

Evans, M. A., Lopez, M., Maddox, D., Drape, T., & Duke, R. (2014a). Interest-driven learning among middle school youth in an out-of-school STEM studio. Submitted to the *Journal of Science Education and Technology*. Manuscript submitted for publication.

Evans, M. A., Duke, R. F., & Jones, B. D. (2014b). *Characterizing youth academic engagement with STEM in an afterschool design studio*. Manuscript submitted for publication.

Evans, M. A., Maddox, D., & Lopez, M. (2014c). *Youth interest in and motivation toward informal STEM education: Two case studies*. Paper presented at the annual meeting of the American Educational Research Association, Philadelphia

Evans, M. A., Won, S., & Drape, T. (2014d). Interest-driven learning of STEM concepts among youth interacting through social media. *International Journal of Social Media and Interactive Learning Environments, 2*, 3–20.

Fraser, B. J., & Kahle, J. B. (2007). Classroom, home and peer environment influences on student outcomes in science and mathematics: An analysis of systemic reform data. *International Journal of Science Education, 29*(15), 1891–1909.

Fusco, D., & Barton, A. C. (2001). Representing student achievements in science. *Journal of Research in Science Teaching, 38*(3), 337–354.

Gattie, D. K., & Wicklein, R. C. (2007). Curricular value and instructional needs for infusing engineering design into K-12 technology education. *Journal of Technology Education, 19*(1), 6–18.

Gerber, B. L., Cavallo, A. M. L., & Marek, E. A. (2001). Relationships among informal learning environments, teaching procedure and scientific reasoning. *International Journal of Science Education, 23*, 535–549.

Gold, M., & Mann, D. W. (1984). *Expelled to a friendlier place: A study of effective alternative schools*. Ann Arbor: University of Michigan Press.

Griffin, J., Brandt, C., Bickel, E., Schnittka, C., & Schnittka, J. (2015). *Imbalance of power: A case study of a middle school mixed-gender engineering team*. Princeton: IEEE Integrated STEM Education Conference.

Gross, L. (2005). As the Antarctic ice pack recedes, a fragile ecosystem hangs in the balance. *PLoS Biology 3*(4), 557–561. Retrieved from http://www.plosbiology.org/article/info%3Adoi%2F10.1371%2Fjournal.pbio.0030127

Gruenewald, D. A. (2003). Foundations of place: A multidisciplinary framework for place-conscious education. *American Educational Research Journal, 40*(3), 619–654.

Hewson, M. G., & Hewson, P. (1983). Effect of instruction using students' prior knowledge and conceptual change strategies on science learning. *Journal of Research in Science Teaching, 20*, 731–743.

Hewson, P. W., & Hewson, M. G. B. (1984). The role of conceptual conflict in conceptual change and the design of science instruction. *Instructional Science, 13*(1), 1–13.

Hidi, S., & Renninger, K. A. (2006). The four-phase model of interest development. *Educational Psychologist, 41*(2), 111–127.

Hmelo-Silver, C. E. (2004). Problem-based learning: What and how do students learn? *Educational Psychology Review, 16*(3), 235–266.

Honey, M., & Kanter, D. E. (2013). *Design, make, play: Growing the next generation of STEM innovators*. New York: Routledge.

Hsu, P. L., & Roth, W. M. (2010). From a sense of stereotypically foreign to belonging in a science community: Ways of experiential descriptions about high school students' science internship. *Research in Science Education, 40*(3), 291–311.

Ito, M., Baumer, S., Bittanti, M., Boyd, D., Cody, R., Herr-Stephenson, B., Horst, H. A., Lange, P. G., Mahendran, D., Martinez, K. Z., Pascoe, C. J., Perkel, D., Robinson, L., Sims, C., & Tripp, L. (2010). *Hanging out, messing around, and geeking out*. Cambridge, MA: The MIT Press.

Ito, M., Gutierrez, K., Livingstone, S., Penuel, B., Rhodes, J., Salen, K., Schor, J., Sefton-Green, J., & Watkins, S. C. (2013). *Connected learning: An agenda for research and design, The digital media and learning*. Irvine: Research Hub.

Jenouvrier, S., Caswell, H., Barbraud, C., Holland, M., Stroeve, J., & Weimerskirch, H. (2009). Demographic models and IPCC climate projections predict the decline of an emperor penguin population. *Proceedings of the National Academy of Sciences, 106*(6), 1844–1847. Retrieved from http://www.pnas.org/content/106/6/1844.full.pdf+html

Jones, B. D. (2009). Motivating students to engage in learning: The MUSIC model of academic motivation. *International Journal of Teaching and Learning in Higher Education, 21*(3), 272–285.

Jones, B. D. (2015). *Motivating students by design: Practical strategies for professors*. Charleston, SC: CreateSpace.

Jones, B. D., & Skaggs, G. E. (in press). Measuring students' motivation: Validity evidence for the MUSIC model of academic motivation inventory. *International Journal for the Scholarship of Teaching and Learning*.

Jones, B. D., & Wilkins, J. L. M. (2013a). Testing the MUSIC model of academic motivation through confirmatory factor analysis. *Educational Psychology: An International Journal of Experimental Educational Psychology, 33*(4), 482–503. doi:10.1080/01443410.2013.785044.

Jones, B. D., & Wilkins, J. L. M. (2013b). *Validity evidence for the use of a motivation inventory with middle school students*. Poster presented at the annual meeting of the Society for the Study of Motivation, Washington, DC.

Jones, M. G., Jones, B. D., & Hargrove, T. Y. (2003). *The unintended consequences of high-stakes testing*. Lanham: Rowman & Littlefield.

Jones, B. D., Osborne, J. W., Paretti, M. C., & Matusovich, H. M. (2014). Relationships among students' perceptions of a first-year engineering design course and their engineering identification, motivational beliefs, course effort, and academic outcomes. *International Journal of Engineering Education, 30*(6A), 1340–1356.

Jones, B. D., Sahbaz, S., & Chittum, J. R. (2015a, April). *Science class motivational beliefs that impact students' science identification and career plans*. Paper presented at the annual meeting of the American Educational Research Association, Chicago, IL.

Jones, B. D., Chittum, J. R., Akalin, S., Schram, A. B., Fink, J., Schnittka, C., et al. (2015b). Elements of design based science activities that affect students' motivation. *School Science and Mathematics, 115*(8), 404–415.

Joseph, B., Shoemaker, C., & Martin, H. J. (2010). How using social media forced a library to work on the edge in their efforts to move youth from "Hanging Out" to "Messing Around". *The Journal of Media Literacy Education, 2*(2), 181–184.

Kanter, D. E., & Konstantopoulos, S. (2010). The impact of a project-based science curriculum on minority student achievement, attitudes, and careers: The effects of teacher content and pedagogical content knowledge and inquiry-based practices. *Science Education, 94*(5), 855–887.

Katehi, L., Pearson, G., & Feder, M. (Eds.). (2009). *Engineering in K-12 education*. Washington, DC: The National Academies Press.

Kitsantas, A., & Dabbagh, N. (2011). The role of Web 2.0 technologies in self-regulated learning. *New Directions for Teaching and Learning, 2011*(126), 99–106.

Lee, O., & Anderson, C. W. (1993). Task engagement and conceptual change in middle school science classrooms. *American Educational Research Journal, 30*(3), 585–610.

Madden, M., Lenhart, A., Duggan, M., Cortesi, S., & Gasser, U. (2013). *Teens and technology 2013*. Washington, DC: Pew Internet & American Life Project.

Maltese, A. V., & Tai, R. H. (2010). Eyeballs in the fridge: Sources of early interest in science. *International Journal of Science Education, 32*, 669–685.

Markowitz, D. G. (2004). Evaluation of the long-term impact of a university high school summer science program on students' interest and perceived abilities in science. *Journal of Science Education and Technology, 13*(3), 395–407.

Merrill, C., Custer, R., Daugherty, J., Westrick, M., & Zeng, Y. (2007). *Delivering core engineering concepts to secondary level students. Proceedings of the American Society for Engineering Education*. Washington, DC: ASEE.

Merrill, C., Custer, R. L., Daugherty, J., Westrick, M., & Zeng, Y. (2009). Delivering core engineering concepts to secondary level students. *Journal of Technology Education, 20*(1), 48.

Moore, T., & Richards, L. G. (2012). P-12 engineering education research and practice. *Advances in Engineering Education, 3*(2). Downloaded from http://advances.asee.org/wpcontent/uploads/vol03/issue02/papers/aee-vol03-issue02-p01.pdf

Moore, T. J., Miller, R. L., Lesh, R. A., Stohlmann, M. S., & Kim, Y. R. (2013). Modeling in engineering: The role of representational fluency in Students' conceptual understanding. *Journal of Engineering Education, 102*(1), 141–178.

Morrison, J. S. (2006). *Attributes of STEM education: The students, the academy, the classroom. TIES STEM Education Monograph Series*. Baltimore: Teaching Institute for Excellence in STEM.

Motto, A., Brandt, C. B., Schnittka, C., Evans, M. A., & Jones, B. D. (2011). *Studio STEM/Save the Penguins: Connecting youth to environmental issues through designbased projects*. Roundtable presented at the American Educational Research Association meeting, New Orleans.

National Academy of Engineering and National Research Council. (2012). *Assuring the U.S. Department of Defense a Strong Science, Technology, Engineering, and Mathematics [STEM] workforce*. Washington, DC: The National Academies Press.

NGSS Lead States. (2013). *Next generation science standards: For states, by states*. Washington, DC: National Academies Press.

Osborne, J. W. (1997). Identification with academics and academic success among community college students. *Community College Review, 25*(1), 59–67.

Osborne, R., & Freyberg, P. (1985). *Learning in science. The implications of children's science*. Auckland/Portsmouth: Heinemann Educational Books.

Osborne, J. W., & Jones, B. D. (2011). Identification with academics and motivation to achieve in school: How the structure of the self influences academic outcomes. *Educational Psychology Review, 23*(1), 131–158.

Osborne, J. W., & Rausch, J. L. (2001). *Identification with academics and academic outcomes in secondary students*. Paper presented at the American Education Research Association, Seattle.

Osborne, J. W., & Walker, C. (2006). Stereotype threat, identification with academics, and withdrawal from school: Why the most successful students of colour might be most likely to withdraw. *Educational Psychology, 26*(4), 563–577.

Piaget, J. (1980). The psychogenesis of knowledge and its epistemological significance. In M. Piattelli-Palmarini (Ed.), *Language and learning* (pp. 23–34). Cambridge, MA: Harvard University Press.

Pintrich, P. R., Marx, R. W., & Boyle, R. A. (1993). Beyond cold conceptual change: The role of motivational beliefs and classroom contextual factors in the process of conceptual change. *Review of Educational Research, 63*(2), 167–199.

Potter, J. (2003). Discursive psychology: Between method and paradigm. *Discourse & Society, 14*(6), 783–794.

Rahm, J. (2008). Urban youths' hybrid positioning in science practices at the margin: A look inside a school – museum – scientist partnership project and an after-school science program. *Cultural Studies of Science Education, 3*(1), 97–121.

Rahm, J., & Grimes, K. (2005). Embedding seeds for better learning: Sneaking up on education in a youth gardening program. *Afterschool Matters, 4*, 33–41.

Reid, K. C. (1981). Alienation and persistent school absenteeism. *Research in Education, 26*, 31–40.

Riegle-Crumb, C., Moore, C., & Ramos-Wada, A. (2011). Who wants to have a career in science or math? Exploring adolescents' future aspiration by gender and race/ethnicity. *Science Education, 95*(3), 458–476.

Reynolds, B., Mehalik, M. M., Lovell, M. R., & Schunn, C. D. (2009). Increasing student awareness of and interest in engineering as a career option through design-based learning. *International Journal of Engineering Education, 25*, 788–798.

Rhodes, J. E. (2004). The critical ingredient: Caring youth-staff relationships in afterschool settings. *New Directions for Youth Development, 2004*(101), 145–161.

Rogoff, B. (1990). *Apprenticeship in thinking: Cognitive development in social context*. New York: Oxford University Press.

Roth, W. M. (2008). The nature of scientific conceptions: A discursive psychological perspective. *Educational Research Review, 3*(1), 30–50.

Sadler, P. M., Sonnert, G., Hazari, Z., & Tai, R. (2012). Stability and volatility of STEM career interest in high school: A gender study. *Science Education, 96*(3), 411–427.

Schnittka, C. G. (2009). *Engineering design activities and conceptual change in middle school science*. ProQuest LLC. 789 East Eisenhower Parkway, PO Box 1346, Ann Arbor, MI 48106.

Schnittka, C. G. (2012). *How Kentucky coal keeps the lights on: Preservice teachers' conceptions about energy*. A paper presented at the University of Kentucky STEM symposium, Lexington.

Schnittka, C. G., & Bell, R. L. (2011). Engineering design and conceptual change in the middle school science classroom. *International Journal of Science Education, 33*, 1861–1887.

Schnittka, C. G. & Ewald, M. L. (2013). *Research results: The Alabama STEM Studio for Afterschool Learning (TASSAL)*. A paper presented at the Auburn University Outreach Symposium, Auburn.

Schnittka, C. G., Brandt, C. B., Jones, B. D., & Evans, M. A. (2012). Informal engineering education afterschool: Employing the studio model for motivation and identification in STEM domains. *Advances in Engineering Education, 3*(2), 1–31.

Schnittka, C. G., Evans, M. A., Drape, T., & Won, S. (2013). Looking for learning in afterschool spaces. *Proceedings of the American Society for Engineering Education*, Atlanta.

Schnittka, C. G., Turner, G., Colvin, R., & Ewald, M. L. (2014). A state-wide professional development program in engineering with science and math teachers in Alabama: Fostering conceptual understandings of STEM. *Proceedings of the American Society for Engineering Education*, Indianapolis.

Schnittka, C. G., Evans, M. A., Drape, T. D., & Won, S. (2015). Looking for learning in afterschool spaces: Studio STEM. *Research in Science Education*. doi:10.1007/s11165-015-9463-0

Seiler, G. (2001). Reversing the "standard" direction: Science emerging from the lives of African American students. *Journal of Research in Science Teaching, 38*(9), 1000–1014.

Senge, P. (1994). *The fifth discipline fieldbook: Strategies and tools for building a learning organization*. New York: Doubleday.

Shapka, J. D., Domene, J. F., & Keating, D. P. (2006). Trajectories of career aspirations through adolescence and young adulthood: Early math achievement as a critical filter. *Educational Research and Evaluation, 12*(4), 347–358.

Sobel, D. (2004). *Place-based education: Connecting classrooms & communities*. Great Barrington: The Orion Society.

Strike, K. A., & Posner, G. J. (1982). Conceptual change and science teaching. *European Journal of Science Education, 4*(3), 231–240.

Tai, R. H., Sadler, P. M., & Mintzes, J. J. (2006). Factors influencing college science success. *Journal of College Science Teaching, 36*(1), 52.

Torp, L., & Sage, S. (1998). *Problems as possibilities: Problem-based learning for K-12 education*. Alexandria: ASCD.

Trusty, J. (2002). Effects of high school course-taking and other variables on choice of science and mathematics college majors. *Journal of Counseling and Development, 80*(4), 464.

Wang, H-H., Moore, T. J., Roehrig, G. H., & Park, M. S. (2011) STEM integration: Teacher perceptions and practice, *Journal of Pre-College Engineering Education Research (J-PEER)1*(2), Article 2. http://dx.doi.org/10.5703/1288284314636

Wigfield, A., & Eccles, J. S. (2000). Expectancy-value theory of achievement motivation. *Contemporary Educational Psychology, 25*, 68–81.

Wiggins, S., & Hepburn, A. (2005). Discursive psychology. *Discourse & Society, 16*(5), 595–602.

Chapter 6
Instrumental STEM (iSTEM): An Integrated STEM Instructional Model

Daniel L. Dickerson, Diana V. Cantu, Stephanie J. Hathcock, William J. McConnell, and Doug R. Levin

Instrumental STEM (iSTEM) is a novel instructional model for science teachers that assists in the incorporation of technology, engineering, and mathematics in ways that are organic, understandable, and replicable. The general premise is that students design, build, and maintain the tools and instruments they need to do authentic scientific inquiry. This model produces relevance for students by requiring the successful design, fabrication, and maintenance of tools and instruments necessary to answer questions they may have during the inquiry process. It also helps build creativity, critical thinking skills, and teamwork as students engage in design process, identify options and provide rationales for materials selection, and work collaboratively with others (e.g. students, teachers, industry, the public, etc.). This process is an organic and authentic one, in that, many practicing scientists must understand, build, modify, and maintain tools and instruments in order to do their jobs. This means that if a science teacher intends to teach a student how to do science like practicing scientists, then the instruction would include building, modifying, and maintaining tools and instruments. Additionally, to help students construct appropriate understandings, the teacher needs to make explicit the technology, engineering, and mathematics concepts involved, not just the scientific concepts. All science disciplines use tools and instruments, so iSTEM is applicable across all science disciplines. This model has been successfully implemented on a small-scale

---

D.L. Dickerson (✉) • D.V. Cantu
East Carolina University, Greenville, NC, USA
e-mail: dickersond15@ecu.edu

S.J. Hathcock
Oklahoma State University, Stillwater, OK, USA

W.J. McConnell
Virginia Wesleyan College, Norfolk, VA, USA

D.R. Levin
Washington College, Chestertown, MD, USA

© Springer International Publishing Switzerland 2016
L.A. Annetta, J. Minogue (eds.), *Connecting Science and Engineering Education Practices in Meaningful Ways*, Contemporary Trends and Issues in Science Education 44, DOI 10.1007/978-3-319-16399-4_6

during a NOAA-funded effort, Project SEARCH. The model and its application are described in this chapter.

The Standards Connection

There is a national push to promote STEM education for K-12 students. In the report, *National Action Plan for Addressing the Critical Needs of the U.S. Science, Technology, Engineering, and Mathematics Education System*, the National Science Board (2007) has advocated for STEM education reform as they believe current and future generations will need basic STEM literacy in order to function in our global society. This concern was echoed in *Rising Above the Gathering Storm, Revisited: Rapidly Approaching Category 5* (NRC 2010), which found that many of the original recommendations made in the initial *Rising Above the Gathering Storm* (NRC 2007), have not been acted upon. Examples include increasing science and technological literacy, and changing education so we are producing more scientists, technologists, engineers, and mathematicians (NRC 2010). President Obama and many states throughout the country have also championed the creation of a STEM literate society and have endorsed legislation and funding (U.S. D.O.E. 2010; NSTCC 2011; CRS 2012) toward meeting this goal. During a speech to endorse the Educate to Innovate Campaign, a national push for STEM education initiatives on November 23, 2009, President Obama's dedication to STEM education was made clear, "I'm committed to moving our country from the middle to the top of the pack in science and math education over the next decade," (The White House 2009). Committees within the federal government (PCAST 2010; NSTCC 2011) and at the state level (National Governors Association NGA 2011; Governor's STEM Task Force 2009) have delved further into what can be done to achieve this goal. Building a STEM literate society has left many to wonder how we are going to change education to cultivate domestic talent in order to remain globally competitive. Some believe reworking standards documents and frameworks may help in achieving this goal.

Individual STEM standard documents have come to suggest a more integrative approach within their standards. For example, *A Framework for K-12 Science Education* (NRC 2012) and the *Next Generation Science Standards* (NGSS Lead States 2013) describe integration of mathematics, technology and engineering within science teaching, and the *Standards for Technological Literacy* (ITEA 2007) provide guidance in the development of content knowledge and skills that bridge standards in other STEM disciplines. However, these integrative STEM approaches and standards present certain challenges as it may not be common practice to create purposeful integrative lessons. Another concern is whether various standards are fully understood in order to create alignment during the creation of an integrative lesson. Standards can continue to be written, however if they are written in such a way that teachers cannot understand or use them, then they will not provide the desired results for which they were written (PCAST 2010). The President's Council of Advisors on Science and Technology (PCAST) (2010) believes that for STEM education to flourish, teacher preparation, school leadership, and high-quality instructional materials are necessary to facilitate the successful use of standards.

Krajcik and Merritt (2012) assert that past science curricula have "present[ed] too many ideas too superficially, often leaving students with disconnected ideas that cannot be used to solve problems and explain phenomena they encounter in their everyday world" (p. 10). They believe the new framework may empower science educators to teach concepts in a deeper and more meaningful way. For science educators, the new science framework helped establish the challenge to include science, technology, and engineering practices in their classrooms and instruction. The committee of *A Framework for K-12 Science Education* believed the addition of technology and engineering into the framework would enhance science education. They state,

> Science, engineering, and technology permeate nearly every facet of modern life, and they also hold the key to meeting many of humanity's most pressing current and future challenges. Yet too few U.S. workers have strong backgrounds in these fields, and many people lack even fundamental knowledge of them. This national trend has created a widespread call for a new approach to K-12 science education in the United States. (NRC 2011, p. 1)

The *NGSS* (2013) authors suggest that the learning of science can be more coherent if there is emphasis on the integration of scientific explanations and practices needed to engage in scientific inquiry and engineering. These documents also look at how knowledge and practice must be intertwined in designing learning experiences in K-12 science education (National Research Council 2012). According to Bybee (2011), the message is not that scientific inquiry is being replaced, however science teaching and learning is being enriched by the expansion of how science will be now be taught. He states, "When students engage in scientific practices, activities become the basis for learning about experiments, data, and evidence, social discourse, models and tools, and mathematics and for developing the ability to evaluate knowledge claims, conduct empirical investigations, and develop explanations" (Bybee 2011, p. 10).

Technology educators have had science and engineering content in their standards document, *Standards for Technological Literacy* (2007), for over a decade. The International Technology and Engineering Educators Association state that the study of technological literacy is important to our everyday lives as technology impacts almost every aspect of our daily living (ITEA 2007). For students to achieve their full potential, it is important that students understand all the components of STEM and gain the necessary knowledge and skills to become global leaders (ITEEA 2009). Cajas (2001) believes there is an emerging relationship between science and technology in which students will be able to learn valuable ideas that enable them to function in a technological world. He states, "the intersection of science literacy and technological literacy is relevant for science education" (p. 727). According to Cajas (2001), both science and technology share common themes such as tools, systems, models, and scale. Roth (2001) believes science and technology also share fundamental themes, which include the production and transformation of representations and the action-oriented language, which describes both the science and technology domain. Both of these themes have been supported in previous attempts of bringing science and technology together (AAAS 1993).

The response to the NGSS (2013) has come with mixed reviews from technology and engineering educators. According to Keller and Pearson (2012), "in addition to providing the vision for a coherent education in science, the framework has great potential to serve as a rallying point for all groups promoting better student learning" (p. 18). However, they caution there may be some issues that should be considered. The first issue is concerned with whether science education teachers will be able to deliver technology and engineering content sufficiently. The second issue addresses the type of support technology and engineering educators may have to provide science educators delivering the new content. The final issue is concerned with how technology and engineering content is practiced within science education for the enhancement of the new science framework (Keller and Pearson 2012). These three concerns speak to the overarching concern of whether educators can create learning environments conducive for learning within their domain and through an integrative STEM instructional approach. These concerns are leading science educators to seek models of effective instructional strategies that can bridge STEM content and standards for effective science instruction.

Instructional Approaches and Strategies

Science, technology, and mathematics disciplines each have their own K-12 standards and framework documents (NGSS 2013; NRC 2011; ITEA 2007; NCMT 2000; NGAC 2010), which impact what educators do in the classroom. Some of these documents are beginning to reflect integrative STEM education initiatives and more specifically are including heavier doses of technology and engineering concepts. These changes have left many educators to wonder how they can approach science instruction and STEM content in their classrooms in ways that will adequately address such standards. Many teachers feel ill-prepared to address standards that traditionally fall outside of the range of most biology, chemistry, physics, earth science, and science education related programs (Weiss et al. 1994). Furthermore, integration of math and science, a long-standing push, has been and continues to be a struggle for many science teachers. The prospect of adding to that technology and engineering may leave many scratching their heads. Comments such as, *'I'm a biology teacher. I majored in biology not engineering. How can I be expected to know and teach all of this stuff too?'* are regularly heard in many school divisions. This frustration comes in part from a lack of instructional models expressly designed to address integrative STEM content.

Current Efforts

Research exists to support the integration of science, technology, and engineering through design, which helps support student understanding of scientific concepts and promotes both scientific and technological literacy in children (Crismond 2001;

Kolodner 2002). For example, Crismond (2001) presented findings that engaged novice and expert students in design activities coupled with inquiry. The students who had more experience with design, or the expert students, were better able to make connections with science concepts than those who were novices in the design process. Kolodner (2002) engaged students in a program called *Learning by Design*™ where students learned science content by engaging in design activities much like authentic scientific inquiry. Kolodner (2002) states,

> Science education that is truly aimed towards scientific literacy focuses as well on learning the practices of scientists-designing and carrying out investigations in a replicable way, accurate observation and measurement, informed use of evidence to make arguments, explanation using scientific principles, working in a team, communicating ideas, and so on. In fact, scientists and designers practice many of the same skills. (p. 9)

Kolodner (2002) asserts learning design during science instruction allows students to make valuable content connections. These findings have also shown that by integrating science, technology, and engineering concepts and standards, student learning can be increased. In addition, they illustrate the need for further research and teacher support for creating conducive learning environments that create student literacy in both science and technology, and ultimately STEM literacy. Issues exist for many science teachers and administrators regarding such curricula, however. Often the curricula are expensive, require considerable teacher training, and do not easily fit into the current enacted science curriculum.

While some administrators and science teachers attempt to address STEM teaching and learning through commercially available, engineering-focused curricula most employ discrete activities that engage students in design and construction. There are numerous examples of such activities that have been used for years, including building thermometers (Sorey et al. 2010), designing and building boats (Schomburg 2008), designing and building bridges (Roth 1995), etc. Issues also exist for many teachers and students regarding such activities, however. Often the activities are implemented in ways that are not explicitly connected to the science and mathematics content. Such activities may take on the nature of crafts rather than a meaningful, integrated STEM learning experience. Additionally, the products that are produced may not be used in an authentic way, such as to answer a student-generated research question or solve a problem in the student's life. Thus, technological artifact design and construction becomes an act of just doing versus an act of bridging STEM concepts and building STEM conceptual understanding.

To help in the achievement of STEM literacy, science educators are going to need instructional models and strategies that can illustrate how to integrate mathematics, technology, and engineering while conducting effective science instruction. They need to understand how science can be enriched by infusing technology and engineering content, which is now required in their classrooms by the NGSS (2013). Roth (2001) has found that students who engaged in the creation of technological artifacts learned science. He suggested that students who engage in technological activities "are deeply involved in creating and transforming representations in the directions of science and technology, arenas traditionally noted as separate" (p. 786). He goes on to state, "technological activities are therefore prime contenders for an integrated approach to teaching" (Roth 2001, p. 787). The revised Blooms Taxonomy now puts creation at the top of the cognitive process dimension (Anderson et al. 2001). Thus artifact-building helps students not only create and design, but also learn science. However, Sidawi (2009) asserts developing a technological artifact requires more than the knowledge of scientific concepts, but rather the appropriateness of using technology and design in a specific context. Consideration of the creation of more meaningful learning activities where integrative STEM can be used is also important. Novak (2002) believes that for students to achieve learning that is meaningful to them, "the construction of new meanings requires that an individual seeks to integrate new knowledge with existing relevant concepts and propositions in their cognitive structure" (p. 557) which he states is "not easy to move science and mathematics instruction from the traditional approaches emphasizing rote memorization to patterns where meaningful learning predominates" (p. 561). When designing technology in a science classroom, careful consideration must be given to the context through which students can apply scientific knowledge and processes. It then becomes important for science teachers to have instructional strategies and models to follow in order to achieve STEM integration.

Roberts and Cantu (2012) proposed three instructional models that can be used in teaching STEM: silo, embedding, and integration. Silo is the traditional model in which each of the STEM domains is taught in isolation. This model runs counter to the recommendations incorporated in the NGSS (2013) and other STEM education standards and frameworks documents. Embedding is characterized as teaching and emphasizing science content while bringing in engineering and technology concepts which are not assessed (Roberts and Cantu 2012). Integration refers to addressing the content of each STEM domain (as described in the NGSS (2013)) in a common context with lesson objectives and related assessments tied to each discipline. This enables a student to see a purposeful connection between each of the STEM disciplines, thus making the learning experience more intrinsic to the learner. Therefore, we advocate for an integrated approach to STEM instruction. Many science teachers feel the same. However finding an instructional model that explicitly addresses content from four different domains in a single context within a single lesson has been frustrating. In response, we sought to design a model that would adequately address pedagogical, content, and logistical considerations – the result was Instrumental STEM (iSTEM).

iSTEM Model

We consider Instrumental STEM (iSTEM) to be an instructional model that can be implemented in formal, non-formal, and informal contexts (Ainsworth and Eaton 2010) for the purpose of enhancing understandings of STEM-related content and processes. The nature of the STEM-related content is integrated and occurs in an authentic scientific inquiry context. Consequently, we acknowledge that the philosophical perspectives that form the underlying assumptions in the model originate from a scientific position consistent with the basic elements of nature of science and inquiry process skills as articulated in guiding standards documents and frameworks, such as the Next Generation Science Standards (NGSS Lead States 2013).

We do not suggest that this is the only or even the best model for integrated STEM instruction for every instance because many variables exist that impact optimal pedagogy for particular content in a given context. That is, pedagogical content knowledge (PCK) (Shulman 1994) may dictate a different instructional model based on the context. For example, effectively teaching students skills may require a more direct teaching approach that does not incorporate high levels of inquiry. We do consider iSTEM, however, to be a powerful tool for curriculum developers, classroom teachers, and students to engage in effective STEM teaching and learning. Much of the strength of the iSTEM model comes from leveraging other empirically supported, best-practice strategies in science education. For example, iSTEM should be implemented using practices that are inquiry-based (e.g. 5E Learning Cycles and Project-Based Learning) (NRC 1996; Llewellyn 2011; NRC 2000; Carin et al. 2005; Fraser-Abder 2011; Koch 2010; Buxton and Provenzo 2007), use appropriate instructional technologies (e.g. implementation of authentic tools of the discipline such as data loggers) (NRC 1996; Llewellyn 2011; Settlage and Southerland 2007; Sherman and Sherman 2004), involve place-based learning (e.g. learning is placed in the context of the students' community) (Tippins et al. 2010; Bodzin et al. 2010), and use authentic, alternative assessments (e.g. formative and performance-based assessments) (NRC 1996; Keeley 2008; Enger and Yager 2001; Liu 2010; NRC 2000).

The iSTEM model is one that can be superimposed over almost any inquiry-based instructional model that engages students in the collection of data. iSTEM does require inclusion of some elements that are not always made explicit or student-centered, such as identifying the tools to be used in a given scientific study. It also requires explicitly addressing additional content, namely technology and engineering concepts. In general, the iSTEM model is operationalized as such:

1. Determine the scientific study in which you will engage
2. As part of developing your study and its methods, identify the tools and instruments commonly used in such studies
3. Learn how the tools and instruments are made, how they work, how they are typically applied, and how they are maintained
4. Based on the nature of the study and instruments to be used, have students design, build, and/or maintain the instrument or apparatus
5. Conduct the study
6. Students provide self-assessment and indicate what they would change about their instrumentation and why and how it impacts their findings
7. Teacher makes explicit content connections regarding the nature of the domains (i.e. science, technology, engineering, and mathematics)
8. Teacher explicitly addresses the content for each domain's standard
9. Have students present their findings, explicitly commenting on the tools used

10. Conduct a new, related study, that incorporates the new knowledge they constructed

A more detailed explanation of teacher and student roles is included in Table 6.1.

It is important for the steps to be taken as indicated (i.e. 1–10). Deletion or rearrangement of the steps tends to weaken the model's impact. For example, if one were to rearrange a 5E learning cycle by switching the 'explain' and 'explore' portions, then the innate inquiry-based nature of the lesson is diminished, if not lost. The content included in this model falls under two categories: (1) natures of the domains (i.e. nature of science, nature of technology, nature of engineering, and nature of mathematics) and (2) domain specific content (e.g. forces and motion, use of prototypes, etc.). Lessons are typically derived from state and national standards documents and frameworks (e.g. Next Generation Science Standards (2013)) and incorporate one or both of the two categories of content always addressed in the iSTEM model. The teacher or school division determines the primary and related standards to be addressed. From these primary and related standards, the teacher often develops objectives for the lesson. Each iSTEM lesson typically has 4 primary and 4 related standards (i.e. 4×4 system) and associated objectives. Each one of the primary and related standards is focused on each of the STEM domains in each of the two content categories. An example is provided in Table 6.2. If the primary standards address the Natures of Domains (NOD), then the related standards would address domain specific content. Conversely, if the primary standards address domain specific content, then the related standards would address NOD. The main point is that standards regarding NOD need to be addressed each time an iSTEM lesson is taught. While this may seem like a lot of standards for a single lesson, it is important to note the lesson duration, which is considerably longer than a single lesson on a single standard. According to our experience, the number of standards addressed in a given time period using iSTEM is not out of the ordinary, particularly when interdisciplinary lessons are taught. Typically one objective is developed from each standard, although at times two or even three may be appropriate. Multiple objectives within the iSTEM model are the exception rather than the rule however due to the many moving parts of the lesson to which the teacher needs to attend. Additionally, a single standard may be used multiple times to address the different domains (see Table 6.2 for example). Using the 4×4 system of primary and related standards, content from each of the STEM domains can be incorporated into a single lesson using an authentic context that allows students an opportunity to experience, in a supported setting, how the different domains of STEM are often incorporated into scientific practice.

The teacher should explicitly and directly address the content (Steps 8 and 9) in ways that explicitly and directly connects to the experiences the students had while conducting their studies. Connecting back to the students' shared experiences is an important element of the iSTEM model. It promotes student interest in the content by providing them context and reason to care about the material. Typically student groups conduct the studies. These groups generally consist of four or fewer students per group. It is critical that if student groups are used, that each member of the group

Table 6.1 iSTEM teacher and student roles

iSTEM component	Teacher actions	Student actions
1. Determine the scientific study	Provide a topic that should be derived from the standards or program goals Next provide a research question, hypothesis, and methods or instruct the students to develop their own research question, hypothesis, and methods, if using complete student-centered inquiry Only answer questions directly if necessary, otherwise use appropriate inquiry-based questioning strategies	Review age appropriate literature regarding the research topic to better understand how the teacher developed the research question, hypothesis, and methods or to inform their own development of the research question, hypothesis, and methods Ask questions about conducting literature reviews and identifying authoritative sources Ask questions about the study design Be creative Record important information
2. Identify the tools and instruments needed	Based on the students' age and level of student-centeredness of the lesson, either tell the students what tools and instruments are to be used or have them identify those that are most appropriate	Record what tools and instruments are needed for their study
3. Learn how the tools and instruments work	Provide the students with direct instruction regarding the tools and instruments you told them would be used or have them explore the literature to learn about the tools and instruments they identified	Record how the tools and instruments are made, how they operate, how they are typically applied, and how they are maintained
4. Design and build the instrument or apparatus	Based on the students' age, level of student-centeredness of the lesson, and understanding of design process, provide the students with an activity that guides them through the construction of a tool or instrument or have them engage in design process to design and fabricate an instrument or apparatus Only answer questions directly if necessary, otherwise use appropriate inquiry-based questioning strategies Monitor for student safety	Follow the instructions to build the tool or instrument and determine how it should be used in the study or engage in design process and design and fabricate the instrument or apparatus Use materials and tools to build an instrument or apparatus that they will actually use to collect data Be creative Ask questions
5. Conduct the scientific study	Facilitate student completion of scientific studies Facilitate student maintenance of the instrument or apparatus to be used in the study Monitor for student safety	Conduct their study including recording their methods, results, and conclusions Maintain an instrument or apparatus to be used in the study Ask questions

(continued)

Table 6.1 (continued)

iSTEM component	Teacher actions	Student actions
6. Student self-assessment of their study	Provide prompt for student self-assessment Visit individual students or groups to ask probing questions	Work individually or in groups to determine how successful their study was and what role their instrument played in that success Identify what they would change next time regarding instrument design and implementation to improve their study
7. Natures of the domains (NOD) (i.e. science, technology, engineering, and mathematics)	Provide explicit instruction regarding elements and processes involved in science, technology, engineering, and mathematics Provide prompt that requires students to reflect on their studies Review student responses for appropriateness	Attend to the explicit instruction regarding the nature of science, technology, engineering, and mathematics Ask questions Work individually or in groups to reflect on how the tenet and/or processes presented by the teacher were exemplified during the course of their studies
8. Domain specific instruction	Provide explicit instruction regarding the content standards dictated by the standards documents in each of the four domains Provide a prompt that requires students to reflect on their studies Review student responses for appropriateness	Attend to the explicit instruction regarding the content standards Ask questions Work individually or in groups to reflect on how the content presented by the teacher was exemplified during the course of their studies
9. Presentation of the scientific studies	Provide instructions for the presentations Facilitate discussions and ask probing questions during the presentations	Develop and deliver presentations on their studies Include all elements typically present in professional scientific presentations Answer audience questions
10. Conduct a new, related study	Facilitate student completion of scientific studies Monitor for the application of their newly developed content understandings Monitor for student safety	Conduct their new, related study including recording their methods, results, and conclusions Use the new language and content they learned from earlier in the lesson Ask questions

6 Instrumental STEM (iSTEM): An Integrated STEM Instructional Model

Table 6.2 iSTEM 4×4 standards matrix: example lesson for middle school students (NGSS – MS 6–8, PS2A: Forces and motion – MS. Forces and interactions)

	Science	Technology	Engineering	Mathematics	
STEM content	Newton's third law. (NGSS PS2.A 2013, p. 59)	Use tools, materials, and machines safely to diagnose, adjust, and repair systems. (Standards for technological literacy, ITEA 2007, p. 130)	Defining and delimiting engineering problems. (NGSS ETS1.A 2013, p. 86)	Write, read, and evaluate expressions in which letters stand for numbers. (Common core, NGAC 2010, 6. EE.A.2, NGSS webpage for forces and interactions, grades 6–8)	
	Science knowledge is based upon logical and conceptual connections between evidence and explanations (NGSS 2013, NGSS webpage for forces and interactions, grades 6–8)	The uses of technologies and any limitations on their use are driven by individual or societal needs, desires, and values; by the findings of scientific research; and by differences in such factors as climate, natural resources, and economic conditions. (NGSS 2013, NGSS webpage for forces and interactions, grades 6–8)	Men and women from different social, cultural, and ethnic backgrounds work as scientists and engineers. (NGSS 2013, Appendix H, p. 6)	Mathematicians reason abstractly and quantitatively. (Common Core (NGAC 2010, MP.2, NGSS webpage for forces and interactions, grades 6–8)	**Nature of discipline**

has work to do that is essential to the success of the project and that over the course of the year, all students have equal access to all components of the study design and implementation. In other words, if it does not take a group to complete the study, then do not use one. Normally, it should however, particularly when using the iSTEM model. This is one way to decrease the amount of time it takes to complete the lesson. There are multiple instances where division of labor will provide each student with an essential task while requiring a shared understanding of the study. However, a student is unlikely to be able to engage with all the various tasks of a given study in isolation, particularly in a realistic timeframe. This is one rationale for the need to implement the iSTEM model multiple times over the course of the school year. Additionally, students need to be able to engage in the model's elements multiple times so students can apply what they have learned. This is one reason that Step 10 is so important. This new, related study does not have to occur on Wednesday if Step 9 was completed on Tuesday. We suggest, however, that the new study occur before the class has moved on to a unit that is not directly related to the content addressed in Steps 1–9. This allows for continuity between the studies and additional opportunities for assessment within the unit.

We purposefully have not included appropriate grade levels because we consider this model to be one that can be implemented across a wide range of ages and abilities. This flexibility comes from the teacher's choice regarding the spectrum of (1) student-centered scientific inquiry and the (2) design and fabrication of the instrument. Younger students and those new to scientific inquiry or design process typically need instruction that falls on the teacher-directed end of the spectrum. As students learn more, they are capable of increased autonomy regarding both spectrums. Dependent on the opportunities the students received to learn and use scientific process skills and engage in designing and building things, they will likely be able to progress across the spectrums at different rates. The iSTEM model requires special attention on the part of the teacher regarding both the science and the design-based spectrums. For example, it is common for students in an AP Chemistry class to be proficient at setting up and conducting a scientific study, but be relatively unfamiliar with design process and even more so with fabrication tools and techniques. Unlike most settings, where differentiation occurs infrequently, superficially, and unsustainably across grade levels, the iSTEM model offers the potential for students to excel in areas where they are already proficient and become proficient in areas where they need significant improvement regarding each of the spectrums.

iSTEM Applied

The iSTEM model was further developed and applied during a 3-year project funded by the National Oceanic and Atmospheric Administration (NOAA) Bay Watershed Education and Training Program (B-WET). The project entitled, *S cience E ducation A dvancing R esearch of the C hesapeake Bay and Its H abitats (SEARCH) (NA09NMF4570008)*, provided urban, rising sixth graders with Meaningful Watershed Experiences (MEWEs) and supported the teachers of those students in implementing the MEWEs by providing materials, in class support, and professional development opportunities. The MEWEs and teacher experiences involved building observation buoys in order to explore their own research questions, field sampling aboard a research vessel, interacting with scientists and environmental lawyers from underrepresented populations, and learning about current local scientific issues from nationally recognized experts and from the data they collected and analyzed. More specifically, the project objectives included: (a) Enhance urban teachers' and students' STEM literacy with an emphasis on their relationship with the Chesapeake Bay system; (b) Increase interest of underrepresented populations in earth/environmental science fields; (c) Need to enhance American competitiveness by increasing students' abilities to conduct authentic scientific inquiry using appropriate technologies; and (d) Increase teachers' knowledge and use of technology-enhanced, inquiry-based instructional strategies.

Development of iSTEM

Initially, the inquiry-based instructional model and strategies we included in our work with teachers and students revolved around the 5E Learning Cycle (Bybee 1997), inquiry-based questioning strategies (Llewellyn 2002), and assessment of and for conceptual understanding (Baker and Piburn 1997). Shortly after the start of Project SEARCH, I (Dr. Dickerson) participated in the Basic Observation Buoy (BOB) II Workshop that was executed at the University of North Carolina – Wilmington Center for Marine Science by SECOORA, COSEE SE, and NC Sea Grant. At the workshop I connected with other science educators, scientists, classroom teachers, and students who were using BOBs, including the original developer Dr. Doug Levin. Additionally, industry representatives (e.g. YSI) provided demonstrations of various instruments. Participants discussed their experiences with various sensors (e.g. Pasco and Vernier), describing the pros and cons of each. Design issues were extensively discussed, including data retrieval and use. We had an opportunity to handle and use various sensors and instruments from multiple vendors. The lessons learned from that workshop and subsequent conversations with Dr. Levin served as the seeds to the development of iSTEM.

Implementation of iSTEM

The primary implementation of iSTEM occurred in the context of the SEARCH Summer Academies. Each year for 4 years (the first 3 years were funded by NOAA and the forth year was funded by Dominion Power), students from Portsmouth Public Schools were engage from 9 am–3 pm in intensive hands-on, inquiry-based STEM activities that challenged them to think about socio-scientific issues related to their local environment (i.e. the Chesapeake Bay). Guest speakers, included: (a) personnel from a local NGO, Elizabeth River Project (ERP), who informed students about ERP and about how ERP could use the data the students collect from their buoy in conjunction with their ongoing programs; (b) a faculty member specializing in industrial technology from ODU's STEM Education Department helped students learn about design process; and (c) an attorney with the Southern Coalition for Social Justice, presented information regarding environmental law, environmental justice, and integrated STEM careers. Students worked together to learn about observation buoys and their roles in helping answer questions about our environment by initially building small-scale buoys. As the week progressed, the teachers and students constructed a full-scale observation buoy outfitted with educational and scientific quality instruments. Additionally, students deployed passive sampler buoys to collect toxicological data. While there were slight changes made from year to year across the iterations of implementation, the general experience occurred as follows.

iSTEM Step 1: Determine the Scientific Study

On the first day of the Summer Academies we introduced students to the environmental challenges the Elizabeth River and their community faced. We provided the topic and overarching research question, "What is the condition of water of quality in the Elizabeth River?" From this prompt and a requirement to use an observation buoy, student groups sought information about what others have found regarding the health of the Elizabeth River. This included visiting many different websites, asking questions of the program leaders, and listening to and asking questions of the invited speakers. Once they felt they had enough information to move forward, they decided what element of water quality they wanted to examine. Common parameters selected included temperature, salinity, dissolved oxygen, and turbidity. Other students also wanted to examine the area for the presence of micro and/or macro organisms. Based on the research question the program leaders posed, we did not require students to attempt to design experiments because we considered other scientific methods more appropriate for the research question and for the students' ages and abilities. Instead students employed case study and naturalistic observation. We stressed that in order for the research question to be viable, it needed to add to the body of knowledge in the field. We explained that one way to do this was to identify a new study context. For example, while dissolved oxygen studies had been performed on the Elizabeth River and those were available, the sample sites did not include locations near the students' homes. By focusing on these locations, the students would be adding to the body of knowledge. They used other studies they read or were told about to inform the development of their own studies. Some used the exact same methods described in the published studies, while others modified the methods for various reasons (e.g. access to research equipment was limited, time was limited, they believed their idea was more creative, etc.).

iSTEM Step 2: Identify the Tools and Instruments Needed

After the students decided on their general methods, they began making a list of the all the instruments and tools they would need in order to conduct their study. They generated the list primarily from the resources they found to complete Step 1. Many students learned that ships, satellites, and buoys are used to collect water quality data. Some discovered that sensors are used and other found out that some scientists use nets to collect organisms.

iSTEM Step 3: Learn How the Tools and Instruments Work

Students used multiple instruments to address a broad range of research subquestions relating to the program leader-provided overarching research question. All of these subquestions and their related instruments cannot be described in this chapter. As such, we will focus on one of the more common instruments used, a thermometer. While many different research subquestions were asked that involved the use of the thermometer they often required different applications of the instrument. For example, some thermometers needed to be above the water line, while others needed to be just below the surface, while others needed to be multiple meters down in the water column. None of this information was explicitly communicated to the students during this Step however. Instead, we provided students with an opportunity to build their own thermometer (California Energy Commission 2006). They also read online about other types of thermometers that exist (e.g. bi-metal, digital, infrared) and how they work.

iSTEM Step 4: Design and Build the Instrument or Apparatus

Students engaged in an open-ended, trail and error design lesson where they competed against their fellow students to build a buoy out of PVC pipe that would hold the most golf balls (Dickerson et al. 2012). After the students completed the activity, in a scaffolded fashion, we provided the students with questions that asked them about placement and rationale of use for different types of thermometers. The first worksheet asked about the thermometer they built in class. The second asked about a commercial glass bulb thermometer and the third asked about a digital thermometer with a data recorder. After completing the worksheets and discussing their responses with their group, students learned that the type of instrument and its placement determines the type of data you will get. This begins the portion of lesson where most groups began an iterative process between Steps 1, 2, and 3 in terms of altering their selection of instruments and apparatus to better address their research subquestion or they altered their subquestion.

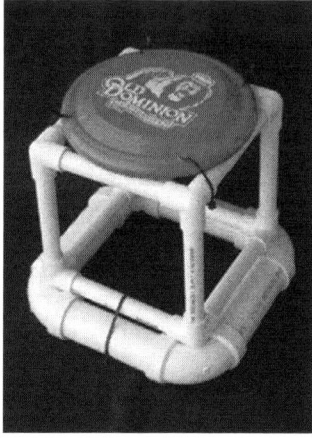 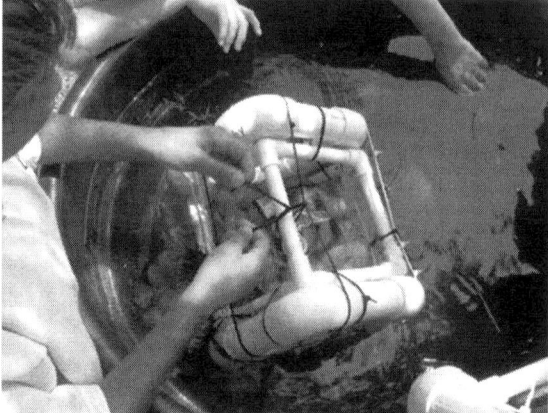

An initial and unstable buoy design

A more stable design supporting many golf balls

iSTEM Step 5: Conduct the Scientific Study

During this Step, the students are applying what they have learned from the other Steps to implement an authentic scientific research study that uses a researcher designed and fabricated, instrument or apparatus. During this Step, we told the students what materials we had and asked them what additional materials were needed. Then we went and bought them if available. For the students in Project SEARCH, this meant using larger pieces of PVC (up to 4" diameter) to create buoys that held educational sensors such as Vernier probeware to professional YSI sondes with telemetry. The buoys outfitted with Vernier probeware had to be checked manually on a regular basis but were sufficient for the research subquestions asked. Other subquestions that required long periods of frequent data collection in remote areas required the YSI instrumentation and telemetry, which sent data via cell signal to a server that housed a web portal at the University of North Carolina at Wilmington. Other subquestions required digital cameras (e.g. GoPro) or sonar (e.g. electronic fish finder).

Large observation buoy with YSI instrumentation and telemetry

Direct, explicit instruction regarding the telemetry

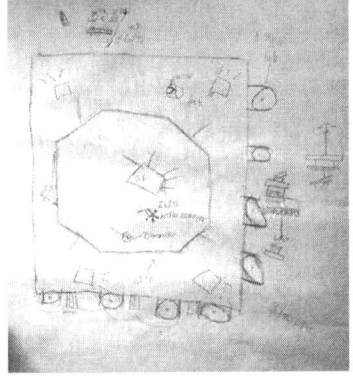
An example of one iteration of student design

Components of the telemetry system

iSTEM Step 6: Student Self-Assessment of Their Study

Students were required to reflect on their findings and how their choices regarding instrumentation impacted their studies. Every buoy design was different even when the subquestions were very similar. This was an interesting element that allowed students to discuss how they thought their designs and builds impacted their findings as well as their fellow students'. Often during this Step students expressed their desire to do another round of data collection or another study because there were elements of their instruments they wanted to change.

iSTEM Step 7: Natures of the Domains (NOD)

During this portion of the lesson, we explicitly addressed a number of NOS misconceptions including the misconception of the single, rigid scientific method. Students worked in groups to map their experiences while conducting authentic scientific inquiry onto a list of NOS and NOD ideas we provided. For example, students correctly identified that they conducted case studies and naturalistic observation and they were, in fact, engaged in authentic scientific practice. Additionally, we asked students how is what they did today like what scientists, technicians, engineers, or mathematicians do? (Akerson et al. 2014)

iSTEM Step 8: Domain Specific Instruction

Domain specific instruction occurred throughout the lesson, although it was student-led and supported by inquiry-based questioning strategies. In this Step, however, we provided a lecture regarding the content in each domain. Specifically, we provided content instruction regarding habitats and point and non-point source pollution. We also provided instruction regarding design process. Students again mapped their experiences of building their small golf ball challenge buoys and their larger observation buoys onto a design process figure. Technology issues related to stainless steel construction as opposed to galvanized or zinc-plated were addressed in relation to their buoy deployments and oxidation. Additionally, we discussed the use of copper and other technologies used to address biofouling. Mathematics concepts addressed explicitly during the lecture portion included issues related to measurement such as use of measuring devices and unit conversion. Precision and accuracy was also addressed. Several algebra and geometry concepts were addressed too, including solving for an unknown and finding the areas and volumes of objects.

iSTEM Step 9: Presentation of the Scientific Studies

Students developed presentations that they delivered to each other and in some case to other students and adults within the community. They incorporated the language and concepts learned throughout the lesson in their presentations. They also included limitations, particularly with regard to their instrumentation, and suggestions for future study.

iSTEM Step 10: Conduct a New, Related Study

Since Project SEARCH was a grant funded effort and the development and refinement of iSTEM took multiple years, sustained experiences including conducting additional studies beyond those completed during the Summer Academy occurred primarily in an after-school capacity. For example, an after school club, SEARCH Club, was started and members from each Summer Academy attended as able. The SEARCH Club offered an opportunity for the students to continue their participation in the studies they had begun as well as beginning new studies in which they were interested. The students continued to follow the iSTEM model but at a much slower pace as the SEARCH Club met only once a month. For example, specific activities included experiences aboard the RV Slover where they were introduced to marine science tools and instruments and how they work (i.e. iSTEM Step 3).

Students learning about marine science instrumentation aboard the RV Slover

Students began to broaden their communication of findings by providing professional presentations regarding their work to other students and adults. For example, through NOAA funding, we were able to provide a stipend for former Summer Academy students to present to subsequent Summer Academy students. Through funding for another separate initiative, (e.g. RiverQuest), that addressed environmental education and environmental science education topics, Project SEARCH students received a stipend to present their work and were included in the movies produced as part of that project. In these ways, students were engaged in iSTEM Step 9 of both the original and subsequent studies.

Bringing iSTEM to Preservice Teachers

We have also very recently implemented the iSTEM model with preservice teachers in the context of a secondary science education methods course and university student chapter of the National Science Teachers Association (NSTA). We followed the same model and similar activities to those presented with the Project SEARCH students. The two primary differences involved the context, which was solely aimed at the use of passive samplers and the direct and explicit incorporation of pedagogical instruction. These students used passive sampler technology to engage in Policy-Ready Citizen Science (Dickerson and Hathcock 2011). That is, through a grant from the Virginia Department of Environmental Quality (VA DEQ), the program leaders completed paperwork that paved the way for the data collected by the students to be used directly in state and federal reports. There is no middleman. The citizen *IS* the scientist and is actually partnering with a state agency in that capacity to collect water quality data regarding environmental toxicants. So rather than building larger buoys that house very expensive instrumentation, the preservice teachers used small quantities of PVC or crab pot buoys, stainless steel cages, and data collection plates which come to less than $100US including paying for the chemical analysis of the plates. Those data then went straight to VA DEQ and then to Congress where they are used in policy-making decisions.

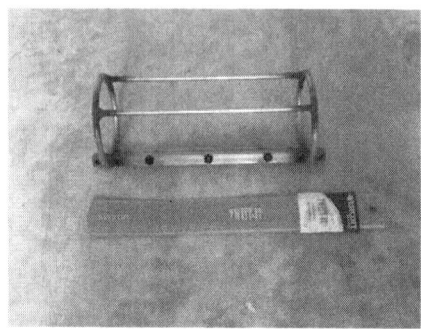 

Passive sampler cage   Teaching about biofouling the best way possible

The preservice teachers have very much enjoyed the challenging, hands-on aspects of the activities (e.g. the Build-A-Buoy Challenge). They also enjoyed working collaboratively with our Industrial Technology students. For example, when making the stainless steel cages, we buy flat and round stock and then use the TIG welder to build the cages. This means that the students need to understand how the film on the plates works and about biofouling in order to know that stainless steel should be used and how the cage should be designed in order to protect the film.

General Strengths

In addition to the basic structure, there exist a number of strengths that make the iSTEM model one that warrants consideration by those interested in implementing an inquiry-based, integrated STEM instructional model. Some of the key issues in the development of the model were congruency with best practice strategies in science education, working within the logistical constraints of the current public education system, and flexibility to support customization by teachers and continued fit with emergent technologies and instructional innovations. The following provide an overview of some of the general strengths associated with the iSTEM model.

Makes Use of Existing Curriculum

There are examples within formal science education contexts where integrated STEM classes are being developed and taught, but this is the exception rather than the rule. Overwhelmingly, science education continues to occur throughout K-12 education as isolated units that address discipline specific content. For example, in most high schools, courses such as biology, chemistry, physics, and earth/environmental science continue to be the norm. In middle schools, some variations of these exist as life science, physical science, and earth science. Elementary school science curricula perhaps comes the closest to integrating the different domains of STEM, when it is taught. This in large part is an artifact of themed instruction (much more common in elementary education) that draws from various domains and disciplines. For example, a unit that uses a school garden may include lessons on Native American culture (social studies), an experiment with worms (life science or biology), and activities regarding soils (earth or earth/environmental science). This same school garden would provide a very rich context for iSTEM lessons. The issue is not opportunity but rather formal, replicable instruction that occurs in a purposeful and authentic way with regard to STEM integration. That is the novel element the iSTEM model brings to the table. Using the iSTEM model, the science teacher does not have to adopt a whole new curriculum or significantly alter what they teach or when they teach it. They can still teach their unit on Native American culture using the school garden by mapping what they do into the iSTEM 4×4 Standards Matrix (Table 6.2) and identifying what tools or instruments would help the students conduct a study. Those tools or instruments are likely to be ones that they are already using, such as magnifying glasses, rain gauges, etc. Additional time will be needed to explicitly address the additional standards, but that time will need to be used toward that end at some point in the year anyway.

Addresses NGSS Standards in a Meaningful Way

The challenges the initial incorporation of the NGSS (2013) entails are significant. Many science teachers view the new standards as a signal that they now are inadequate in terms of content expertise. Their undergraduate science degree is now no longer sufficient to teach all of what they are being asked to teach. Additionally, they must rethink when and how these new standards will be incorporated into their curriculum. The iSTEM model provides a clear and simple way to insure that these newer standards are adequately addressed through the use of the 4×4 Standards Matrix (Table 6.2). For a given iSTEM lesson, the teachers can map their current state science content standards, a state math-related standard, an ITEA (2007) technology-related standard, and an NGSS engineering-related standard. It does not matter where the standards come from so long as they are meeting the needs of the students and teachers. The Matrix also insures that standards related to nature of science are also included, along with standards that address elements of the natures of the other three domains. Many states include within their standards documents, one or more standards directly related to nature of science and science process skills in response to the former national standards document, the National Science Education Standards (NSES) (1996). The NGSS (e.g. NGSS Appendix H) and other relevant standards documents and frameworks also include standards that address elements of the other domains. Understanding the natures of the domains (NOD) is essential to being able to differentiate between them thereby recognizing domain-specific strengths and applications that informs the user's capacity to effectively use each. Employing a model that assists teachers in delivering instruction that includes NOD serves as an additional strategy to promote the inclusion of NOS as a thread rather than a lesson or two at the beginning of the school year. All of the standards are included in a single lesson that typically lasts several days. By implementing several iSTEM lessons throughout the course of the school year, all the integrated STEM standards in the NGSS can be appropriately addressed.

Teacher Preparedness

Of course, appropriately addressing the standards through curriculum mapping is only part of the story in impacting student achievement in STEM. There are many elements, not the least of which is teacher preparedness. The vast majority of teacher education programs in the US train teachers to be domain and discipline specific. Although instructional models and strategies that include traditional integration approaches (e.g. math and science team teaching, reading across the curriculum, etc.) are often addressed in science education methods and other teacher education courses, the same cannot be said for STEM integration (as defined in this work). The result is teachers in science classrooms who do not feel prepared to teach integrated STEM. The iSTEM model helps to mitigate this effect by drawing on the strengths of the science teacher, namely implementing hands-on, inquiry-based science activities. Science teachers who graduate with a base pedagogical knowledge and skill level that includes understanding and using inquiry-based science instruction will have little difficulty implementing an iSTEM lesson. This is primarily because the iSTEM model is based upon the same foundational elements of other inquiry-based instructional models and strategies. Even the components that need to be added (e.g. the design and fabrication of scientific instruments) are potentially familiar science activities with a slightly altered focus. For example, building a thermometer (e.g. California Energy Commission 2006) is an established science activity that can be employed in an iSTEM lesson. With the iSTEM model, teachers can still use the best-practice strategies and activities they love and have found to be effective for them and their students.

Supports the Use of Best Practice Strategies

The iSTEM model can be superimposed on just about any reform-based curriculum in science education. The primary reason rests with the fact that many of the reform-based efforts in science education over the last 40 years are inquiry-based to varying degrees. Instructional inquiry, or pedagogy that facilitates student-generated questions, is at the heart of iSTEM. As such, iSTEM shares foundational elements with instructional models like Learning Cycles (e.g. 3E, 4E, 5E) and problem-based learning. Specifically, the iSTEM model emerged from a constructivist perspective (Baker and Piburn 1997). We also consider conceptual change theory (Posner et al. 1982) to be an appropriate paradigm to approach the teaching and learning of science. As such, we have taken a Piagetian tact towards dissatisfaction with one's current ideas (Baker and Piburn 1997) and employ the use of discrepant events, particularly as they relate to the design, fabrication, and implementation of instruments in a scientific study (i.e. iSTEM Model Steps 4, 5, and 6). During these steps students often are surprised to observe that they are not getting the results they anticipated and that their instrumentation has everything to do with that. These discrepant events prompt student-generated questions, which by definition, constitutes inquiry-based pedagogy. They go on in subsequent steps to modify the instrument or apparatus to collect data that are meaningful.

The iSTEM model also includes the incorporation of NOS instruction that ways that make it hardwired into its structure. This instruction is intended to be explicit, reflective, and contextual adhering to current best practice strategies regarding NOS instruction. Elements of the natures of mathematics, technology, and engineering directly identified within the NGSS and other standards and frameworks documents are also included. These ideas are taught using the same best practice strategies for NOS. Collectively we refer to the natures of the domains as NOD. We consider instruction regarding NOD to be a critical element for the success of our students. Without a complete and appropriate understanding of NOD, it is unlikely that students will be able to make complete and appropriate connections across domain. For example, if a student does not know what science is or how it operates, it is unlikely that the student will fully understand how math is used within the scientific community, even if they hold complete and appropriate understandings of the nature of mathematics because the context of application is the domain of science. Additionally, words may hold different meanings across domain and common elements such as argumentation may be expressed very differently (e.g. NGSS Appendix L 2013). Students are left to fill in the blanks on their own when teachers do not explicitly address both the similarities and differences across domains. This is the driving argument behind the inclusion of NOD within the iSTEM model.

Best practices in assessment are crucial to a successful instructional model. In iSTEM, conceptual, formative assessment is the primary focus. This means that the use of non-standardized test items is preferred (e.g. drawings, portfolios, interviews, essays, presentations, open-ended items, etc.). We view assessment as a teaching and learning tool. Students' ability to explain their thinking is essential to their individual and collective learning. Due to the nature of the activities, it is very important that the students receive feedback in an appropriate timeframe, which will vary among students, but can sometimes be on the order of minutes or seconds. Micro-interviews using inquiry-based questions are well suited for this task. Often simply providing a response to a student's question, with "Can you test that?" will allow the student to move forward. Other times guiding questions may be needed to help move a group in gridlock over an idea that cannot be tested within the class setting. Through the course of these micro-interviews, the teacher can build an understanding of what particular students think. This assessment is important to inform additional questions from the teacher to assist the student in learning the standards-based content. The presentations in Step 7 allow for student-to-student interactions and promote critical thinking. The teacher using formal observations can assess both the questions and the responses from students in order to gauge learning. While conceptual assessments are preferred and are critical to the students' understanding of integrated STEM content and processes, students need to be able to pass standardized science tests. We would argue that the focus on conceptual understanding adequately prepares students for such a challenge; however, we recognize the reality that many teachers and administrators insist upon the use of standardized test items within science assessments. The rationale is often that engaging with these items helps students practice science vocabulary and build familiarity with the format and strategies associated with standardized tests. Such assessments can be introduced at multiple times using the model. Typically an iSTEM lesson will last several days. Teachers often have activities and structures that are logistical and support classroom management that are employed concurrently with other instructional models (e.g. taking roll, bell ringers, exit slips, etc.). Exit slips, or quick assessments at the end of each class, are easy ways for teachers to include sample test items from standardized state tests.

The use of conceptual-based assessment that requires students to explain their thinking facilitates differentiated learning in the science classroom. Once the teacher has evidence regarding a student's mental model, he or she can move forward to directly and explicitly address misconceptions and begin building from a point of solid conceptual foundation. Differentiated instruction is further supported through the use of group work that requires individual responsibility and completion of tasks and the rotation of those tasks over multiple implementations of iSTEM lessons. For example, students that need additional support regarding scientific method have the opportunity to continue to build on their skills and understanding in this area, while other students who may need additional time to master design and fabrication skills and concepts can focus on these roles more. The key to successful differentiated learning for iSTEM students is sustained instruction. Sustained instruction is recognized as an important element in building complete and appropriate conceptual understanding and serves as a key fixture among ideas and materials in science education (e.g. spiral curricula, parallel curricula, learning progressions, etc.). Step 10, Conduct a New Related Study, is critically important in building students' understandings of integrated STEM content and processes. Students need a chance to apply what they have learned in a novel context to test their own understandings, build confidence in their abilities by excelling at elements at which they are already proficient and becoming proficient at elements at which they need improvement, explore what elements of STEM they enjoy, find out what parts of STEM they do not like, and experience integrated STEM in multiple contexts thereby increasing the likelihood that they find a high degree of relevance in a scenario. All of these things take time to develop and sustained engagement in iSTEM lessons provides the time, context, and support to enhance student achievement.

Increases Authenticity

The iSTEM model is an organic means of integrated STEM instruction because it makes use of authentic contexts. While not every scientist makes his or her own instruments, some do and most must contend with issues of maintenance, placement, tolerances, associated apparatus, interaction effects among instruments, etc. Understanding and manipulating instrumentation is key to scientific process for most practicing scientists. Requiring students to engage in science rather than just 'act like scientists' is important and that means requiring students to not only manipulate, but also understand, the instrumentation used. One of the best means of constructing this understanding is by building their own instruments. Building instruments and apparatus is not as difficult as it may seem. There are many 'how to' guides in print and on the web. With the advent of the 3D printer and their growing popularity and presence within schools, great potential exists for ever more complex builds. Many of the designs are available for free online and are as easy to fabricate as following prompts and pushing buttons. For example, using a MakerBot Replicator 2X, and a design discussed in a Popular Science article (Grushkin 2013), a science teacher can make a centrifuge very cheaply. More and more designs are being shared online daily. These designs can be altered using software that exports in .stl (e.g. Inventor, TinkerCad, etc). Another example of an instrument design that is readily available in print and online is a thermometer (e.g. California Energy Commission 2006). Many elementary and secondary science teachers have made these with their students using a plastic bottle, straw, water, alcohol, food coloring, and modeling clay. No software is required. Options abound for low-tech to hi-tech designs that teachers can use without modification. An example of the novel design component is to have students figure out how to attach or place the instrument to collect the data they want using student-designed apparatus. So, even though the students may not design the instrument themselves, they still engage in design process when they have to attach or place the instrument on the apparatus to collect the desired data.

Challenges to Implementation

Of course there are challenges to implementing iSTEM, as with any instructional model. Consideration of the issues below is warranted for those thinking about implementation. Although we identify several challenges, the strengths far outweigh them. Additionally, we offer some suggestions for mitigating the challenges.

Resistance to Inquiry Means Resistance to iSTEM

Using a foundation of best practice can be considered a double-edged sword regarding implementation. For teachers currently using best practice strategies, iSTEM fits nicely into their practice and serves to enhance what they are already doing in synergistic ways. However, for those teachers who are not currently using best practice strategies, the iSTEM model may prove to be a challenging prospect. This is not always a matter of pedagogical knowledge, but rather an issue of philosophical position. We have worked with teachers, both in-service and pre-service, who subscribe to constructivist philosophies and either regularly use, occasionally attempt, or anticipate implementing constructivist-based, best practice strategies. Regardless of their level of pedagogical knowledge and years of experience, these teachers are as receptive to iSTEM as any other best practice strategy they encounter in a science methods course or through professional development. Conversely, we have worked with teachers, both in-service and pre-service, who are resistant to constructivist-based, best practices. These teachers demonstrate a similar resistance to iSTEM. It appears to us that the challenges associated with delivering professional development regarding reform-based efforts (e.g. inquiry-based instruction), will also be present for those conducting professional development regarding iSTEM.

Need for Pedagogical and Content Knowledge

Pedagogical knowledge of reform-based, best practices is essential. In other words, it will be extremely difficult, if not impossible, for a teacher who does not use best practice strategies to effectively implement iSTEM. As such, teachers need to have been prepared through pre-service or in-service professional development in the areas of inquiry-based instruction, conceptual-based assessments, differentiation, safety, NOS instruction, and authentic scientific inquiry. There are instances where content knowledge, or lack thereof, can negatively impact implementation as well. In particular, discipline-specific, authentic scientific inquiry practices, NOS content, NOD content, and design process stand out as areas in which many teachers need increased content knowledge. The iSTEM model can operate with a basic understanding of best practice pedagogies, experimental design, NOS and NOD elements, and design process, however, its impact grows along with increased pedagogical and content knowledge. The good news is that professional development is being rapidly deployed nationally to address the NGSS and in particular enhanced attention to design process. This is occurring concurrently with ongoing professional development efforts aimed at enhancing reform-based practices.

Logistics of Using Inquiry-Based Instructional Practices

One issue we have experienced is logistical problems in implementing iSTEM lessons. Planning is essential. iSTEM lessons address multiple standards and are designed to last several days, which may or may not be consecutive. Planning ahead as one would with any multi-day project or lesson is the easiest way to mitigate this potential problem. Another issue, depending on the level of student-centeredness, is gauging the amount of time it will take students to complete the lesson. This can be challenging the first few times. We have found anticipating logistics to be very similar to experiences reported by those implementing highly student-centered, inquiry-based activities (e.g. a problem-based learning context or science fair).

Materials

Doing science as opposed to talking about it is a materials intensive endeavor. Building the instrument or apparatus adds to this issue. Many times the materials needed are ones that teachers already have available (e.g. building a thermometer), however this means that larger quantities of those materials will be needed. Often donations of materials to build instruments and apparatus come easier than cash. The reason appears to be that the materials needed for the instruments (e.g. screws, old tools, wire, scrap wood, metal, and plastic) are those that people are looking to get rid of anyway when cleaning out their garages. We approach this as an opportunity to be environmentally responsible and reuse items rather than them going to a landfill. Of course the teacher will need to review all materials first for safety reasons, but often such donations can go a long way to building a fabrication shop area within a science classroom. In addition to consumables, tools may be needed to build or fabricate the instrument or apparatus students plan to use. Career and technical education (CTE) teachers within a school or school division can be wonderful resources and are often times willing to partner with you. The shops in some school divisions are amazingly well equipped to handle building items from wood, metal, or plastic. These teachers are also trained to supervise student use of tools that may be unfamiliar to some science teachers (e.g. TIG welder). Many contain emergent and state of the art production technologies (e.g. 3D printers and CNC machines). The partnership possibilities are considerable and the products produced are ones that are truly functional. While challenges to implementation do exist, as with all instructional models, they are not insurmountable.

Conclusion

The iSTEM model holds considerable potential as an effective instructional model specifically designed for authentic, integrated STEM learning from a scientific perspective. We are just now beginning to conduct studies to collect empirical data regarding efficacy of the model. Data collected from the grant-funded projects indicate that the iSTEM model may be effective, but because the model was in the process of being developed and the original evaluation and research questions focused on other elements, the data are largely antidotal or tangential. There exists a need for studies to examine the efficacy of iSTEM and other integrated STEM instructional models across diverse populations and contexts. Science teachers are in need of models that they can realistically implement in their classrooms to address the new challenges they face from the NGSS (2013). They need models that go beyond just using traditional discrete activities and instead integrate the STEM domains in ways that are authentic, meaningful, and sustained. We cannot expect our science teachers to become mathematicians or engineers, but we can expect them to become better scientists and science teachers. The iSTEM model provides a viable pathway to improvement. At a time when science teachers are begging for concrete ways to teach their students STEM, the development of the iSTEM model is a pragmatic response that can help move this critical conversation forward.

This work was supported by funds from NOAA, VA DEQ, Dominion Power, and the Beazley Foundation.

References

Ainsworth, H. L., & Eaton, S. E. (2010). *Formal, non-formal and informal learning in the sciences*. Calgary: Onate Press. (ERIC Document Reproduction Service No. ED511414).

Akerson, V. L., Weiland, I. S., Nargund-Joshi, V., & Pongsanon, K. (2014). Becoming an elementary teacher of nature of science: Lessons learned for teaching elementary science. In *Science teacher educators as K-12 teachers* (pp. 71–87). Dordrecht/New York: Springer.

American Association for the Advancement of Science (AAAS). (1993). *Benchmarks for science literacy*. Retrieved from http://www.project2061.org/publications/bsl/online/index.php?intro=true.

Anderson, L. W., Krathwohl, D. R., & Bloom, B. S. (2001). *A taxonomy for learning, teaching, and assessing: A revision of Bloom's taxonomy of educational objectives*. Boston: Allyn & Bacon.

Baker, D. R., & Piburn, M. D. (1997). *Constructing science in middle and secondary school classrooms*. Boston: Allyn & Bacon.

Bodzin, A. M., Klein, B. S., & Weaver, S. (2010). *The inclusion of environmental education in science teacher education*. New York: Springer.

Buxton, C. A., & Provenzo, E. F., Jr. (2007). *Teaching science in elementary and middle school: A cognitive and cultural approach*. Los Angeles: Sage.

Bybee, R. W. (1997). *Achieving scientific literacy: From purposes to practices*. Portsmouth: Heinemann.

Bybee, R. (2011). Scientific and engineering practices in K-12 classrooms: Understanding a framework for K-12 science education. *Science and Children, 49*(4), 10–15.

Cajas, F. (2001). The science/technology interaction: Implications for science literacy. *Journal of Research in Science Teaching, 38*(7), 715–729.

California Energy Commission. (2006). *Science projects: Make a thermometer*. Retrieved from http://www.energyquest.ca.gov/projects/thermometer.html

Carin, A. A., Bass, J. E., & Contant, T. L. (2005). *Teaching science as inquiry*. Upper Saddle River: Pearson/Merrill Prentice Hall.

Congressional Research Service (CSR). (2012). *An analysis of STEM education funding at the NSF: Trends and policy discussion* (R42470). Retrieved from http://www.fas.org/sgp/crs/misc/R42470.pdf

Crismond, D. (2001). Learning and using science ideas when doing investigate and redesign tasks: A study of naive, novice, and expert designers doing constrained and scaffolded design work. *Journal of Research in Science Teaching, 38*(7), 791–820.

Dickerson, D. L., & Hathcock, S. (2011, October). *Project SEARCH – buoys, non-traditional STEM fields, and policy-ready citizen science*. Presented at the annual meeting of the North American Association of Environmental Educators, Raleigh.

Dickerson, D. L., Hathcock, S., Stonier, F., & Levin, D. (2012). The great build-a-buoy challenge. *Science and Children, 50*(4), 62–66.

Enger, S. K., & Yager, R. E. (2001). *Assessing student understanding in science: A standards-based K-12 handbook*. Thousand Oaks: Corwin Press.

Fraser-Abder, P. (2011). *Teaching budding scientists: Fostering scientific inquiry with diverse learners in grades 3–5*. Boston: Pearson.

Governor's STEM Task Force. (2009). *Investing in STEM to secure Maryland's future*. Retrieved from http://www.marylandpublicschools.org/NR/rdonlyres/443201CC-D7DA-43E8-8FFE-191296AD7F5F/29748/STEMTaskforceReportAug20092.pdf

Grushkin, D. (2013). Spin master: Turn a Dremel tool into a lab-grade centrifuge. *Popular Science, 283*(2), 76.

International Technology and Engineering Educators Association (ITEEA). (2009). *The overlooked STEM imperatives*. Reston: Author.

International Technology Education Association. (2007). *Standard for technological literacy: Content for the study of technology*. Reston: Author.

Keeley, P. (2008). *Science formative assessment: 75 practical strategies for linking assessment, instruction, and learning*. Thousand Oaks: Corwin Press/NSTA Press.

Keller, T. E., & Pearson, G. (2012). A framework for K-12 science education: Increasing opportunities for students learning. *Technology and Engineering Teacher, 71*(5), 12–18.

Koch, J. (2010). *Science stories: Science methods for elementary and middle school teachers*. Belmont: Wadsworth.

Kolodner, J. L. (2002). Facilitating the learning of design practices: Lessons learned from an inquiry into science education. *Journal of Industrial Teacher Education, 39*(3), 9–40.

Krajcik, J., & Merritt, J. (2012). Scientific practices: What does constructing and revising models look like in the science classroom? Understanding a framework for K–12 science education. *Science and Children, 50*(3), 6–10.

Liu, X. (2010). *Essentials of science classroom assessment*. Los Angeles: Sage.

Llewellyn, D. (2002). *Inquire within: Implementing inquiry-based science standards*. Thousand Oaks: Corwin Press.

Llewellyn, D. (2011). *Differentiated science inquiry*. Thousand Oaks: Corwin Press.

National Council of Teachers of Mathematics (NCMT). (2000). *Principals and standards for school mathematics*. Retrieved from http://www.nctm.org/standards/content.aspx?id=16909

National Governors Association (NGA). (2011). *Building a science, technology, engineering and math education agenda*. Retrieved from http://www.nga.org/files/live/sites/NGA/files/pdf/1112STEMGUIDE.PDF

National Governors Association Center for Best Practices & Council of Chief State School Officers. (2010). *Common core state standards for mathematics*. Washington, DC: Authors.

National Research Council. (1996). *National science education standards*. Washington, DC: National Academy Press.

National Research Council. (2000). *Inquiry and the national science education standards*. Washington, DC: National Academy Press.

National Research Council. (2007). *Rising above the gathering storm: Energizing and employing America for a brighter economic future*. Washington, DC: The National Academies Press.

National Research Council. (2010). *Rising above the gathering storm, revisited: Rapidly approaching category 5*. Washington, DC: The National Academies Press.

National Research Council. (2011). *A national framework for K-12 science education: Practices, crosscutting concepts, and core ideas*. Washington, DC: National Academies Press.

National Research Council. (2012). *A framework for K-12 science education: Practices, crosscutting concepts, and core ideas*. Washington, DC: The National Academies Press.

National Science and Technology Council Committee (NSTCC). (2011). *The federal science, technology, engineering, and mathematics (STEM) education portfolio: A report from the federal inventory of STEM education fast-track action committee on STEM education* (pp. 111–358). Washington, DC: US Department of Education.

National Science Board. (2007). *National action plan for addressing the critical needs of the U.S. science, technology, engineering, and mathematics education system (NSB-07-114)*. Washington, DC: National Academies Press.

NGSS Lead States. (2013). *Next generation science standards: For states, by states*. Washington, DC: The National Academies Press.

Novak, J. D. (2002). Meaningful learning: The essential factor for conceptual change in limited or inappropriate propositional hierarchies leading to empowerment of learners. *Science Education, 86*(4), 548–571.

Posner, G. J., Strike, K. A., Hewson, P. W., & Gertzog, W. A. (1982). Accommodation of a scientific conception: Toward a theory of conceptual change. *Science Education, 66*(2), 211–227.

President's Council of Advisors on Science and Technology (PCAST). (2010). *Prepare and inspire: K-12 education in science, technology, engineering, and math (STEM) for America's future*. Washington, DC: Author.

Roberts, A., & Cantu, D. (2012). Applying STEM instructional strategies to technology and design curriculum. In T. Ginner, J. Hallstrom, & M. Hulten (Eds.), *Technology education in the 21st century* (pp. 111–118). Linköping: LiU Electronic Press.

Roth, W. M. (1995). Inventors, copycats, and everyone else: The emergence of shared resources and practices as defining aspects of classroom communities. *Science Education, 79*(5), 475–502.

Roth, W. M. (2001). Learning science through technological design. *Journal of Research in Science Teaching, 38*(7), 768–790.

Schomburg, A. (2008). The better boat challenge. *Science and Children, 46*(2), 36–39.

Settlage, J., & Southerland, S. A. (2007). *Teaching science to every child: Using culture as a starting point*. New York: Routledge.

Sherman, S. J., & Sherman, R. S. (2004). *Science and science teaching: Methods for integrating technology in elementary and middle schools*. Boston: Houghton Mifflin Company.

Shulman, L. S. (1994). Those who understand: Knowledge growth in teaching. In B. Moon & A. S. Mayes (Eds.), *Teaching and learning in the secondary school* (pp. 125–133). London: Routledge.

Sidawi, M. M. (2009). Teaching science through designing technology. *International Journal of Technology and Design Education, 19*(3), 269–287.

Sorey, T., Willard, T., & Kim, B. (2010). Make your own digital thermometer! *Science Teacher, 77*(3), 56–60.

The White House. (2009). *Remarks by the President on the "Education To Innovate" campaign* [Press release]. Retrieved from http://www.whitehouse.gov/the-press-office/remarks-president-education-innovate-campaign

Tippins, D. J., Mueller, M. P., van Eijck, M., & Adams, J. D. (2010). *Cultural studies and environmentalism: The confluence of ecojustice, place-based (science) education, and indigenous knowledge systems*. New York: Springer.

U.S. Department of Education (USDOE). (2010). *ESEA reauthorization: A blueprint for reform*. Retrieved from http://www2.ed.gov/policy/elsec/leg/blueprint/index.html

Weiss, I. R., Matti, M. C., & Smith, P. S. (1994). *Report of the 1993 national survey of science and mathematics education*. Chapel Hill: Horizon Research.

Chapter 7
Robotics Education Done Right: Robotics Expansion™, A STEAM Based Curricula

Anthony J. Nunez

Robotics as a Method to Interest Children in STEM (Why It Can Work)

Robotics is an exciting and effective method to interest children in Science, Technology, Engineering, and Math (STEM), but it requires trial and error, an effective means to implement the child's input and most importantly an accurate depiction of what it takes to make a robot. Recently, Robotics has been introduced to educational programs looking to increase children's interest in the Engineering (E) portion of STEM. The concept has been to show children that learning about engineering can be fun and exciting by building programs around something that moves and obeys a child's commands. Working as a large group, students build a robot from a pre-packaged kit. The students piece together these robotic kits' physical parts and then work with a computer to connect blocks on the screen so that the robot can now do something. The packaged kits include all the pieces to hold everything together as well as the motors and sensors already assembled in order to facilitate construction. These kits have only a few set ways things can go together and all the pieces that are premade for them to fit nicely together. There are typically only a few ways these pieces can adjoin and, therefore, which eliminates the potential to make mistakes. The computer work, often mistakenly referred to as coding, is simply a series of dragging and dropping pre-labeled items. If the dragging and dropping on the screen or assembling the connecting rods into connecting holes of the physical structure becomes too challenging for a child, a set of instructions to do all of the above is provided.

A.J. Nunez (✉)
Infamous Robotics LLC, Arlington, VA, USA
e-mail: anthony.j.nunez@gmail.com

The children are excited. They have just taken block building from childhood to the next step. They have learned to add moving pieces and even sensory pieces. They also enjoy the addition of learning new skills on the computer. The excitement is correlated to excitement about STEM, and the hope has been that the children now have an accurate introduction to engineering through the construction of this pre-packaged block kit; however that is not the case. Using this method may show short-term STEM excitement and interest, but it does not foster long-term learning or the STEM skills that students will require to continue in the sciences after high school graduation. Our goal is to develop STEM proficiency and engagement so that children grow into adults with STEM careers in order to foster a healthy economy. Our research shows that pursuit of this goal should begin at the beginning of the child's education—at the elementary school level.[1] The journey must begin there and be approached in such a way as to accomplish the long-term goal.

Robotics can be a method to interest children in STEM as stated by Barker and Ansorge (2007), but it must be done accurately. A successful method includes not only the upside of having fun and the excitement of accomplishing something, but also the more challenging downside of a project not working right away and having to go back and determine where the error or mistake was. It is only in this personal journey that each child fully engages in the scientific process. The idea that robotics, engineering, science or math is as simple as putting two pre-made and perfectly fitted pieces together is not an accurate one. It is rather the opposite most times. Pieces you thought would work together usually do not the first time. Every time a student makes a mistake in any one of the STEM subjects, it is often from trying something new or experimenting with a process they just learned. I often tell my students that I can remember my mistakes more than my successes from building robots. The reason is that I learned something valuable each time I made one of those mistakes.

Let us take a closer look at the pre-packaged kits recently made available to teach children about robotics. First, however, when discussing robotics as part of STEM, it must be acknowledged that more than just the "E" of "Engineering" is involved because robotics is more than just engineering. In robotics, many different aspects of engineering and concepts that encompass the various STEM fields are involved. The building of a true robot structure involves mechanical engineering, physics, math, and, speaking from personal experience, some elbow grease, sweat and tears. That experience is withheld from the child with these pre-packaged kits.

What Robotics Truly Is at Its Core

At its core, robotics is a field that brings together multiple STEM disciplines. For science, robotics has a major science portion to it that is often overlooked. In robotics we often hypothesize how a solution we are developing will work. We have a problem and need to solve it. Sometimes we are correct and other times, we prove ourselves wrong, the approach we had would not have worked. It is a process by which we must experiment often with the different variables available to us for a solution. As is often stated in the Next Generation Science Standards (NGSS), while experimenting and analyzing this data often leads us to a solution, it is frequently only after many unforeseen lessons in the process. As a roboticist, you often must notice the relationship and interaction two items you are working with undergo. An example would be a wheel. Most people think this is an easy selection, choose a wheel, put it on the robot, and go. First, however, we have to ask important questions such as what type of surface will the robot be travelling over, or how fast do we want the robot to go? These are all interactions and relationships with the robot itself and how it will operate as well as interactions with the outside world.

The technology portion of the STEM equation is evident for those who follow the robotics industry and its ever-changing landscape. The technology created in other industries often finds an application within robotics. One such example would be the touch screen, which came from the telecommunications industry. The creation of a screen you can touch to issue commands has already been integrated into robot design. People's familiarity with it lends to ease of use, and its availability due to production for telecommunications are two reasons this technology will be quickly integrated into the robotics field. There is more to technology than just incorporating other devices, however. More technology invented for use in robots will go mainstream in the future for other uses. This has already happened with the invention of the robotic arm, which was originally intended for robots that would assist our exploration of outer space. That same technology can now be used for prosthetics to help humans with disabilities.

The engineering behind robotics is most evident as much of the attention placed on robots is given to the software engineering that creates the code and to the mechanical engineering that creates robotic movement. The engineering is important[2] for the structure of the robot, which will hold everything together and allow a robot to accomplish its mission. Military robots are a good example of why structure is important in robotics. In this case, the robot needs to be thrown into a room via a window, survive the fall, turn itself upright, and then proceed to its mission. In this example you can see that if the design of the structure does not survive the fall, the robot is nothing more than a decorative paperweight. As another example, robots that fly, such as a new class of robots called Micro Aerial Vehicles (MAVs) rely on their weight, shape, and design to navigate the air space. All the code in the world could not make a MAV fly without the proper structural design. I recall working on a robotic project where the robot was overheating and when it got beyond a certain temperature the robot became erratic. The root cause of the problem was the design of the circuitry and was easily solved making some changes to allow for the electronics to cool down. The point being that, in robotics, everything is tied together.

The math component of STEM appears in robotics in everything from taking the measurement of a robot and all its various pieces to calculating the amount of necessary current to charge it. There are many applications of math necessary to build a robot and make it work. This is not complicated math like differential equations or calculus, but fun math like multiplication, addition, and even geometry. In our Robotics Expansion programs, we frequently have students asking to do more math problems. Yes, it is true.

Students of Infamous Robotics LLC applying math to make their own design (© Copyright Infamous Robotics LLC, 2014)

Current Options for Children Learning about Robotics in STEM

Up until the development of the Robotics Expansion™ programs by Infamous Robotics LLC, there were two basic options for children to learn robotics. The first option involves purchasing a high cost kit that comes with plenty of components designed to wow the eye. Parents have come to believe that these kits are the only way children can learn about robotics and therefore purchasing them is necessary for their children's robotics education. There are two versions currently on the market, and they are very similar. Both offer structural components that have pre-determined placements, leaving little to no room for innovation or creativity on the part of the child. The kits are not all inclusive, contain a limited amount of components, cost $1200.00 dollars, and still you are not able to build a decent, fully autonomous robot. In order to accomplish this, additional pieces must be purchased from the company. These additional pieces will still only provide enough parts for one robot.

Much of the emphasis for this pre-packaged kit is on the 'programming code' for the robot. The companies have designed a product that requires children to spend the majority of the time building their robot sitting in front of a computer. Where there is no innovation possible in the structural design, there is some available in choosing the robot's code. Though the kits come with completely finished sample code, both versions offer software packages that will allow a child to either drag blocks of finished code or use a drag and drop menu to "write" code. These pre-packaged kits though made with slightly different designs (one metal and the other plastic) are the exact same at their core. They each offer lesson plans heavy on following instructions, but light on accurately reflecting what it takes to build a robot.

The cost of these kits is a significant issue. The cost is so high that organizations or schools that use them find themselves putting students in large teams of six to eight students for one robot. Team work is important in robotics, but it is also important that each child has a position on the team that is equal in workload to others. At this level of their education, putting them in a large group where they have limited access to equipment and limited exposure to the material, then it is not accurately depicting what it takes to make a robot as part of a team. The bottom line is children do not learn equally about fundamental lessons they need to understand when using one of these kits.

Some smaller companies saw an opportunity in the expensive kits and bought toy robotic kits from overseas and put together their own robotics class. Unfortunately, the companies had no expertise in engineering or robotics. They took information from the instruction booklet for the toys and used it for their descriptions. With 'robotic sounding' phrases and a price that appeared more reasonable than the earlier pre-packaged versions, a short-lived excitement ran through the robotics educational community. This option, known as the robot toy kits, offered children the opportunity to build a robot and take it home. The robots, however, were not robots at all, but simply toys. The brief elation of taking a new toy home was short lived and ended where the child's STEM potential did, up on the shelf. This course, deceptively entitled Robotics, is a class that focuses solely on the construction of the robot toy and playing with it in a thoughtlessly designed mockery of a robot competition. The injustice for STEM education has proliferated into additional programs as the never ending list of imported robot toys grows. The robot toys do not teach the children about core concepts, rather throw a few words their way that have important meaning, but no focus is put on them and no foundation of understanding is established. We often have children from this type of class in our Robotics Expansion programs who would not be able to answer three questions about their robot toys. It is disheartening for us to see children so put off the topic by these toy kits, and we are trying to undo the damage done by such a false representation of robotics. The toy kits offer no actual programming language that robot builders use, no concepts of robotics construction, no circuit design instruction, no foundation of understanding in any topic, and most of all no math that pertains to robotics. The fact remains, the focus for both the pre-package kits and the toy kits is on the E in STEM, though even the E is not done justly for the students. If we want more engineers, more people with technical skills to invent and create new business to develop this sector of our economy, then there has to be a better way. We at Infamous Robotics LLC have been working on developing a solution to this problem for 7 years now, and we call it Robotics Expansion™.

Where Each of Them Has Short Comings

The current standard for introducing children to robotics is inadequate, gives children a false representation of robotics, and has many limitations. The pre-packaged kits offer up an array of false representations for students wanting to learn more about robotics. The construction of a robot is never pre-determined. This means that you are not always guaranteed to have two things fit together perfectly. When it comes to how a robot fits, how it adjoins, and how it supports—all key features any mechanical engineer must work with—these may look terrific on paper, but not so good in execution. So why do we teach children robotics in a way that avoids these learning features? They do so for ease of execution. It is far easier to avoid those lessons and frustrations than to have students go through the actual process of engineering a robot structure. The lessons a child will devise for themselves in the structural components of a robot are important because this makes up one major piece of robotics. In bypassing this lesson with these pre-packaged kits, we are not preparing the students properly or introducing them to what they may find to be an exciting and rewarding aspect of robotics. Another false representation from these robotics kits is one we have repeatedly seen in with their students that come to our programs, and that is the frustration of not understanding programming. Programming a robot is like learning a new language; there is no short cut. The pre-packaged kits will have you believe programming is drawing a line between two blocks on a computer screen that tells the child what they are, or they give children an expensive software that offers features such as drop down menus.

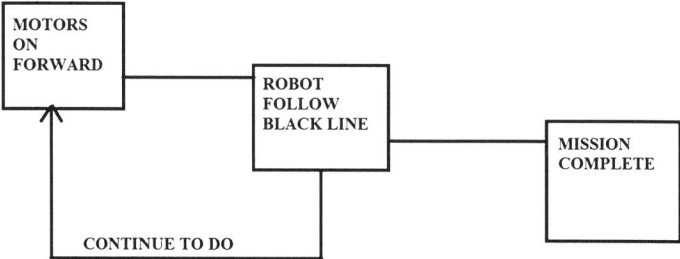

Block programming from prepackaged kits does not provide an accurate depiction of robot programming. As shown here there is no code, just connecting lines (© Copyright Infamous Robotics LLC 2014)

A computer programmer will tell you that this is a false representation of what it is like to write code, but as a roboticist, I can assure you this is again an inadequate way to educate students, and we can do better. Writing a program to control a robot is a process that has many speed bumps along the way, but it is in those speed bumps that children will discover what programming a robot is all about. It is about learning that sometimes the smallest thing you put incorrectly in your code will affect many other parts of your code. Discovering that small missing piece is a journey, sometimes in that journey students can get frustrated and confused, but if they are instructed correctly, they can work through that, refer to their education in the topic, and continue on. At the end of that journey to find this error in the code, that missing piece, is a joyful moment for a child. The sense of accomplishment and the deeper respect the child holds for the subject are un-paralleled.

7 Robotics Education Done Right: Robotics Expansion™, A STEAM Based Curricula

A major shortcoming in the pre-packaged kits is the complete lack of circuit building they offer. This is a significant inadequacy with these programs that cannot be ignored, overlooked, or brushed to the side. Circuit building is a major component of robotics and leaving this out of a robotics education is difficult to explain. I often see the result of this absence on robotics students who have to resort to seeking fundamental circuit advice in online forums, pleading for information to help them complete the last part of their robot, only to become very discouraged when they realize the help they need does not come in the form of a simple answer. Circuit building is hands-on learning, but also a result of establishing an understanding of the topics surrounding it. Pre-packaged kits leave this out, and it is to the detriment of students in the field.

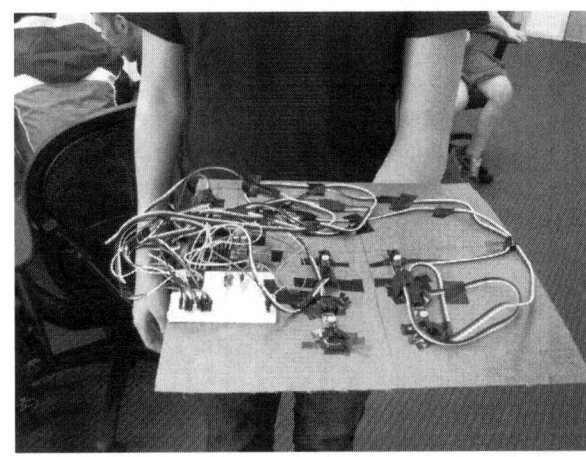

Location system for a robot circuit built 100 % by elementary students of the Robotics Expansion programs at Infamous Robotics LLC (© Copyright Infamous Robotics LLC, 2014)

The toy robot kits simply do not suffice, summarize, or equate to anything that should be associated with the STEM education initiative, particularly and most definitely the robotics education of children. The pre-packaged kits misstep and miss the mark on robotics education, and the robot toy education effort actually set children back a few steps. The lack of any room for intuition at even a small level, any room for design, or any lessons outside of putting something together is a disastrous mix for inspiring students about STEM and Robotics. Students of these programs are instructed on portions of material as it relates to the toy kit, and the child is left to fill in the blanks. Students of robot toy kit classes differ from pre-packaged kits in that the emphasis is on the building rather than the programming as the latter. There is not code, nor engineering, nor design, nor trial and error, and definitely no pathway for the child following the one or two classes of robot toy building available. Not setting a clear path for the child following the robotics education program will result in poor results for obtaining a strong STEM based workforce. The toy kits place an image in the student's mind that robotics, engineering, and math simply require that you plug in the pieces you are shown, and then you are done. That is not engineering nor robotics, but rather a lesson in following instructions.

Robotics Expansion™-How It Came About

Robotics Expansion™ is influenced from my time spent working in the field as an engineer, with a focus on electrical engineering and software engineering. In the field, I was fortunate to work with an older group of engineers, the ones with decades of experience, applying things that had been passed down to them. I became determined to carry on the legacy of what they knew and that seemed to me was being lost. The lessons I learned and how they were taught to me would become major components of Robotics Expansion™. Establishing my first robotics company more than 7 years ago was not my first exposure to a robotics project. Over 4 years prior to that, I worked on two Defense Advanced Research Projects Agency (DARPA) projects with the goal of designing a robot car capable of driving autonomously. I also built a robot for George Mason University under a stipend from the Entrepreneurship program while building the robot for my senior design project. It is important to note here, none of these robot projects used pre- packaged kits. They were all from the ground up, and for the most part, built as a single member team.

One exception was the work on the DARPA projects, which was part of a larger team that was comprised of smaller segments of team groups. The smaller teams operated as individual groups with their own set of responsibilities. Often we divided the workload and met regularly to discuss progress or issues. There are many types of skills acquired from working in large groups and working in small groups around robotics, much of it is unique to robotics because of the diversity of disciplines involved in the field. Looking at the Next Generation Science Standards (NGSS), topics for grades 3–5 such as 3-PS2, 3-LS1, 4-PS3, 4-PS4, 5-PS2 and 5-ETS-1 all cover topics related to robotics that have been incorporated in the Robotics Expansion™ programs we developed from the aforementioned experience. See table below for further information.

NGSS unit	Application in robotics
3-PS2: Motion and Stability: Balance	Balance of a mobile robot in a competition
3-LS1: Developing and using models	Modeling your idea in robotics and its importance
4-PS3: Energy	Importance of energy in circuits
4-PS4: Waves and their Applications in Technologies for Information Transfer	How certain sensors operate and transfer information
5-PS2: Motion and Stability: Forces and Interactions(gravity)	Understanding the effects of gravity on robots that leave the ground
5-ETS-1: Engineering Design	Applying constraints to a problem and coming up with a solution

The research encompassed a major component for the development of the successful Robotics Expansion™ programs. Understanding where the pitfalls would be in trying to execute the lessons designed and how the material would be received by the students was a major piece. Before that could come about, we had to evaluate the previous programs from a roboticist point of view, referencing not only examples of real world applications but also leaning on my previous experiences in robotics projects. Two of us went to a NASA facility for a robotics education class where one of the pre-packaged kits would be used to help train teachers to implement a class at their schools. The goal of the program was to build a robot to traverse above a terrain using a motor and to take measurements of the various structures below as quickly and as effectively as possible. We began by going over some of the pre-packaged lessons provided in the box as a group. We were then handed the pre-packaged kits and told to begin. Laptops were set up for us to drag and drop blocks of pre-cooked code and connect them with lines to make the robot do something. We stepped away from the others to study the course and sketch out a design on paper to pushing the limits of what could be done with this pre-packaged kit and then got down to work.

The instructors at NASA, most experience engineers or physicists, walked around observing our designs. Our design and approach was so different that we were told it could not work. In fact, the competition culminated with our robot finishing first and being the most efficient. We learned first hand that day the serious disadvantage the pre-packaged kits put the other students in. We knew it could be done better as our approach in this challenge referenced concepts and an understanding from lessons plans we had already developed. Most of these involved a kind of thinking outside the box that came from having a greater understanding of the various fields in robotics that come into play in any solution. We could not help but notice the group of teachers and their difficulties in understanding the material and how that would reflect on administering the material to their students. This would be a critical component for students to effectively understand the material as the teachers would need to have an understanding of the material in order to effectively teach it. We have designed our training manuals for teachers and administers of the curriculum from a unique perspective that they can relate to and follow.

Robotics Expansion™ has evolved since its inception, improving every step and constantly updating as the ever changing robotics field grows. The application of the material was consistently studied and adjustments were made throughout its early stages. We wanted this to be the best material for students and felt that it could be, as long as we stayed true to the end goal, getting students to pursue STEM fields beyond the classroom. Improvements were made only to enhance the performance and execution of the material for students and those administering the material.

New Approach-High Level Overview of Robotics Expansion Programs

We have developed a multi-tiered approach in which a student enters into the Robotics Expansion™ program, and upon completion of the introductory levels, the student has a strong interest in the subject, a solid and accurate understanding of STEM fields as applied in robotics, a clear sense of which field of study within robotics they desire to pursue, and the ability to complete Robotics-STEM

(R-STEM) exercises independent of the classroom. The first step in a program such as this is establishing interest of the child. Children are already excited about robotics, so let us not ruin it by misleading them or giving them false representations of robotics with robot toy kits or these pre-packaged block programming kits. Often once a child has been introduced to robotics in such a false way, it is an uphill challenge to explain core concepts and re-teach them about robotics in the real world. Our initial levels attract new children to the field, excite those already into the subject, and build excitement among groups of children as to the endless possibilities in the field. Children feel the need to explore and understand their environment, and robotics is no different. The initial levels are critical, getting the student interested in the field through instruction followed by hands-on activities so they can know exactly what they are doing and be able to explain it in their own words. We introduce certain R-STEM topics in these interest levels in such a manner that children are left hungry for more answers.

Children are inquisitive and getting them started correctly in robotics is key if we are going to keep them interested in STEM according to NGSS (2013). Once the interest is established and the students have a sense of topics to be covered at their next steps, we begin with the foundation. We introduce students to the ways that the various STEM topics are encompassed in robotics and teach them through instruction and hands-on learning. Students engage in a range of activities with topics that vary across the broad spectrum of robotics topics to learn these items and relate them to items in their daily lives. Once this is done, the excitement truly begins as going beyond this level students apply what they have learned in a series of upper levels that educate to innovate.

Children cannot simply be taught a subject through lecture, and often incorporating hands-on components is only a short- term goal. We have noticed through our studies, applications, lessons, and programs that much more is required, and we have discovered a way to do it. Children need to be given a stake in the game, be able to take ownership on what they are doing. The child's input in R-STEM subjects is critical. Take for example the pre-packaged kits for sale. By the time the student has finished, there is no sense of ownership because the student merely followed instructions to build. They may have had an idea, but they were confined to a box to build it. We must set the child free from this box to design what they want to; otherwise, they really have not made any input.

Having the freedom to design something with no limitations is one way it is accomplished in R-STEM subjects. Another way is in the actual building process of a robot. We establish the interest, build for them a strong foundation of understanding, and give them a stake in the game as Banilover, Smith, Malzahn, Weis and Campbell have recognized is important. Now let us help them find their place in robotics. Each student is different than the next. Student A may gravitate more towards programming while another may enjoy the mechanical building process more. Furthermore, student C may enjoy the aesthetic appearance and how it relates to the robot accomplishing it tasks. I am a firm believer that there is a place for every student in robotics, showing them their options in the right way and giving them the option to choose where their path forward is what Robotics Expansion™ programs do. We bring them to the fork in the road and rather than turn them loose blind, we demonstrate throughout the levels what can be found down each path and what to expect and the challenges in that field so they may set their own goals in R-STEM subjects.

How Robotics Expansion™ Displays the Core of Robotics

The Robotics Expansion™ programs display the core of robotics in a way that is easy to learn and provides an accurate depiction of the STEM fields. Merely understanding the fields involved will not guarantee successful delivery of the material, it requires many other items to be taken into account, some of which I will mention here. Teaching engineering concepts learned at the collegiate level can be a challenge, especially when the students are in elementary school. The concepts need to be articulated in such a way that a student can understand. If the teacher begins by using words such as "differentiating" or "end-effector," confusion will set in and frustration will follow. I recall watching a Doctoral candidate trying to explain a simple concept in robotics to a young child. I watched as the child's face turned from excitement about learning to confusion and to a face of disgust. The child walked away in anger, now convinced robotics was not for him. We must not do this to our children; we must open the doors rather than close them.

Communicating the concepts and lessons in an easy to learn way is only one piece to this complicated puzzle. Each of the paths for learning a subject like robotics has further paths because of all the subjects involved. The R-STEM topics must be introduced in a fun and exciting way. Making the exercises hands-on is only one part of this structure. Hands- on is exciting and fun, but the goal is to have students pursue a career in one of the STEM fields and hands-on alone does not suffice. This is a major misstep in the pre-packaged kits and robot toy kits. The topics have to be fun and exciting by themselves. By this I mean that without hands-on exercises a teacher must be able to instruct the students of core concepts so they are interested and engaged. This is where our programs thrive. We do so in a variety of ways at calculated steps in the program. The students are given information and then a robotics problem, as seen in 3-5-ETS1-2 (Generate and compare multiple possible solutions to a problem based on how well each is likely to meet the criteria and constraints of the problem) of NGSS and instructed to utilize what they have learned in the program to solve it.

The catch is in robotics any problem you have can be multi-layered, and so the students often begin by giving answers to the problem with little to no planning only to realize that the solution will only be found through following the steps.

They must take into account multiple variables and issues through the problems they are given. The end result is students or student teams devise solutions through a rigorous trial and error process that challenges them. Students like and need such challenges as these when it comes to their education as mentioned by Clifford (1990). The NGSS will provide a route for this. Our programs see this all the time.

When incorporating math problems that relate to robotics, our students are always asking for harder and more challenging problems. I know this sounds unbelievable, not math, impossible you might be saying, but the truth remains, students want to be challenged in the setting we provide.

The final component here is accurately promoting STEM and Robotics. In the world outside of school, you are never given the answer along with the problem, neither are you shown instructions that outline the step-by-step method to complete a project. Scientists do not make breakthroughs or discoveries without experimenting. A scientist will learn concepts that apply to the solution, practicing and applying the concepts into examples that will give known results. That will give the scientist something to compare to, something to fall back on. The same thing applies in robotics and the use of the STEM fields in robotics. It is not accurate to give the child a kit, whether it be a toy or a pre-packaged one and expect them to accurately learn the processes of R-STEM. The consequences of doing so are that we will continue to miss the mark and have the 'shock factor' when students reach the University level of their engineering education and drop out at a rate of 60 % in their first year alone, Marcus (2012).

Why Is the Core of Robotics Important in Creating the Next Generation Workforce of Engineers and Roboticists

The Carnegie Foundation commission says "the nation's capacity to innovate for economic growth and the ability of American workers to thrive in the modern workforce depend on a broad foundation of math and science learning, as do our hopes for preserving a vibrant democracy and the promise of social mobility that lie at the heart of the American dream" (Carnegie Corporation, Institute for Advanced Study 2013). This goes beyond simply getting children excited about STEM subjects. This excitement is simply the first step and not the end as it is for pre-packaged robot kits and robot toy lessons. The pre-packaged robot kits will have you believe that participating in their programs will lead to further STEM excitement beyond the class but that is not necessarily the case. What happens when a child wants to continue the excitement at home? The cost for one of their boxed cookie cutter kits is $300+, and that is just the introduction kit. If a child has an imagination and wants to move beyond the box by any measure, a parent or organization will be spending close to $1000.00 more by the time all is said and done. Unfortunately, that does not get a child far in robotics. The child knows only what was within the confines of the box, which can never truly prepare them for the vast and wide world that is R-STEM.

The issue becomes evident when the child seeks to venture off beyond the pre-packaged box kit and into the real world of robotics only to realize they were ill prepared and quickly become frustrated and upset that the limited lessons of their confinement will not suffice to accomplish what they desire. This is what I call the boxed robot bait and switch.

Very few if any, by sheer will, forge their own path to a STEM career. Those lost along the way or left behind by the limitations that the pre-packaged kit creators placed upon them are forced to find another career and lose their passion for the STEM subjects. The Robotics Expansion program is the best option for accomplishing the commission goals, not only for those who are naturally drawn to the STEM subject, but also for those who have yet to find their passion. The Robotics Expansion early levels focus on a proper introduction to the R-STEM fields and hooking a child's interest in these subjects. Using that excitement for robotics and then translating that into studying STEM subjects establishes the necessary connections for students to find successful careers in the STEM fields.

An important step into the STEM fields takes place in the first 2 years of College or University level education. It is here that many make decisions for their future careers that define not only their jobs, but their own American dream so eloquently mentioned by those at the Carnegie Foundation. The status quo of pre-packaged robot kits is more of a disservice at this stage than a contribution. Students of their programs are led to believe that robotics, engineering, and science are all about plugging and playing, that everything fits the first time you try it, that help boxes and completed sample programs you can chop and mix with other sample programs to create something slightly different is exactly how it is in the real world. The downfall of their programs becomes evident at this stage. The pre-packaged robot kit students attempt to enter a STEM program, realizing the difference within the box that they were taught in versus the engineering and science classes of College or University is so large, many experience a 'STEM shock.' While it may be too early to link this directly to those programs, the fact remains, children are not pursuing STEM careers in the numbers that we need them. The students of Robotics Expansion™ programs at this stage in the game would have a clear direction of which field they want to pursue as a career, but also have a good idea of what to expect and how it applies to what they strive to accomplish. This is a major reason the Robotics Expansion™ programs are the way of the future for R-STEM education, as they can reduce that 'STEM shock' factor and bring more children to STEM careers.

A contributing factor to the 'STEM shock' is that the pre-packaged robot kits place a heavy emphasis on only one field in robotics, and that is programming. This is a mistake, as the students become besieged with the notion that building robots means sitting in front of a computer creating endless code. They then tie this image to STEM and inadvertently sever any chance of the non-technically inclined student exploring the possibility of STEM fields. Robotics, as the pre- packaged robot kits would have you believe, is sipping a dark caffeinated beverage for long hours alone with your computer, occasionally grunting at the screen and drained of the last ounce of creativity. Robotics as shown in our Robotics Expansion™ programs, on the other hand, is a passion to give life to an idea that students conceive in their own imaginations.

An example of the benefits of the Robotics Expansion™ programs lies in the issues currently faced in the field. Given the multiple level approach in material and method by which Robotics Expansion™ is instructed to children, current engineering challenges can be presented to students. Students can use the strong foundation of understanding they have in the field to further pursue a solution to these problems in ways that may be overlooked by current experts. One such example is the limitations robotics now faces due to power source size, strength, and longevity. It is a barrier that holds back innovation for smaller and more flexible robots to conduct tasks that humans need for either safety or security. Presenting these real world problems as part of the class curricula creates a rich learning experience and provides an opportunity for students to create actual solutions that may 1 day translate into products.

Age Ranges Robotics Expansion™ Serves and Why

Robotics Expansion™ programs have focused on the elementary school level ages for the purpose of exposing children early to STEM. Though we begin at Kindergarten, our programs have instructed children as young as age 3; thus successfully planting the seed of STEM curiosity. It is never too early to begin children on the road to STEM (Stubbs et al. 2009), it merely needs to be done effectively. At this stage, we use a combination of methods in conjunction with a practical approach to capture the students' interest. The early ages are key (Watkins and Mazur 2013), as this is where the battle is won or lost for interest, where poor impressions by a subject often reflect poorly throughout the child's academic life.

There is much emphasis in this chapter on accurate depiction, it cannot be more important than at this beginning level. One factor in the success of the approach at this level is outlining the path forward for the student and through the beginning levels our students see this clearly. Another effective tool is setting the bar high but providing the support—both educational support through the material and psychological support through encouragement and recognition. The educational support is done through the design of the Robotics Expansion™ curriculum. The curriculum incorporates various avenues within its design for encouragement and recognition. Reward students for getting intellectual answers in STEM, and treat their ability to solve scientific, math or engineering problems in the classroom as important as any sport victory on the field. That positive reinforcement and public recognition for their efforts can do more for their intellectual growth. This is done to some extent now, but the major piece lacking was in the educational material, until now. The Robotics Expansion™ utilizes these tools to complement and enhance the STEM education process, as well diversify it. Diversify it in every means—more women engineers, more minorities, and more children who might not otherwise go into STEM.

The psychological effect on STEM perception must not be overlooked. The factors we have successfully integrated into the curricula work and lead to an excellent side effect when paired with our curricula, and that side effect is confidence in R-STEM subjects. We have been seeing this effect in our students for some time. The material inspires them, gives them the foundation they need to build upon and the encouragement they receive pushes the desire to be recognized in the classroom for getting the answer correct, solving the problem, coming up with a new idea, all of which are interwoven in engineering (Bybee 2010).

Student presenting their robot from the Infamous Robotics LLC first level summer camp (© Copyright Infamous Robotics LLC, 2014)

No More "Cookie Cutter" Robot Competitions

These cookie cutter robot competitions are not only boring and dull, but also quietly undermine the global initiative to strengthen our STEM workforce. The companies host these 'cookie cutter robot competitions' with large groups of student teams that do little more than help promote their products in the long run. I have seen first-hand how a typical student views these cookie cutter robot competitions and demonstrations, and the word is uninterested. Where is the innovation? The answer is: not here.

The approach has to change and that is something we have done at Infamous Robotics. We focus on the technical knowledge in classes through exciting, fun and hands-on exercises and place the emphasis on creativity, which has been shown to correlate with positive attitudes for technical subject matters. Similar attitudes are recognized as important by Oh, Lee & Kim. We have done this by creating a new club based on the sole concept that children are naturally creative and innovative. It is only in becoming an adult that many loose that creativity and innovativeness from people telling us "you can't do that," "it won't work," or "that is wrong." These statements are banned from our programs in order to highlight the paradigm shift our programs are making from the current status quo for robotics programs.

The new club, that is part of our Robotics Expansion™ programs, will revolutionize innovation in this country and bring back the long lost art of creativity in robotics. We call it the Robotics Inventors Club™ (the RIC). The RIC is a groundbreaking concept that puts the creativity in the child's hands and truly takes them through the engineering design process, unlike any pre-stamped prepackaged kit ever could dream of. The engineering design process entails core components of building a model or prototype of the design and then improving the design as necessary. There are other steps, but these two are critical to creativity.

Students working on their robot in the Robotics Inventors Club (© Copyright Infamous Robotics LLC, 2014)

Going Beyond STEM into STEAM

The recent acknowledgement of adding the arts in STEM to create STEAM is something we have known and adopted in our Robotics Expansion programs long before it had an acronym. The Arts play a large role in innovation related to robotics that is 100 % overlooked and ignored by the current pre-packaged approach. The Art in STEAM is important (Miller and Knezek 2013) and implemented in the Robotics Expansion™ programs for three reasons. The first reason is that if you give the child input into what they are doing, they take ownership. The project no longer becomes the class robot, it is now their robot. They name the robot, give it a personality, and seek to improve it beyond the classroom. The beyond the classroom improvement feeds the passion of learning in a personal way, and the material and Robotics Expansion™ programs effectively address this aspect.

The second reason Robotics Expansion™ is a STEAM based curricula is because it helps us casts a wider net. A wider net means the program catches the attention of more children with a wider variety of interests. Our approach has been to incorporate freedom of interest early on in our programs so children recognize it. A student may ask 'can I do this' and our response is "try it and find out." This statement is powerful because it ties directly back to the core of the engineering design process. In fact, it applies far beyond robotics and into other fields such as science where trial and error are a major component of discovery and innovation. Robotics Expansion™ programs see the bigger picture of STEAM subjects and their application in Robotics.

The final reason bringing the Arts into robotics is important is because they help to attract and retain more young women (Cooper and Heaverlo 2013). We have noticed young women's overwhelming acceptance of our programs versus the pre-packaged programs. The young women of our programs enjoy the artistic components we integrate with robotics. Our programs began with teaching the Girl Scouts about robotics in a unique way, and it has always been one of our main goals to inspire more young women into the engineering fields. We have had programs where the ratio of young women to young men students in our class is 4:1. It is exciting to witness these technical classes where there is a major shift like this. I recall being a student in engineering classes and having maybe one young woman in a class. The Robotics Expansion™ programs will change that in a positive way and has already begun to do so.

References

Banilower, E., Smith, P., Weiss, I., Malzahn, K., Campbell, K., & Weis, A. (2012). *Report of the 2012 national survey of science and mathematics education*. Retrieved November 12, 2013, from http://www.nnstoy.org/download/stem/2012%20NSSME%20Full%20Report.pdf

Barker, B. S., & Ansorge, J. (2007). Robotics as means to increase achievement scores in an informal learning environment. *Journal of Research on Technology in Education, 39*(3), 229–243.

Brophy, S., Klein, S., Portsmore, M., & Rogers, C. (2008). Advancing engineering education in P-12 classrooms. *Journal of Engineering Education, 97*(3), 369–387.

Bybee, B. (2010). Advancing STEM education: A 2020 vision. *Technology and Engineering Teacher, 70*(1), 30–35.

Carnegie Corporation, Institute for Advanced Study. (2013). *Transforming mathematics and science education for citizenship and the global economy*. Retrieved October 12, 2013, from http://opportunityequation.org/uploads/files/oe_report.pdf

Clifford, M. M. (1990). Students need challenge, not easy success. *Educational Leadership, 48*(1), 22–26.

Cooper, R., & Heaverlo, C. (2013). Problem solving and creativity and design: What influence do they have on girls' interest in STEM subject areas?+. *American Journal of Engineering Education (AJEE), 4*(1), 27–38.

Haynes, K. T., Cannata, M., & Smith, T. M. (2013). *Reaching for rigor by increasing student ownership and responsibility*

Marcus, J. (2012). *High dropout rates prompt engineering schools to change approach*. Retrieved December 5, 2013, from http://blogs.ptc.com/2012/08/06/high-dropout-rates-prompt-engineering-schools-to-change-approach/

Miller, J., & Knezek, G. (2013). STEAM for student engagement. *Society for Information Technology & Teacher Education International Conference, 2013*(1), 3288–3298.

NGSS Lead States. (2013). *Next generation science standards: For states, by states*. Retrieved January 12, 2014, from http://www.nextgenscience.org

Oh, J., Lee, J., & Kim, J. (2013). Development and application of STEAM based education program using scratch: Focus on 6th graders' science in elementary school. In *Multimedia and ubiquitous engineering* (pp. 493–501). Dordrecht: Springer.

Rockland, R., Bloom, D. S., Carpinelli, J., Burr-Alexander, L., Hirsch, L. S., & Kimmel, H. (2010). Advancing the "E" in K-12 STEM education. *Journal of Technology Studies, 36*(1), 53–64.

Stubbs, K. N., Yanco, H. A., Sathianarayanan, M., Chauhan, M., Saha, S. K., Kumar, S., & Krovi, V. (2009). STREAM: A workshop on the use of robotics in K-12 STEM education. *Robotics & Automation Magazine, IEEE, 16*(4), 17–19.

Watkins, J., & Mazur, E. (2013). Retaining students in science, technology, engineering, and mathematics (STEM) majors. *Journal of College Science Teaching, 42*(5), 36–41.

Chapter 8
Designing Serious Educational Games (SEGs) for Learning Biology: Pre-service Teachers' Experiences and Reflections

Meng-Tzu Cheng and Ying-Tien Wu

Introduction

It goes without doubt that we are living in a digital era where technology is shaping the way we live, think, and learn. Websites are becoming more popular information resources because of its convenience, and we now have online access to a multitude of learning materials and activities. Today more than half of the parents believe that videogame play provides mental simulations and that it is a positive part of child's life (Entertainment Software Association 2013). As a result, various methods have been created to harness the power of technology to support our education. The use of video games in training and learning environments, known as Serious Games (SGs) or Serious Educational Games (SEGs) (Annetta 2008), is one of the increasingly relevant trends which transforms our education because new digital innovations has significantly changed our pedagogical perspectives. Supporters of SEGs claim that video games have huge potential as a vehicle for learning and research evidence also shows its positive impact on students motivation, engagement, and learning outcomes, such as conceptual understandings and science process skills (e.g. Cheng and Annetta 2012; Clark et al. 2011; Connolly et al. 2012; Echeverría et al. 2011; Gee 2003a, b; Giannakos 2013; Lim 2008; Paraskeva et al. 2010; Prensky 2001; Sánchez and Olivares 2011). Although there remain some debates about the educational potential of video game play, the idea becomes clear that well-designed serious gaming do promote some educational goals as long as they can be done right. The type of video games and the desired ends of learning are particularly the issues that should be addressed.

M.-T. Cheng (✉)
National Changhua University, Changhua, Taiwan
e-mail: mengtzu@gmail.com

Y.-T. Wu
National Central University, Taoyuan, Taiwan

The use of SEGs is particularly important to science education, as many scientific concepts which are invisible in the real world and generally abstract and difficult to grasp can be portrayed in the virtual environment. In addition, scientific inquiry ability and problem-solving skills often require long-term cultivation and repeated practices. The complex structure of science, the trouble of reasoning about abstract concepts, and the challenges that arise in problem solving and scientific inquiry often cause students to have a sense of anxiety and difficulties in learning science compared to other subjects (Halff 2005). However, SEGs which combine game characteristics with science content not only motivate and absorb students in the embedded science learning activities, but also increase the probability of bridging virtual reality into reality in numerous dimensions. Thus they can provide students with authentic learning, an instructional approach focuses on learning through experimentation and real-world problem solving, wherein they are allowed to repeatedly experience things that are impossible in the real world without worries of real life consequences (Cheng et al. 2011).

After making a comprehensive survey of literature, we see that most of the available evidence focuses on students' science learning through SEG play; however, research that emphasizes pre- and in-service teachers' perceptions and implementations of using SEG or their professional development through designing an SEG is sparse. People, especially teachers, consider creating a game-based learning environment to be expensive and arduous. Moreover, although many governments worldwide have invested money in developing SEGs that facilitate science learning in elementary and secondary settings (e.g. http://www.fas.org/programs/ltp/games/), accessible resources of SEGs in Taiwan or projects which are funded by Taiwan's government endeavoring to create and develop SEGs are relatively deficient. For example, there are not many researchers doing the research related to serious gaming or not many SEGs or SEG-based instructions available for use in middle schools. All of these make it become more challenging and difficult for Taiwanese teachers to integrate SEGs into science classrooms.

Therefore, in fall semester 2012 a two-credit, 18-week-long course, entitled *Computers in Teaching and Learning Biology*, was delivered to 12 students who were enrolled in a teacher education program (pre-service teachers). In this course, students learned Adobe Flash™ and programming of ActionScript 3.0 and were asked to develop SEGs for biology learning by themselves. They were required to present their SEG idea and script (SEG prototype) for the midterm and demonstrate their SEG as the final exam. In-depth interviews with every pre-service teacher and instructor were conducted and recorded after the conclusion of the semester to collect data regarding feedback and comments towards this course, as well as the challenges and difficulties encountered from the perspective of students and instructor respectively.

This chapter consists of three sections. The first section discusses the theoretical framework underpinning this study. This is followed by a brief introduction of the details of the course, including how it was designed and implemented. Lastly, the major part of this chapter which focuses on the obtained results about the perceptions and challenges encountered in this course from the perspective of instructor, and pre-service teachers who completed this course and finished designing their own SEGs. Implications derived from the results, recommendations and suggestions for future implementation are also provided.

Teacher Education Courses and Technology Integration

Research indicates that teaching with technology supports student science learning in many aspects, such as the facilitation of conceptual understanding and the improvement of problem-solving abilities (Lee et al. 2010; Vogel et al. 2010). In views of the educational potential of technology use, many governments including Taiwan have developed plans to intensify their investments in constructing educational settings wherein instructors and students are encouraged and expected to teach and learn, using technology. Policy making has also responded to these acts. As a result, government institutions in charge of education worldwide have all placed a lot of effort into integrating technology (or information and communication technology; ICT) into national curriculum standards or guidelines. The newly released *Next Generation Science Standards* (NGSS), based on the *Framework for K-12 Science Education* which has its roots deeply in the most current research on science and science education, clearly identifies the importance of scientific and technological literacy for a well-educated society (http://www.nextgenscience.org/). A major commitment of the initiative is to integrate engineering/technological design into the structure of K-12 science education, in order to engage our next generations to become well-prepared citizens in the twenty-first century society who are capable of solving the major societal and environmental challenges they will face. Likewise, the *K-9* and *high school curriculum guidelines* published by Taiwan's Ministry of Education also aims to bring up K-12 students in Taiwan as individuals with knowledge and skills to deal with information and solve problems with technology. In recent years, they further encouraged the use of innovations in the teaching of all subjects to increase teaching quality by integrating technology and pedagogy (Ministry of Education 2008).

Despite the best intentions of administrators endeavor to increase technology access in educational settings, the most challenging issue remains, are teachers competent or well-prepared for teaching with technology? It is obvious that only when teachers are competent of carrying out the task of teaching with technology, do the integration of educational innovations succeed. So if the answer is that they are not capable of doing so, how can we expect students to acquire ICT knowledge and skills, and to learn with the acquired knowledge and skills, and furthermore, to design and create using those knowledge and skills? Research shows that over 90 % of teachers had access to one or more computers or other technological facilities in the classrooms every day. However, less than 50 % of teachers reported that they or their students actually use computers in the classrooms during instructional time on a regular basis (Gray et al.2010). Even if teachers do use the equipped technologies, they are likely to employ them merely for administrative support rather than instructive support, or mainly for informative or expressive purposes of supporting their existing practices instead of engaging and facilitating students in higher-order thinking activities (Wozney et al. 2006). It has been reported that teacher self-efficacy, confidence to perform specific tasks, significantly affects the extent, as well as the way, teachers use technology for everyday instructional practices in classrooms (Paraskeva et al. 2008).

Teacher education had always played a major role in preparing pre- and in-service teachers with knowledge, attitudes, and skills required to teach effectively in the classrooms, and the lack of properly integrating technology into classrooms can be seen as a reflection of the inadequacy of teacher preparation programs provided by teacher education institutes. The National Education Technology Standards for Teachers (NETS•T) require effective teachers to be capable of designing, implementing, and assessing relevant learning experiences which incorporate digital tools and resources to facilitate and inspire student learning and creativity (http://www.iste.org/standards/nets-for-teachers). However, teacher preparation courses related to effective teaching with technology offered by teacher education institutes in Taiwan are relatively few. Moreover, almost all the technology literacy-related courses are elective, hence pre-service teachers might not have to take any course empowering them to succeed in technology integration prior to graduation. Pre-service teachers even felt that many experiences and resources in teacher preparation programs are insufficient and not helpful for technology-integrated teaching (Singer and Maher 2007). Unfortunately, there remains a gap between what is taught in the teacher preparation programs and how teachers use technology effectively in the real classrooms.

Technological Pedagogical Content Knowledge (TPACK)

Obviously, successful and effective integration of technology into instructions is never as simple as merely using innovations for administrative purposes or supporting the existing practices in the classrooms, teachers are actually required to have sufficient pedagogical content knowledge (PCK) and technological knowledge. In other words, the approaches related to teaching with technology have transferred from techno-centric, which merely focuses on technology and the knowledge and skills to use various technologies, to techno-pedagogical integration, which places much more emphasis on putting both pedagogy and technology into practice in the integration process (Yurdakuli et al. 2012).

In order to prepare pre-service teachers for their future teaching career (which likely requires the integration of technology), teacher education programs have to help pre-service teachers to construct their own technology-supported pedagogical and technology-related classroom management knowledge and skills. The Technological Pedagogical Content Knowledge (TPACK) is a model that provides directions for teacher education programs to evaluate the effectiveness of their courses and prepare pre-service teachers as qualified educators with the ability to integrate technology into pedagogical strategies and content representations (Chai et al. 2011). TPACK, an expansive framework based on Shulman's (1986) concept of pedagogical content knowledge (PCK), aims to make three aspects, content, pedagogy, and technology into a whole to describe the required knowledge of using technology in a way which is contextually authentic and pedagogically appropriate in the educational settings for an effective teacher.

Representing the intersections among knowledge of pedagogy, content, and technology, the framework of TPACK includes seven dimensions of professional knowledge, namely, the Pedagogical Knowledge (PK), Content Knowledge (CK), Technological Knowledge (TK), Pedagogical Content Knowledge (PCK), Technological Pedagogical Knowledge (TPK), Technological Content Knowledge (TCK) and Technological Pedagogical Content Knowledge (TPACK) (Mishra and Koehler 2006). We introduce these seven dimensions as below (Abbitt 2011; Chai et al. 2011):

1. Pedagogical Knowledge (PK): Knowledge about the nature of processes and practices or methods of teaching and learning (e.g. instructional strategies, classroom management, etc.)
2. Content Knowledge (CK): Knowledge of the actual subject matter that is to be learned or taught (e.g. biology, physics, etc.)
3. Technological Knowledge (TK): Knowledge and skills required to operate particular technologies for information processing, communication, and so forth.
4. Pedagogical Content Knowledge (PCK): Knowledge of using different strategies and teaching practices to represent and formulate a given subject matter.
5. Technological Pedagogical Knowledge (TPK): Knowledge of the affordances and constraints of using technology for facilitating pedagogical approaches.
6. Technological Content Knowledge (TCK): Knowledge of using technology for representing or exploring a given subject content.
7. Technological Pedagogical Content Knowledge (TPACK): Knowledge of appropriate integration among content, pedagogy, and technology for facilitating students learning.

The TPACK framework has been widely used as a lens through which to observe and think about teacher knowledge and practice of teaching with technology in research and many evaluation studies, as most of the current research has made a lot of effort to develop valid and reliable tools and methods for assessing the obtained knowledge of teachers in terms of evaluating teacher preparation experiences. However, teachers' understandings of TPACK should also be embedded in their created learning environment. In addition to considering TPACK as a framework for evaluating teacher skills, it might be more interesting to see TPACK as a framework for teachers to design digital learning environments, such as games and simulations (Gibson 2008).

Gibson (2008) argues that in order to produce an SEG, teachers are required to improve their understandings of content, technology, and pedagogy and consequently integrate these understandings into a highly interactive innovation which the audience interacts with. Even though engaging teachers in designing an SEG is time- and effort- consuming and the process of development is complex, it is worthwhile as the whole process powerfully situates teachers in an authentic and meaningful context where personal motivation and relevance are much more increased. Also thinking of how to produce an effective SEG would require teachers to deeply consider: (a) the prior knowledge students bring to the game; (b) logical progression of the content; (c) effective scaffolding of student thinking/decision-making; and (d) ongoing formative assessment. All of this has been known for decades to be crucial for effective teaching and learning. In other words, to design and develop an SEG which provides players with a highly interactive experience not only benefits the audiences, the teachers who create the SEG do actively learn through the whole process of making SEG. Hence, the requirement of designing an SEG facilitates pre-service teachers to actively construct their own TPACK and in turn embeds their TPACK into the created SEGs. In short, the courses offered by current teacher education programs do not seem to prepare teachers with competence for teaching with technology and the process of designing an SEG might be helpful in facilitating teachers to construct their own TPACK which is required for effective integration of technology into pedagogical practices.

Project-Based Learning

Because of the aforementioned issues the course *Computers in Teaching and Learning Biology* was offered with an aim to provide pre-service teachers with an experience of project-based learning (in this case, project refers to the development of an SEG). Project-based learning has its roots in Dewey's (1938) idea of learning by doing. It is an instructional approach and offers a contextualized learning activity wherein learners are presented with problems to solve or product to develop. It is defined as "a model that organizes learning around projects" and these projects are "complex tasks, based on challenging questions or problems, that involve students in design, problem-solving, decision making, or investigative activities; give students the opportunity to work autonomously over extended periods of time; and culminate in realistic products or presentations" (Thomas 2000, p. 1). Thomas (2000) further suggests five criteria for defining an exemplary project of project-based learning:

1. The project is the central teaching strategy, not peripheral one to the curriculum. In project-based learning, students encounter and learn the central concepts of disciplines and construct understanding via the project.
2. The project is focused on questions or problems which are so ill-defined that "drive" students to encounter (and struggle with) the central concepts and principles of a discipline.
3. The project involves and engages students in a goal-directed, constructive investigation including inquiry, knowledge building and resolution. The project is quite different from an exercise as it cannot be easily carried out by students merely with the application of already-learned information skills.

4. The project is student-driven to some significant degree. It allows a great deal of student autonomy and doesn't have a predetermined outcome or path.
5. The project is realistic, rather than school-like. It mainly focuses on authentic problems or questions and where solutions have the potential to be implemented.

The idea of learning by doing is consistent with the perspective of constructivist learning theory which provides a philosophical view on how people come to understand. Constructivism has influenced the practice of teaching and the design of learning environment greatly since it considers our understanding as being contextualized in our interactions with the environment, and also that learning is stimulated and results from individual's cognitive conflict or puzzlement and knowledge evolves through social negotiations (Savey and Duffy 1985). It turns out that project-based learning is a constructivist approach which creates a learning environment supporting engagement in problem-solving situations where students actively construct their own knowledge. Research has identified many positive effects of project-based learning, including the development of positive attitudes towards learning as well as the improvement of abilities on problem-solving, critical thinking, collaboration and so forth. Moreover, it results in better learning outcomes and turns students into active problem solvers rather than passive knowledge receivers (Gülbahar and Tinmaz 2006). It is a systematic teaching method concentrating both on the end-product and the experience of the process. In terms of our case, the use of project-based learning focuses on not only the SEGs created, but also the process of creating the SEGs.

However, how does one implement an effective project-based instruction? Barron et al. (1998) have identified four important design principles for reaching this tough goal. The first principle is that educators have to clearly define learning-appropriate goals that lead to deep understanding of the how and why of a project in advance. Then, suitable scaffolds of providing a series of problem-solving activities and contrasting cases need to be offered before projects are really carried out. The third design principle is the provision of frequent opportunities for formative assessment and revision, which allows both students and instructors to realize what is and isn't being learned so that the instructions can be adjusted accordingly and immediately. Finally, social organizations that promote participation and support active, collaborative learning should be encouraged.

The Course *Computers in Teaching and Learning Biology*

Computers in Teaching and Learning Biology is a two-credit, elective course of the Department of Biology at the National Changhua University, Taiwan for undergraduates who are majoring in biology and are enrolled in teacher education program. The course is of particular importance in the entire teacher education curriculum offered by the Department of Biology because it is the only course in the curriculum which aims to foster skills of pre-service teachers in designing digital learning environments and practically integrating technology with biology teaching and learning. The instructor, who is a science education researcher as well as experienced computer programmer, has many years of experience in game development. With the help of two science education experts whose research interests have focused on educational technology, the instructor designed the course to be project-oriented and design-based in such a way that students would construct their own knowledge and skills by collaborating with their group members to design and develop their own SEGs. Five learning objectives were addressed:

1. To enhance Information Communications Technology (ICT) competences and technological/engineering literacy.
2. To improve Technological Pedagogical Content Knowledge (TPACK).
3. To develop proficiency in logic/analytical thinking.
4. To cultivate abilities of creative thinking and problem solving.
5. To foster skills of communication and collaboration.

Although there are many tools which allow entry-level novices to easily create a game without programming (e.g. GameMaker http://www.yoyogames.com/game-maker/studio), the less flexibility for expansion of those tools/engines doesn't allow game creators to take as much control as pure coding would. The benefit of making games without programming soon becomes a disadvantage because students don't really experience what real-life game programmers/engineers do. Hence, we finally decided to employ ActionScript 3.0™ as the programming language taught in this course. There are five reasons that ActionScript 3.0™ was chosen instead of other programming languages (Agarwal 2010; Brimelow 2008):

1. Adobe Flash™ is one of the widely used tools for e-learning, and ActionScipt 3.0™ is designed to be primarily used for the development of Web-based games and rich Internet applications with streaming media targeting Adobe Flash Player™ platform.
2. It is an object-oriented programming language with reusable code bases. The visual design of ActionScript 3.0™ is more accessible and comprehensive.
3. It includes strictly debugging and troubleshooting functionality allowing for easier error checks.
4. Programming structure of ActionScript 3.0™ is on the same level as writing in other higher-level languages like Java and C$^\#$, which makes it easier for students to get into more advanced programming someday.
5. Work in ActionScript 3.0™ leads directly to portability among other Adobe technologies (e.g. Adobe Integrated Runtime™ (AIR)), which allows singular experience to be delivered across multiple devices.

The course was a two-credit, 18-week-long course and the course syllabus is presented as Table 8.1. There were a total of 12 students registered in this course. They were finally divided into four groups (2–4 individuals/group) to carry the project out by group collaboration. In this course, students were taught basic principles

Table 8.1 Course syllabus of Computers in Teaching and Learning Biology

Weeks	Topics/tasks	Laboratory assignments
1	Introduction to Adobe Flash™ and ActionScript 3.0™ Object-oriented programming Introducing flowcharts of game programming	Dividing students into groups (2–4 students/group) Discussing SEG script
2	Timeline, layer and frame Event and function	Development of flowcharts
3	Variables, objects/classes, movie clip properties Path and the framework of programming	Development of storylines Finish initial idea of SEG script
4	Playing with text Loops	Assignment 1 (loops)
5	Statements	Assignment 2 (statements)
6	Keyboard event – events for keyboard	Assignment 3 (button)
7	Arrays	Assignment 4 (collision with motion tweening)
8	Add sounds and audio effects Add videos	
9	Midterm: presenting SEG prototype	
10	Presenting game sample-collision detection	Assignment 5 (collision with motion tweening)
11	Demonstrating game sample-random	
12	Demonstrating game sample: group 1	Discussion and practice of programming that group 1 needs
13	Demonstrating game sample: group 2	Discussion and practice of programming that group 2 needs
14	Demonstrating game sample: group 3	Discussion and practice of programming that group 3 needs
15	Demonstrating game sample: group 4	Discussion and practice of programming that group 4 needs
16	Demonstrating game sample	Review of ActionScript 3.0™
17	Final exam-ActionScript 3.0™ Final project Q&A	
18	Final project showcase	

of ActionScript 3.0™ programming so that they can use Adobe Flash Player™ as a platform to demonstrate their created SEGs.

The course schedule can be divided by midterm into two parts. Before midterm (week 1–8), the instructor placed much more emphasis on basic concepts and fundamentals of ActionScript 3.0™. After midterm (week 10–16), the instructor in turn introduced specific programming which each group needs according to their SEG script. Two presentations and one paper-and-pencil test were required. Each group had to present game idea and script (SEG prototype) in the midterm (week 9) and demonstrate the SEG (end-product) they created in the end of the semester (week 18).

Table 8.2 Criteria for assessing student performance

Criteria	Percentage
Participation	10
Homework assignments	30
Final exam (paper-and-pencil test)	10
Midterm presentation of SEG prototype	20
Final demonstration	30

Moreover, there was an exam assessing what they had learned about ActionScript 3.0™ in the final (week 17). In addition to in-class practices, five homework assignments were also distributed to ensure that students did learn the programming, which were taught. Although the 18-week lectures mainly emphasized the development of programming skills, each group had to regularly discuss their SEG idea and script with a science education expert at times out of classes to ensure the validity of scientific content and pedagogical methods embedded in their games.

For the midterm presentation of SEG prototype and final demonstration of the created SEG, students were required to clearly address the below questions:

- What is the main idea of creating the SEG? What is the originality of the SEG?
- What are the learning objectives?
- What are the scientific concepts embedded?
- How does the art design appear? (prototype)
- What are the programming needed? (prototype)
- Presenting the whole game script (including storylines, scenes, characters, user interface, etc.) (prototype)
- Demonstrating the created SEG (final product)

The entire course was graded according to the criteria provided by the instructor (Table 8.2).

Research Design

To explore the pre-service teachers' experiences and reflections on designing SEGs for learning biology, several tape-recorded in-depth interviews with every pre-service teacher and instructor were conducted after the semester. The pre-service teachers were asked to answer several leading questions. However, a semi-structured method was employed, which allows conversational, two-way communication between the interviewer and the person being interviewed to probe for details. These leading interview questions are presented as below:

1. What was your motivation for taking this course?
2. What have you learned from this course?
3. Are there any distinctions between your expectations and the actual practices of this course? If yes, please tell me about the distinctions?

4. What are your perceptions about the learning processes in this course?
5. Have you ever felt frustrated during this course? If yes, how did you overcome the frustration?
6. What are your suggestions for the instructor on his teaching practices in this course?
7. What are your recommendations on the arrangement of the course?
8. Do you think that this course is helpful for your instructional practices in the future?

Moreover, the instructor of the course was also interviewed in this study with the following leading questions:

1. What were the difficulties for you when teaching the pre-service teachers to make serious educational games?
2. According to your observation on the pre-service teachers' learning in making serious educational games, what was the most difficult part for them?
3. After teaching these pre-service teachers to make serious educational games, how will you modify your course design in the future?

Each interview with each interviewee lasted about 15–20 min. During the interview, each pre-service teacher or instructor was required to provide their feedback and thoughts towards this course and challenges and difficulties encountered in the course. We collected data and heard different voices from the perspective of the students and the instructor. With interviewee permissions, all the interviews were transcribed verbatim into transcriptions for data analysis. These transcriptions were first separated into narrative segments that expressed a specific idea/concept or described a particular experience, and then these narrative segments were again read repeatedly by researchers to find emerging categories. Recurring and qualitative distinct themes, conclusions, and explanations were drawn from these categories. There is one thing which should be noticed. The participants were required to provide their response which they thought was the most important for most of the questions, so most of the response categories only have 12 total responses. Although we can obtain the most important factors affecting participant learning in this course, other potential important data might likely be lost, which should be acknowledged in the future work.

Results

Pre-service Teachers' Motivation for Taking the Course

As shown in Table 8.3, about half of the pre-service teachers mentioned that they took this course because it teaches them to design and make SEGs, which they could use in their biology teaching. For example, Pre-service Teacher #12 mentioned, "I took this course because I could design and make a Serious Educational

Table 8.3 The pre-service teachers' motivation on taking the course (n = 12)

Motivation	n
1. Can design and make an educational game, and use it in biology teaching	5
2. Curious about how to use educational game in teaching	3
3. Can learn how to make animations with flash	3
4. It seems fun	1

Game by myself. It seems very interesting. I could also use it in my biology classes". Moreover, three of the pre-service teachers took this course due to curiosity about the use of SEGs in biology teaching. For instance, Pre-service Teacher #11 mentioned, "I took this course because I was curious about how to use Serious Educational Games in biology teaching." However, it should be noted that three pre-service teachers in this study mentioned that they took this course for the reason that they could learn how to make animations with Flash. For example, Pre-service Teacher #3 mentioned that "I took this course because I wanted to learn Flash in this course."

It seems that most of the pre-service teachers in this study had some basic understanding regarding SEGs. As a result, the basic understanding about SEG motivated them to take this course focusing on designing and making SEGs.

What Did the Pre-service Teachers Learn from This Course?

Learn from the Course

According to Table 8.4, only two pre-service teachers mentioned that they learned how to design SEGs. Pre-service teacher #5 mentioned that "I learned that if I wanted to design a Serious Educational Game, what I should take into account. For example, I should consider what my students could learn from playing the serious educational game. Also, I learned about how to make a simple Serious Educational Game." Moreover, a pre-service teacher (Pre-service Teacher #1) mentioned that "I learned how to transfer content knowledge of a specific topic into a Serious Educational Game in this course", and another pre-service teacher (Pre-service Teacher #2) pointed out that "In this course, I learned about how to work collaboratively with others." These descriptions stated by the aforementioned pre-service teachers are exactly in alignment with the learning outcomes expected by the instructor.

However, half of the pre-service teachers reported the skills for programming with Flash as the learning outcome derived from taking this course focusing on designing and making SEGs. The Pre-service Teacher #6 mentioned that "I learned how to program with Flash, and became familiar with using computers." Moreover, four of the teachers mentioned that they learned about designing and making games

Table 8.4 The pre-service teachers' self-reported learning outcomes derived from taking the course (n = 12)

Learning outcome	n
1. How to program with flash	6
2. Designing and making games	4
3. How to design an serious educational game	2
4. Transferring the domain knowledge into a serious educational game	1
5. How to collaboratively work with others	1

Table 8.5 The distinctions between the pre-service teachers' expectations and the actual practices of this course (n = 12)

Distinction	n
1. More efforts should be paid during taking this course	5
2. Making a game is not so easy	4
3. Programming with flash is difficult	3
4. The need for collaboration in making a serious educational game	1

throughout the course. For instance, the Pre-service teacher #3 mentioned that "I learned a lot in this course. At first, we learned how to design a game. Then, we learned how to program the modules with Flash, and made a whole game."

It is surprising that most of the pre-service teachers may have placed their focus on learning to program with Flash or making games, rather than designing and making "Serious Educational" games. This may be due to the fact that the pre-service teachers in this study lacked relevant knowledge or ability in programming with Flash, hence their insufficient prior knowledge in programming may have distracted their attention during their learning processes. Consequently, they paid most of their attention on programming rather than integrating educational purposes into the games they designed and made.

The Distinctions between the Pre-service Teachers' Expectations and the Actual Practices of This Course?

Only 2 of the 12 pre-service teachers expressed that the actual practices of the course were almost the same as what they expected before taking the course. However, the other teachers mentioned various distinctions between what they expected and the actual practices of the course. Table 8.5 summarizes the distinctions that the pre-service teachers mentioned. Most of the pre-service teachers, such as Pre-service teacher #7, mentioned that "The teacher told us that lots of efforts would be needed in developing a game. However, the efforts I put in during this course were much more than I expected." Besides, some pre-service teachers also

Table 8.6 Pre-service teachers' perceptions of their learning processes (n = 12)

Perception	n
1. Experiencing interesting and meaningful learning	5
2. I spent a lots of time in coding	2
3. Experiencing student-centered instruction	1
4. The loading increased; however, there was insufficient time	1
5. More detailed explanations from the instructor will be helpful	1
6. I could not follow the teacher's instruction	1
7. Lots of homework to be done after school	1

mentioned that making a game was not so easy. Pre-service Teacher #6 expressed that "At beginning, I felt that it would not be too difficult to design and make a Serious Educational Game. However, I was able to design a game but was not able to finish making it." Other pre-service teachers, such as Pre-service Teacher #10 mentioned, "I spent lots of time in programming with Flash. However, it was still very difficult for me."

It seems that the workload of this course was too heavy for the pre-service teachers. In particular, for the pre-service teachers without prior knowledge in programming with Flash, as it took them substantially longer periods of time to complete their game design.

Pre-service Teachers' Perceptions of Their Learning Processes

As revealed in Table 8.6, five of the pre-service teachers perceived their learning processes as "interesting and meaningful." For example, Pre-service Teacher #3 mentioned, "I felt that this course is interesting." Another one, pre-service Teacher #4, mentioned, "In traditional classes, it is always teacher-centered; however, in this course, we experienced learner-centered instruction in class."

Yet, it is worthy to note that six course participants did not report the learning process was either interesting or student-centered. In fact, they expressed less positive perceptions regarding this course and perceived the course as very effort- and time- consuming because they had to spend much time in coding and homework assignments. For example, Pre-service Teacher #1 expressed that "The loading of the course gradually increased; however, there was insufficient time for me to study it." Pre-service Teacher #8 mentioned, "I felt lots of homework to be done after school. However, I did not have sufficient time."

The explanations offered by the instructor might be another important issue as one participant needed more detailed explanations from the instructor and one felt the instruction offered by the teacher is hard to be followed. Pre-service Teacher #2 mentioned, "I was not so good in programming. Therefore, I always felt that more detailed explanations from the instructor would be needed and helpful." Pre-service Teacher #8 mentioned, "I felt lots of homework to be done after school. However, I

Table 8.7 Pre-service teachers' frustrations during taking this course (n = 12)

Frustration	n
1. Programming and coding	8
2. Completing the homework	1
3. How to integrate what I have learnt into the game	1
4. How to implement our design	1
5. Low quality of the game	1

did not have sufficient time." It turns out that the major challenge for these pre-service teaches probably lies on their lack of prior knowledge of programming, suggesting the very need for modifying the design and arrangement of this course in the future.

Pre-service Teachers' Frustrations During Taking This Course

The tape-recorded interviews in this study also explored what the pre-service teachers felt frustrated during taking this course. Only Pre-service Teacher #12 did not feel frustrated during taking this course, while the others mentioned various frustrations. However, it is interesting to find Pre-service Teacher #12 got the lowest score on this course, although he said that he did not encountered any frustration in this course.

As shown in Table 8.7, most of the teachers mentioned that they felt frustrated in programming and coding, and one of them (Preservice Teacher #1) also mentioned that she felt frustrated because she was always not be able to complete the homework on time. Compared with aforementioned frustrations, some teachers expressed their frustrations were caused by further personal commitments. For example, Pre-service Teacher #5 expressed that "I felt frustrated when I tried to integrate what I have learnt into the Serious Educational Game we designed and developed." Pre-service Teacher #10 said, "I felt frustrated. When we had finished the game design (prototype), we could not implement our game design (end-product)." Also, Pre-service Teacher #6 mentioned that "I felt so frustrated because the quality of the Serious Educational Game we made is low." It seems that the more efforts were made by these pre-service teachers in this course, the more frustrated they might be oriented to feel. These frustrations might be a result from the pre-service teachers' insufficient experiences of mastering in designing and making SEGs.

The Ways that the Pre-service Teachers Overcome Their Frustrations

Table 8.8 shows that discussing with peers or teachers in classes is the most common way for these pre-service teachers to overcome the frustrations they had. Besides, some pre-service teachers tried to overcome their frustrations by using

Table 8.8 The ways that the pre-service teachers overcome their frustrations (n = 12)

Way of overcoming frustration	n
1. Discussing with team members and then discuss with the teacher in classes	12
2. Making use of other resources, such as Google, YouTube, other books	4
3. Practicing with the instructional materials provided by the instructor	1
4. Asking the instructor after school	1

other online resources, such as the tutorials provided on Google™ or YouTube™, to find out how to do programming with Flash™. Also, they might read books about programming. Pre-service Teacher #3 also stated, "when I felt frustrated, I would practice again with the instructional materials provided by the instructor."

It is admirable that most of the pre-service teachers spent much time discussing with peers or teachers in classes, yet only one tried to seek assistance from the instructor after classes. Pre-service Teacher #4 mentioned, "I always discussed with the instructor on Facebook after school when I did not know how to solve the problems." The reason might be due to the issue that these pre-service teachers didn't know the instructor none the less provides assistance anytime even after classes, so they felt hesitated to ask. It suggests that it would be much more helpful if more scaffolds, such as available resources or synchronous and /or asynchronous interactions after school, were provided in the future.

Pre-service Teachers' Suggestions on Instruction

The pre-service teachers in this study provided three major suggestions for the instructor:

1. More basic instruction in programming and coding would be helpful: For instance, Pre-service teacher #2, #3, and #6 mentioned that "If the instructor could provide more basic instruction on programming and coding, we might be able to progress in programming and coding step by step."
2. More instructional time would be needed: Five pre-service teachers (#4, #5, #9, #10, and #12) stated that more instructional time would be needed because both designing and making Serious Educational Games were required in this course.
3. Classroom videos and more detailed handouts for students would be helpful: Pre-service Teacher #7 and #11 advocated for classroom videos and more detailed handouts for students. Then, the students could practice and rehearse by themselves after school.

Pre-service Teachers' Recommendations on the Arrangement of the Course

The pre-service teachers in this study also provided three major recommendations on the arrangement of the course:

1. The instructional time of the course should be extended: Half of the pre-service teachers suggested that the course should be extended to a two-semester course.
2. The loading of homework should be reduced: Five pre-service teachers mentioned that if the loading of homework is too heavy, the students may feel frustrated.
3. The participants of the course should be limited: a pre-service teacher also advocated that the class should be a small-sized one. Then, every student will have more time discussing with the instructor in classes.

Pre-service Teachers' Attitudes towards the Usefulness of the Course for Their Instruction in the Future

Except for Pre-service Teacher #12, all pre-service teachers expressed positive attitudes towards the usefulness of the course for their biology teaching in the future. Through the process of creating their own SEGs, they got a better understanding about how to visualize the scientific concepts which are abstract into concrete representations. They also gained a clearer idea about what it is meant by authentic, meaningful learning and how to implement it in their biology classes to provide their students a context wherein they can apply the learned knowledge to solve problems in novel situations. These are perceived helpful for their biology teaching in the future.

However, it is interesting to find that two felt positive toward the course, yet low desire to further develop another SEG in the future was found. Pre-service teacher #6 mentioned that "I think taking this course is helpful for my instruction in the future. However, I will not develop a Serious Educational Game by myself in the future. I would like the Serious Educational Game developed by others, and I could use it in my classes." Pre-service Teacher #10 also expressed a similar perspective. Some teachers also mentioned that they obtained basic experiences in designing and making a Serious Educational Game, and these experiences would help them to use a Serious Educational Game in the classrooms more effectively (the Pre-service Teacher #3, 4, 11). Again, the programming issue seems a major challenge and stress for these pre-service teachers, which decreases their inclination to create new SEGs in the future.

We thought the most encouraging result is the inspiration of their conceptual ideas about creating an SEG. These participants did learn about how to situate their scientific knowledge into a game format and also knew how to write a creative game script. We believe that they are still willing to create their own SEGs as long as appropriate assistance can be provided, such as cooperation with game designers and programmers who can help them transfer their conceptual ideas into end-products.

The Instructor's Reflection

In this study, the instructor was also interviewed to provide directions for the improvement of the course. The instructor was asked to reflect on the difficulties he encountered when teaching pre-service teachers on SEG design and development. He mentioned that "As beginners in programming and coding, these pre-service teachers faced huge challenges in programming and coding with their insufficient ability. They also faced the challenges in integrating their professional knowledge into the Serious Educational Games." In other words, the major challenge that the instructor encountered while teaching these pre-service teachers to design and make SEGs was about how to help these pre-service teachers obtain basic ability in programming and coding, and guide them to integrate their professional knowledge in biology and biology teaching into the games.

During the interview, the instructor also stated the difficulties for the pre-service teachers in making SEGs. He mentioned that "Basically, the pre-service teachers did not have any problem in programming or coding in classes. However, they often lacked sufficient practices after school. Without continuous practices, these pre-service teachers shortly forgot what they have learned in classes and accordingly, they felt more difficult in programming and coding as an increasing content should be mastered. When I divided the instruction of programming into several parts, students were not able to make connections among these parts by themselves. That is, it is not easy for them to construct an integrated understanding regarding the computer program for an SEG. However, the most difficult thing for students was to have the insights on the logic of how a computer program is executed." The above-mentioned results indicated that for those who are with only professional knowledge in biology and biology teaching, programming and coding will be the major obstruction for their success in developing SEGs. Finally, the instructor also proposed possible ways in helping pre-service teachers design and make SEGs. He advocated that if the instructional time can be extended, the students could have more time practicing in classes. Also, it will allow students to complete their homework in classes. Then, instructors could provide immediate feedbacks to students, and students' frustrations could be reduced and the quality of the created SEGs as well as students' learning outcomes could be improved.

Discussion

Researchers argue that teacher education has a primary, inherent goal, which is to enable pre-service teachers to effectively transfer what they learn in teacher preparation courses to their future teaching (Howard 2002). Especially since we are now living in a digital era where technology is closely and inseparably connected to our daily lives, today's educators have to put all of their effort to not only reach these two goals, but also achieve the two goals through the integration with technology. Since the use of SEGs in science education has gradually grabbed much attention and its positive impact (e.g. improving knowledge acquisition, increasing learning motivation, enabling problem solving, encouraging collaboration, and so forth) on science education now has been widely evidenced, we thought it might be feasible to reach the two primary goals by engaging pre-service teachers in the activity of developing SEGs. We believe the act of creating SEGs, which provides players with an authentic context and a highly interactive experience, not only benefits the audiences but also allow the pre-service teachers to actively learn through the whole process of designing and creating SEGs. Therefore, the course *Computers in Teaching and Learning Biology* was offered as a project-based, design-oriented, and student-centered course with the expectation that pre-service teachers can construct their own knowledge and skills of TPACK through the processes of designing and making SEGs and embed the TPACK into the created SEGs.

However, the course *Computers in Teaching and Learning Biology* had been suspended for several semesters and was only re-offered since the fall of 2012. As it was the first time the course was delivered as a project-based experience which requires pre-service teachers to create their own SEGs, it was a wholly new and unfamiliar experience for both the students and the instructor alike. Hence, many challenges and difficulties were encountered while implementing this course as was expected. We interviewed both pre-service teachers and the instructor in-depth with a main focus on investigating the challenges and difficulties in carrying out this course. In so doing, we hope to provide a wide-ranging set of contextualized findings to support further research and it is also expected that the conditions associated with successful implementation of such a project-based learning (learning by designing SEGs) for pre-service teacher education can be somewhat delineated. By further examining the above-mentioned results derived from this study, we found these pre-service teachers basically had a positive attitude towards this course. They generally agreed that the whole learning experience of taking this course is interesting, student-centered, and meaningful, and they also thought this course is helpful in their future biology teaching as long as they are not required to do programming by their own. They also did learn some basic ideas about the integration of technology into pedagogy that they don't see elsewhere in their curriculum. However, four major issues: time, programming, course loading, and transfer/integration, were also revealed regarding challenges and difficulties of the implementation of this course, from both the perspective of the pre-service teachers and the viewpoint of the instructor. The descriptions of the four issues are provided as below.

Time Issue

The insufficiency of instructional time seems the major deficit of this course as mentioned by both students and the instructor. In fact, the curriculum of teacher preparation programs in Taiwan (unlike most countries) has a rather tight schedule, so that teacher preparation courses related to effective teaching with technology offered by teacher education institutes are relatively few. Moreover, even when the courses are offered, almost all of them are elective and two-credit only. Hence, the time issue becomes a dilemma for teacher educators in Taiwan. On the one hand, if the course is offered as a three- or four- credit course, the redundant credit(s) might be not able to be counted into the required credits for graduation. On the other hand, if the course remains two-credit, students might think the difficulty is too great for a two-credit course. Both of these situations will significantly decrease students' motivations of taking this course, which clearly reflects the inadequate arrangement of the current teacher preparation courses for improving the technological literacy of pre-service teachers.

Programming Issue

Almost all of these pre-service teachers mentioned that they felt frustrated in programming and coding. However, according to the instructor's reflections, these pre-service teachers appeared to not have any problem in programming and coding in classes. The major problem was they usually lacked sufficient practices after school and didn't effectively construct an integrated understanding. This is actually a serious problem which has to be overcome if the course still has to be offered in the future. A very important reason why ActionScript 3.0™ was used as the programming language to be taught in this course instead of using other existing tools which allow novices to easily create a game without programming was that, we hoped the analytical and logical thinking of these pre-service teachers could also be improved by learning how to program, a worthy byproduct perhaps. It was believed that while it might be difficult for these pre-service teachers to learn programming, once they are familiar with programming and learn the logical thinking process behind it, it would be very helpful for them in solving real-life problems and for making decisions in the future. Also, being familiar with the programming structure of ActionScript 3.0™ will make it much easier for them to get into more advanced programming someday (if they need). But indeed, it is impossible for pre-service teachers or others to gain mastery within a short time period, particularly when the skill in question is complex programming. Substantially more time is required to allow repetitive practice in order to construct an integrated understanding of the execution of computer programs, so that mastery of programming can be gained. Some pre-service teachers expressed that they felt their ability of analytical and logical thinking had improved after receiving the training of programming in this course.

Course Loading Issue

Another issue regarding challenges and difficulties is course loading. The criteria for assessing student performance in this course include participation (10 %), homework assignments (30 %), final paper-and-pencil exam (10 %), midterm presentation of prototype (20 %), and final demonstration (30 %) (as shown in Table 8.2). Despite the midterm presentation of prototype and final demonstration of SEGs that were group work, the five homework assignments and final paper-and-pencil exam required pre-service teachers to finish individually. The assignments and exam were basically designed with an aim to forcing pre-service teachers to practice the taught programming during the time outside of classes and to ensure that they learned to program. Although these pre-service teachers were required to hand in these assignments individually, they could discuss with their group members, peers, and the instructor, as well as work together to figure out how to finish the assignments. However, as mentioned earlier that these pre-service teachers are undergraduates who are majoring in biology and are enrolled in teacher education program, meaning that they have to take responsibilities for not only the assignments in this course, but also the other requirements of the department of biology. Needless to say, the students with biology major would have the tightest course schedule compared to students with other majors since they need to carry out many laboratory experiments. Their feeling that the course workload was too heavy was therefore understandable. Not enough instructional time to allow these pre-service teachers to have sufficient practice in classes again becomes the major issue. Later, we will propose some solutions and suggestions that might be helpful.

Transfer/Integration Issue

The transfer/integration issue is difficult, to unravel but also important to consider. However, It was frustrating, but not surprising to us, to find that some pre-service teachers still have difficulties in transferring what they have learned into games or completing their games, so they felt the quality of their games was quite low. According to the results, we can see that there might be two transfers/integrations that needed to be taken into account. First is the "transferring/integrating" of their professional knowledge in biology and biology teaching into the game format (integrating scientific concepts, educational objectives, and instructional strategies with game features), and the other is "transferring/integrating" the design of prototype into a real game product. These pre-service teachers showed fewer difficulties in the first transfer/integration after regularly discussing with the science education expert and their group members, which ensures the content validity of their game scripts. However, they were not able to properly transfer the design of prototype into a real game product, even though they might be able to write a very good game script which appropriately integrates scientific concepts with game features and develop a sound prototype. This might be because these pre-service teachers were not familiar with the design process and had insufficient design thinking skills. According to Razzouk and Shute (2012), the design process is iterative, exploratory, and sometimes chaotic. Besides, to properly transfer the design of prototype into a real game product, these pre-service teachers should have design-thinker characteristics as revealed in the study by Razzouk and Shute (2012). These design-thinker characteristics include having learner-centered concern, ability to visualize, predisposition toward multifunctionality, systemic vision, ability to use language as a tool, affinity for teamwork, and avoiding the necessity of choice. These pre-service teachers likely did not experience the design process before taking the course, and they might lack of training in design thinking and skills. Moreover, as mentioned earlier in this chapter that one of the learning objectives of this course was to improve students' TPACK. However, in reviewing the framework of TPACK, we found that it represents the intersections among knowledge of pedagogy, content, and technology, so that seven dimensions are included. For these pre-service teachers, they might have sufficient knowledge in pedagogy and content because the course was offered for juniors who have had took many related courses to obtain a certain extent of content and pedagogical knowledge, yet they still have not gained mastery of technology (programming). The lack of programming skills sooner becomes the major obstacle for their success in developing SEGs. Besides, these teachers might not have sufficient experiences in playing digital games or not familiar with instructional design. Their lack of understanding of game and/or instructional design also hindered their success in developing SEGs. In other words, the insufficient knowledge of technology, lack of understanding of game, or instructional design resulted in immature construction of TPACK, and this immature TPACK is eventually revealed in the final game product. That is the major reason that the pre-service teachers felt frustrated in implementing their game design (prototype) and integrating what they have learned about the programming into the game construction.

What Has Been Learned?

From this experience, a model representing students' learning from the processes of designing and making SEGs (Fig. 8.1) has emerged. Transforming an original/conceptual idea into an end-product (in this case, SEG) requires iterations of design/redesign cycle (unfortunately, the course as taught did not provide enough time to do). The design/redesign cycle describing the prototype of design has to be iteratively modified based on feedback from continuous tests. Only by taking great pains to perform the design/redesign iterations, can a valuable end-product finally being created. The iterations of design/redesign cycle have a reciprocal interaction with students' design thinking/skills, TPCK, communication, and collaboration. Namely, students have to properly employ their design thinking/skills, TPCK, as well as communication and collaboration in order to run the cycle well, and reciprocally,

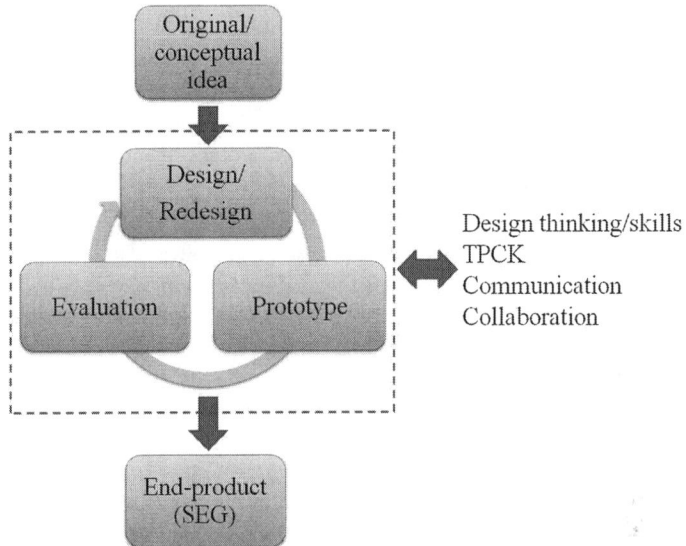

Fig. 8.1 The proposed model showing students' learning from designing SEGs

their design thinking, TPCK, communicative and collaborative skills are further improved through several design/redesign iterations.

In addition to the improvement of TPCK, one of the most valuable parts of the research is the enhancement of students' design thinking. Design thinking is a mindset about our faith of being able to creatively, independently, and resourcefully design meaningful and innovative solutions for making positive impact on the world. It can be characterized as empathy-driven, human-centered, collaborative, optimistic, and experimental (IDEO 2012). Because design thinking needs the full integration of empathy into solutions, students finally realize there is no a perfect answer and a single problem can be addressed in different ways. The importance of design thinking in education has nowadays attracted much attention, as design thinking requires students to actively find effective solutions by looking at a problem from different perspectives and supports the use of outside resources for learning and problem solving. Design thinking inspires changes and is highly relevant to today's workplace; therefore, many studies have now much encouraged researchers and educators to embed design thinking throughout the curriculum (Beckman and Barry 2007).

Future Work

Several suggestions for the successful implementation of learning by designing SEGs have emerged from the current work.

Cooperating with Other Professional Departments

One suggestion is to offer the course in cooperation with other professional departments, such as the department of computer science and information engineering or management. In fact, in the real world most of the games that are currently available also result from teamwork among experts with specialty in different fields. If there can be collaborations between different departments to offer this course, there would be students with different majors taking it. Consequently, the student groups in this course can be heterogeneous as suggested by much of the literature. For example, a group might consist of different students with major in biology, education, science computer, and/or information engineering. It turns out that students with biology majors can contribute content knowledge of science and pedagogical knowledge of biology teaching, and students from the department computer science or information engineering/management give ideas regarding programming and technological issue. This kind of heterogeneous grouping is an enhancer of group work because within the group, everyone learns from everyone else, and students are given more opportunities to participate in classes, just as Vygotsky's advocacy that students' zone of proximal development (ZPD) can be significantly improved through the teamwork within heterogeneous groups.

Providing More Scaffolds and Social Organizations for Helping Student Learning in This Course

The provision of more scaffolds is absolutely necessary. As students had mentioned in the interviews, more detailed handouts and useful resources, such as books and websites related to the tutorials or teaching of ActionScript 3.0 programming, should be provided as appropriate scaffolds. The use of exemplary cases is also highly encouraged. For an act or an event, there should be many different methods of programming. If the exemplary cases of programming for the same act/event can be provided, then the pre-service teachers or students can analyze and compare the differences and similarities between two or more examples. Moreover, a large number of websites that provide resources with open codes should be suggested. In so doing, it might be much more helpful for pre-service teachers and students in coming up with their own logic and method of programming. In addition, with today's technological advances, digital facilities must be utilized in a more proper way. Some pre-service teachers stated that the in-class instructions should be recorded and saved as tutorial videos. These videos then can be uploaded onto the web so that students could practice and rehearse repeatedly after school. Furthermore, educators and instructors should facilitate interactions between group members or students and instructors after school and assist them to build on-line social organizations wherein synchronous and asynchronous discussions between or within groups can be easily carried out.

Administrating Appropriate Number of Formative Assessments for Self-Diagnosis and Instruction Adjustment

When teaching design, formative assessment provides students feedback for their design work, and is critical for the success in design work. To evaluate student learning process and outcomes, it is important to assess the levels and various iterations in design. Although some participants argued the course workload was too heavy, we still recommend appropriate number of formative assessment should be administrated during the implementation of the course. However, the way it is administrated could be slightly modified. For example, it could be conducted as a format of self-assessment on-line that students decide when and how many times they would like to carry out these assessments. Or the assessments and assignments can be worked on through teamwork instead of being finished individually. Although it is up to the instructors or educators to determine if the performance of these assessments/assignments should or should not be counted into the criteria of final scores, the most important thing is that the results of these formative assessments and homework assignments benefit both students and instructors by allowing them to realize what is and what is not being learned. The instructions can then be adjusted accordingly and immediately. Moreover, it will also help students develop metacognition which includes the ability to plan their learning, monitor their own understanding, and to find resources and create solutions when necessary.

Whether having teachers learn to create SEGs by their own is feasible or desirable becomes a critical question raised from the current study. From the experience we have learned, we still admire and encourage the efforts of offering the courses in the future. We think the challenges can be overcome with appropriate strategies as previously suggested, and what can be learned for the pre-service teachers through the whole process of learning by creating an SEG is more than the investment required.

Acknowledgements This work was funded by the National Science Council (NSC), Taiwan, under grant contract no. NSC 101-2511-S-018-004-MY3 and NSC 101-2628-S-008-001-MY3. All the supports are highly appreciated.

References

Abbitt, J. T. (2011). Measuring technological pedagogical content knowledge in preservice teacher education: A review of current methods and instruments. *Journal of Research on Technology in Education, 43*(4), 281–300.

Agarwal, Y. (2010). Why use ActionScript 3.0 in Flash based learning Development. Retrieved from http://www.upsidelearning.com/blog/index.php/2010/03/22/why-use-actionscript-3-0-in-flash-based-elearningdevelopment/

Annetta, L. A. (2008). *Serious educational games*. Amsterdam: Sense Publishers.

Barron, B. J. S., Schwartz, D. L., Vye, N. J., Moore, A., Petrosino, A., Zech, L., & Bransford, J. D. (1998). Doing with understanding: Lessons from research on problem- and project-based learning. *Journal of the Learning Sciences, 7*(3-4), 271–311. doi:10.1080/10508406.1998.9672056.

Beckman, S. L., & Barry, M. (2007). Innovation as a learning process: Embedding design thinking. *California Management Review, 50*(1), 25–56.

Brimelow, L. (2008). Six reasons to use ActionScript 3.0. Retrieved from http://www.adobe.com/devnet/actionscript/articles/six_reasons_as3.html

Chai, C. S., Koh, J. H. L., Tsai, C.-C., & Tan, L. L. W. (2011). Modeling primary school pre-service teachers' Technological Pedagogical Content Knowledge (TPACK) for meaningful learning with Information and Communication Technology (ICT). *Computers & Education, 57*(1), 1184–1193. doi:10.1016/j.compedu.2011.01.007.

Cheng, M.-T., & Annetta, L. (2012). Students' learning outcomes and learning experiences through playing a serious educational game. *Journal of Biological Education, 46*(4), 203–213. doi:10.1080/00219266.2012.688848.

Cheng, M.-T., Annetta, L. A., Folta, E., & Holmes, S. Y. (2011). Drugs and the brain: Learning the impact of methamphetamine abuse on the brain through virtual brain exhibit in the museum. *International Journal of Science Education, 33*(2), 299–319.

Clark, D. B., Nelson, B. C., Chang, H.-Y., Martinez-Garza, M., Slack, K., & D'Angelo, C. M. (2011). Exploring Newtonian mechanics in a conceptually-integrated digital game: Comparison of learning and affective outcomes for students in Taiwan and the United States. *Computers & Education, 57*(3), 2178–2195. doi:10.1016/j.compedu.2011.05.007.

Connolly, T. M., Boyle, E. A., MacArthur, E., Hainey, T., & Boyle, J. M. (2012). A systematic literature review of empirical evidence on computer games and serious games. *Computers & Education, 59*(2), 661–686. doi:10.1016/j.compedu.2012.03.004.

Deway, J. (1938). *Experience and education*. New York: Collier Books.

Echeverría, A., García-Campo, C., Nussbaum, M., Gil, F., Villalta, M., Améstica, M., & Echeverría, S. (2011). A framework for the design and integration of collaborative classroom games. *Computers & Education, 57*(1), 1127–1136. doi:10.1016/j.compedu.2010.12.010.

Entertainment Software Association. (2013). *Essential facts about the computer and video game industry*. Retrieved from http://www.theesa.com/facts/pdfs/ESA_EF_2013.pdf

Gee, J. P. (2003a). High score education: Games, not school, are teaching kids to think. *Wired, 11*(5). Retrieved from http://www.wired.com/wired/archive/11.05/view.html

Gee, J. P. (2003b). *What video games have to teach us about learning and literacy*. New York: Palgrave.

Giannakos, M. N. (2013). Enjoy and learn with educational games: Examining factors affecting learning performance. *Computers & Education, 68*, 429–439. doi:10.1016/j.compedu.2013.06.005.

Gibson, D. (2008). Make it a two-way connection: A response to "connecting informal and formal learning experiences in the age of participatory media. *Contemporary Issues in Technology and Teacher Education, 8*(4), 305–309.

Gray, L., Thomas, N., & Lewis, L. (2010). *Teachers' use of educational technology in U.S. public schools: 2009*. Washington, DC: National Center for Education Statistics, Institute of Education Sciences.

Gülbahar, Y., & Tinmaz, H. (2006). Implementing project-based learning and e-portfolio assessment in an undergraduate course. *Journal of Research on Technology in Education, 38*(3), 309–327.

Halff, H. M. (2005). *Adventure games for science education: Generative methods in exploratory environments*. Paper presented at the 12th International Conference on Artificial Intelligence in Education, Amsterdam.

Howard, J. (2002). Technology-enhanced project-based learning in teacher education: Addressing the goals of transfer. *Journal of Technology and Teacher Education, 10*(3), 343–364.

IDEO. (2012). *Design thinking for educators toolkit*. Retrieved from http://designthinkingforeducators.com/

Lee, H.-S., Linn, M. C., Varma, K., & Liu, O. L. (2010). How do technology-enhanced inquiry science units impact classroom learning? *Journal of Research in Science Teaching, 47*(1), 71–90. doi:10.1002/tea.20304.

Lim, C. P. (2008). Global citizenship education, school curriculum and games: Learning Mathematics, English and Science as a global citizen. *Computers & Education, 51*(3), 1073–1093. doi:10.1016/j.compedu.2007.10.005.

Ministry of Education. (2008). *The philosoply of K-9 curriculum standards*. Retrieved August 8, 2013, from http://teach.eje.edu.tw/99CC/discuss/discuss2.php

Mishra, P., & Koehler, M. J. (2006). Technological pedagogical content knowledge: A framework for teacher knowledge. *Teachers College Records, 108*(6), 1017–1054.

Paraskeva, F., Bouta, H., & Papagianni, A. (2008). Individual characteristics and computer self-efficacy in secondary education teachers to integrate technology in educational practice. *Computers & Education, 50*(3), 1084–1091. doi:10.1016/j.compedu.2006.10.006.

Paraskeva, F., Mysirlaki, S., & Papagianni, A. (2010). Multiplayer online games as educational tools: Facing new challenges in learning. *Computers & Education, 54*(2), 498–505. doi:10.1016/j.compedu.2009.09.001.

Prensky, M. (2001). *Digital game-based learning*. New York: McGraw-Hill.

Razzouk, R., & Shute, V. (2012). What is design thinking and why is it important? *Review of Educational Research, 82*(3), 330–348. doi:10.3102/0034654312457429.

Sánchez, J., & Olivares, R. (2011). Problem solving and collaboration using mobile serious games. *Computers & Education, 57*(3), 1943–1952. doi:10.1016/j.compedu.2011.04.012.

Savey, J. R., & Duffy, T. M. (1985). Problem based learning: An instructional model and its constructivist framework. *Educational Technology, 35*(5), 31–38.

Shulman, L. S. (1986). Those who understand: Knowledge growth in teaching. *Educational Researcher, 15*(2), 4–14.

Singer, J., & Maher, M. (2007). Preservice teachers and technology integration: Rethinking traditional roles. *Journal of Science Teacher Education, 18*(6), 955–984. doi:10.1007/s10972-007-9072-5.

Thomas, J. W. (2000). *A review of research of project-based learning*. Retrieved August 8, 2013, from http://www.ri.net/middletown/mef/linksresources/documents/researchreview-PBL_070226.pdf

Vogel, B., Spikol, D., Kurti, A., & Milrad, M. (2010). *Integrating mobile, web and sensory technologies to support inquiry-based science learning*. Paper presented at 2010 6th IEEE International Conference on Wireless, Mobile and Ubiquitous Technologies in Education (WMUTE), 12–16 April 2010. Kaohsiung, Taiwan.

Wozney, L., Venkatesh, V., & Abrami, P. (2006). Implementing computer technologies: Teachers' perceptions and practices. *Journal of Technology and Teacher Education, 14*(1), 173–207.

Yurdakuli, I. K., Odabasi, H. F., Kilicer, K., Coklar, A. N., Birinci, G., & Kurt, A. A. (2012). The development, validity and reliability of TPACK-deep: A technological pedagogical content knowledge scale. *Computers & Education, 58*(3), 964–977. doi:10.1016/j.compedu.2011.10.012.

Part III
Preparing Teachers for the Grand Challenges…Exemplary Professional Development Practices

Chapter 9
Language of Design Within Science and Engineering

Nicole Weber and Kristina Lamour Sansone

In order to meet the Next Generation Science Standards, educators are charged to develop the content knowledge, pedagogical skills, and habits of mind that allow them to meet the demands for improving science learning for students, they must also develop their ability to effectively communicate and transition between multiple means of representation within their respective disciplines. In an attempt to help educators develop and enhance these communication skills while deepening the core content knowledge and transdiciplinary understanding within their classroom, we have developed a research-based approach, integrating the natural sciences, engineering, and graphic design. Here we will define the need for this transdiciplinary work, share the core components within a graphic design multi-sensory form of communication, how the graphic design process is broken down and connects to scientific and engineering practices, and demonstrate how multi-sensory communication can become evident and deepen the learning experience within science and engineering education within educator classrooms.

Defining the Problem

In the development of the Framework for K-12 Science Education and Next Generation Science Standards, the three historical science education camps (of the how to, the crosscutting reasoning, and the disciplinary silos) sat down, discussed,

N. Weber (✉)
Graduate School of Education, Lesley University, Cambridge, MA 02138, USA
e-mail: nweber@lesley.edu

K.L. Sansone
College of Art and Design, Lesley University, Cambridge, MA, USA
e-mail: klamour@lesley.edu

and developed a more holistic approach to teaching and learning science and engineering. Now it is the challenge of educators and researchers to do the same, and look beyond their bookshelves to begin a dialogue with neighboring ideas within their community. Inside every school there lies this opportunity, and here at Lesley University we have begun the deconstruction of disciplinary principles of *graphic design* and *science education* to see where a natural transdisiplinary support can be provided for our learning and teaching.

Creating curriculum connections within and across STEM disciplines is a strategy that has been underutilized when researching local ill-structured issues, even though natural connections are evident and can reinforce important key concepts (e.g., using scientific experimentation to generate data to support and inform engineering design decisions) (NAEE 2011; NAE and NRC 2009). Professional development opportunities for engineering are few and far between, with all in-service initiatives using existing curricula (*not focused on long term support that promote teacher learning*), and no pre-service initiatives are seen that significantly support qualified teacher development (NAE and NRC 2009). As stated in the NRC's (2012) Framework for Science Education, "*Science, engineering, and technology permeate nearly every facet of modern life, and they also hold the key to meeting many of humanity's most pressing current and future challenges… This national trend has created a widespread call for a new approach to K-12 science education in the United States (p. 1).*" Within the global context, there is also the need for an interdisciplinary and collaborative approach in developing innovative solutions to societal problems (NRC 2009; Musante 2011), and a push to both develop and implement core competencies that develop an interdisciplinary understanding of core content related to it's real world applications, while improving communication skills at the same time (Musante 2011).

Inspired by the Framework for K-12 Science Education initiative, the Science in Education program at Lesley University took on the challenge to redesign the masters in education program. Historically the masters program had one of the few engineering courses available to teachers. The new design expands the engineering thread throughout the program, engaging the recommendations of the National Research Council (2012). Faculty developed new courses that highlight the STEM components, while incorporating the science *and* engineering practices, making the connections more transparent.

As engineering education has historically received little attention in K-12 (NAE and NRC 2009). The first big step in implementation is to address the shortfall of qualified Math and Science teachers in the classroom (National Center for Best Practices 2011). As research indicates, elementary-school teachers continue to teach science and math in isolation, the causes of which include barriers of insufficient content knowledge to better integrate disciplines, issues of inadequate instructional time, and limited access to or awareness of curriculum resources that blend disciplines (Strobel et al. 2011). In re-thinking the science education framework as a form of "creative capacity building", where the science and engineering positions itself to bring the different forms of knowledge together to create integrated real life "creative capacities" or innovations, and therefore "creative pedagogies" to develop these skills (McWilliam et al. 2008).

We are beginning this process here at Lesley University, engineering now provides the context to learn other science content (Strobel et al. 2011) within each course (see Example 9.1; in grey box), with local environmental sustainability as the authentic learning thread. This course focuses on science concepts within the Life Sciences core discipline, specifically a sun vs. shade plant experiment looking at natural selection (LS4.B), adaptation (LS4.C) cause and effect (CC 2), and science practices (1–8) of the NRC (2012), here we push the educators out of the content area into the application of their knowledge within an authentic context. In our next step, as teachers develop the content knowledge, pedagogical skills, and habits of mind that allow them to meet the demands for improving science learning for students, we want to provide them with methods to also develop their ability to effectively communicate (visually and textually) and transition between multiple means of representation within their respective disciplines. In an attempt to help teachers develop and enhance these skills and improve their core content science and engineering knowledge, we look beyond the science and engineering education framework to find a more holistic language to better communicate.

Communication Through Pictures and Words

There's a reason why elementary school teachers use "show and tell" in their classrooms, as different parts of the brain deal with language and vision, which are stored and sorted separately as memories; meaning, if information is presented both visually and verbally, there is a better recall of the information (Kosslyn 2007). Integrating pictures and words as a visual expression is referred to as graphic design, and can help a student access content and demonstrate knowledge. Graphic design is already in our mathematics and science education classrooms, however, many users do not fully utilize the communication potential of the graphic image. For example, a teacher who creates a PowerPoint presentation or handout with text alone misses an opportunity to integrate pictures and words and thereby enhance student comprehension. As we are born with a powerful integrated set of tools and capabilities that help us read the world, the visual is often dominant, critical, logical, relevant and hard-wired (Dahaene 2009). The graphic design process provides a control of text and image elements, providing "perceptual organization" (Kosslyn 2007) integrating the visual into the learning arena.

At Lesley University, *Language of Design* is the foundation studio course for the Design program within the College of Art and Design, and we feel the core concepts developed here are applicable to teacher education of all disciplines. In the Science in Education program we hope to translate pedagogy traditionally held in a professional art and design school and bring it into the School of Education, integrating a graphic design language of pictures *(photographs, illustrations, color, texture)* and words *(messages, typestyle choices, choosing lower or upper case)* is uncommon in most teacher preparation and professional development programs *(which tend to privilege the textual)*, yet the understanding of this integrated language can expand the bandwidth of teaching and learning, and bring access and engagement through a visual vocabulary and improve communication across the learning community.

Learning to connect the unique graphic design methodology to teaching communication is becoming more important for teachers, as they develop content for their classrooms, building the ability to read and become more critical of information within our technology based society. Within the *Language of Design* learning experience, the use of observational skills, the construction of meaning, the use of multi-modal languages, and considerations for communication *(relationship between the designer and user)* are introduced. Students in this course look at content as if for the first time considering the potential for meaning construction and communication. Here, we will walk you through the progression of a student within an undergraduate course that holds lessons for science educator classrooms. Yoselin builds a visual communication on her quest to understand bird sounds, with the objective to present her findings using a graphical language. As we present the case study of Yoselin, we will relate the ideas to their overall application within the language of design that stimulate interest in the target content, increase the level of involvement, and engage the learner in higher-order thinking within science and engineering learning (Van Meter and Garner 2005).

Language of Design: Core Concepts

When a student can translate data or text to solve problems (e.g., a graph into a narrative or a oral story into a picture), they are assessed as functioning at a higher level than a student that struggles with these activities (Van Meter and Garner 2005). However, science classrooms often move from topic to communication quickly, not allowing observation and research to inform the learning that visual language integration can support. At first, this learner may have difficulty working with more than one format of information, as the integration of verbal and nonverbal offer a higher complexity, so establishing support in visualizing information early is important for establishing learner independence (Van Meter and Garner 2005).

Within our science classrooms, we often teach as if creativity is not important, and as if science deals only with structured issues with a single solution (DeHann 2011; Mc William et al. 2008). Yet, in the field, scientists deal with messy problems all the time, with many possible paths and solutions, needing to approach these issues through creative problem solving to find a solution that is novel, useful and economically viable (DeHann 2011). An idea or concept is essential to the goals in Language of Design. Without an idea that pulls the learner in, visual communication will have no legs to stand on, and students will be moving elements around arbitrarily.

The Language of Design course, at Lesley University, uses the concepts of observation, construction and communication as broad concepts to unpack the design process. Language of design starts by using a stimulating environment *(in this case, the Museum of Science and the New England Aquarium)* to help students identify a topic of their own interest to communicate. This is not unlike when science teachers ask students to identify a topic to visually present on for the science fair on a poster board, the two worlds we want to bridge. Here we will offer strategies to facilitate that integration, and here are some questions one considers as they begin the process:

- Does the image draw viewers in?
- Does it make them think?
- Does it challenge them?
- Does it promote conversation?
- Does it "feel" like a new idea or invention, or is it something you have seen before?

Here we will walk through the following Language of Design learning concepts: (1) the use of observational skills or visual research, (2) the construction of meaning or idea generation, and (3) the use of text and language (multi-modal languages) to communicate and reflect on the design, including the considerations for communication based on the relationship between the designer and user. Here, Yoselin will walk us through the progression of her learning within the Language of Design course, in building a visual communication on her quest to understand bird sounds.

Observation: Visual Research

Observation is to see and visually research content as if for the first time with a communication style or design form used to capture the essence of the complex information present. Research has shown when students have illustrated their understandings while exploring science content; they were highly engaged and more motivated to learn when compared to the more conventional teaching style (Hackling and Prain 2005, Ainsworth et al. 2011). The concept of *observation* easily connects to many areas within the Next Generation Science Standards (2013), as a foundation within the framework of disciplinary core idea progressions, where the developmental progression "is designed to help children continually build on and revise their knowledge and abilities, starting from their curiosity about what they see around them and their initial conceptions about how the world works" (p. 40, NGSS 2013). Another area is found within the cross cutting concept of *patterns*, where "observed patterns of forms and events guide organization and classification, and they prompt questions about relationships and the factors that influence them" (p. 79, NGSS 2013). Lastly, this is the origin to begin the science and engineering practice of asking questions and defining problems (SEP.1), where one begins with the phenomenon and begins to ask questions to define the issue at hand (NRC 2012).

Yoselin first began to visually examine a diverse array of birds at the museum, through this observation her interest in the topic of bird sounds emerged, using her senses (not a literal linear logic) to determine and define her quest. She began by observing the physical bird forms within the museum collection, then looked at the video stills of a female communication display of the Lyre Bird, followed by video stills and literature on the moon walking bird acoustic communication. After viewing the Lyre Bird video she writes, *"The images show the sequence of the bird's movement when signing or making a call. This was another video that helped me understand the birds' singing because it showed the literal aspect of bird communication. In the video, the male Lyre sings the most complex songs to impress the female."* As she moved on to the "moon walking bird" she stated that the more she researched the system of bird language, the more she understood. As she studied she realized that birds not only use their beak to make sounds, but they also use their wings. She noted that the birds performed this behavior so fast that the human eye cannot notice it, and with the video still images she was able to learn the sequence of the wing snapping. Within the walls of the library, Yoselin decided to continue her research in bird communication through a species comparison within the "Ecology and Evolution of Acoustic Communication" by Kroodsma and Miller (1996). Further observing the process in which birds learn their songs, the sound frequencies and duration of different species, and how this may affect communication.

Construction: Idea Generation

Within this stage, the challenge is to make meaning using text and image (multimodal), through ideation and personal coding or clarification. As Scientists and Engineers frequently encounter ill-structured problems that can have multiple paths to multiple solutions, to approach such a problem creative thinking and a "higher-order" of mental operations (e.g., analysis, synthesis, and abstraction) are key (Dehann 2011). This is a messy stage used to figure out and make sense of things, and often is only understood by the designer, because its about sensing the visual logic of the clues and connections in front of you, not something you can predict. This can directly be connected to NRC's (2012) disciplinary core idea of engineering, technology, and applications of science (ETS) and within the science and engineering practices (SEP). Within the ETS, the ideation stage is found within the engineering design process (ETS1.A and B), which is commonly referred to as brainstorming solutions and creating a prototype. In addition, within the SEP.2, the practice of developing and using models (NRC 2012) is also a great equivalent.

Here Yoselin first decided to construct the notion of sound, by recording the sound of the birds at the museum exhibit itself, where she captured both bird song and shortened calls of different species, then went on to interpreted them visually through a mark-making exercise using ink and newsprint. Here she visualized the wood duck alarm call and the hairy woodpecker song (Fig. 9.1). As research has shown that students often disengage from science, as learners are forced to take on passive roles within the traditionally taught rote/lecture pedagogy (Ainsworth et al.

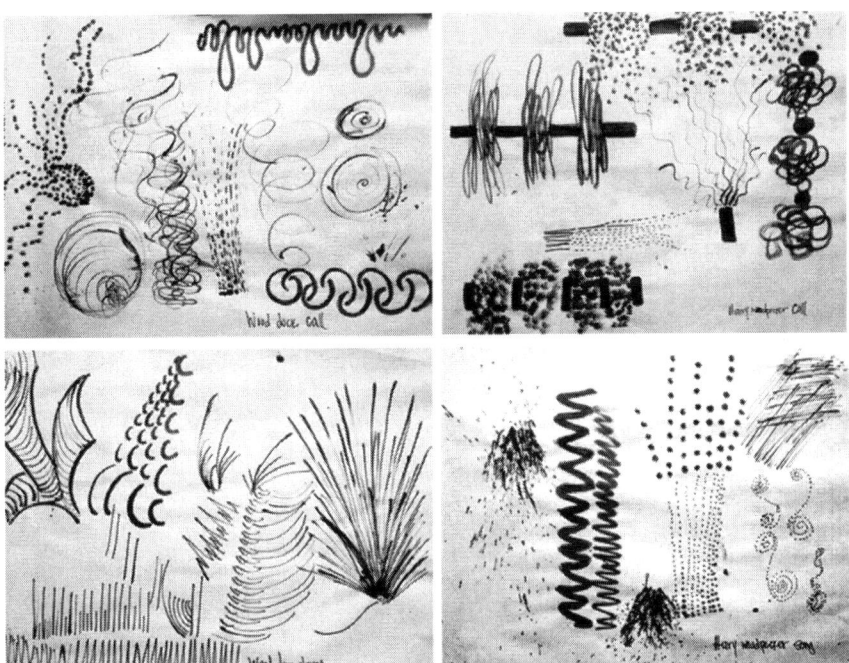

Fig. 9.1 Construct the notion of sound: wood duck call *(left)* and the hairy woodpecker song *(right)*

2011). Emerging research suggests that drawing be recognized as a key component of science education, as students need to learn how scientists use multiple literacies to construct and record knowledge, and learn to generate their own representation (Ainsworth et al. 2011).

In the second phase of this process, students map their varying visual and textual data *(in the class referred to as brown paper mapping)*. Students visually piece together their work to better understand the big picture of their process. Here students are asked to work on a large piece of paper to find themes and connectors. Yoselin worked to define her learning by laying everything down on the table and tried to find connections, and the process allowed her to pinpoint the interesting ideas within her topic of bird communication. She needed both the words and pictures to learn more about her topic. Through this experience she was able to realize how much she had learned in this process on birds and how science was part of the design, with a research question in hand. The draft of her question read, "What are the differences between bird and human language?" The final research question was, "How can bird sounds be presented visually?" As this fits beautifully under the disciplinary focus of Life Science of the NRC (2012), within understanding the complexities of biodiversity and adaptation within communication (LS4: Biological Evolution: Unity and Diversity), this experience can reach into all science disciplines, given an authentic context that provides space for learners to take more ownership over their role within the learning experience.

Fig. 9.2 Black and white symbol studies

The third phase of this process is to connect meaning with form, in the black and white symbol studies (Fig. 9.2). After coming up with the science content area to focus on, the research question was defined to reach her goal. The question was a way of making sure the symbol was conveying the message. To Yoselin, this was the most important part of the process, because it gave my bird symbol meaning. She sketched different symbols that could be an answer to the research question, in an effort to design a symbol that can be place in a different context and fulfill it's utility.

Communication: Reflection

Every visual communication has an audience whether it be broad or specific, and it is important to use your audience as a filter when designing. The audience is able to let you know if your design is usable, age appropriate, and if the viewing distance supports the reader in context. This can directly be connected to NRC's (2012) disciplinary core idea of engineering, technology, and applications of science (ETS) and within the science and engineering practices (SEP). Within the ETS, the optimizing the design solution is found within the engineering design process (ETS1.C), which is commonly referred to as the testing phase. In addition, within the practice of constructing explanations and designing solutions (SEP.6) and within the practice of obtaining, evaluating, and communicating information (SEP.8; NRC 2012) are also equivalent.

At this stage, the designer tests their design through *iteration*, re-evaluation of the *design criteria*, and the connection to the *audience* to identify and a successful concept through critique to assess if the following criteria are met:

Fig. 9.3 Narrowing of form, based on connection to audience

1. Strength in concept and interest – *selecting a set of forms that best represented the research question at this stage.*
2. Connection to audience – *a series of critiques, to help choose best representation (see circled form in Fig. 9.3).*
3. Appropriate choice of elements – *the components of the visual project that support the goal (text style, pictures, backgrounds, and color).*
4. Clear visual hierarchy/reading path – *the form is treated as a visual sentence (deliberate organization of the parts to help our viewer "read" the visual in a particular order).*
5. Efficient use of elements/editing – *do any elements distract from the main idea, and if so how can they be eliminated, replaced, or redesigned?*

For Yoselin, in the final editing, after the second critique of symbol iterations, the base structure of the dotted symbol changed to be the first image on the top left of Fig. 9.3. She then noted that the contour line symbol needed to be simplified through the use of line weight. Here she has put together the images of the editing stages of each symbol leading up to the final version (Fig. 9.4). Lastly, Yoselin took this form a step further in doing a color study of her final form. The goal of this particular project was to capture a research question in a graphical language. In the beginning of the process, Yoselin asked, *"How can bird sounds be presented visually?"* Given that Yoselin was unable to use movement in the static image, she was still able to capture the movement of sound visually. Yoselin went on to explore the connection of text within her structural form, eventually developing a similar relationship within the text created (Fig. 9.5). Yoselin's goal was to find a visual relationship between the expressive logic of her pictorial elements (dots and line language) with varying letterforms and words.

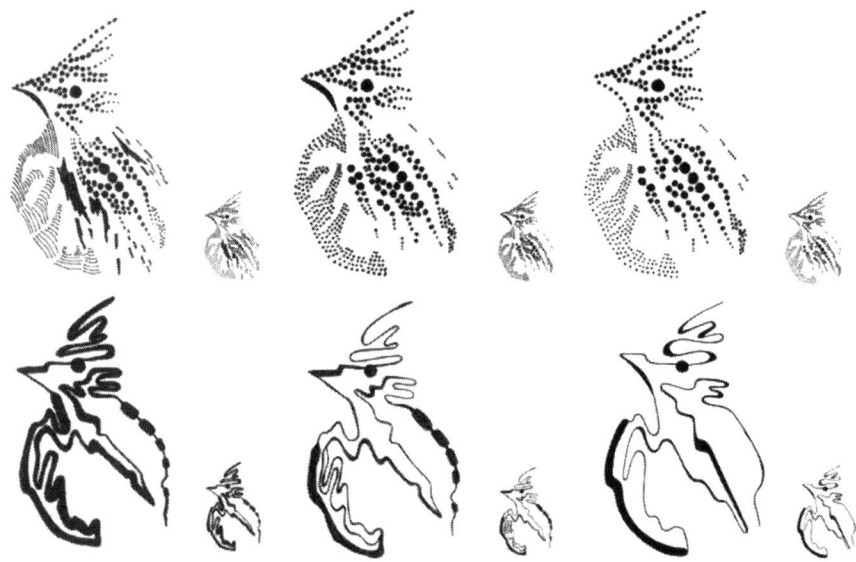

Fig. 9.4 Final stage of editing

Fig. 9.5 Yoselin's text and bird image relationships

Language of Design for Science and Engineering

Emerging research suggests that drawing be recognized as a key component of science education, as students need to learn how scientists use multiple literacies to construct and record knowledge, and learn to generate their own representation (Ainsworth et al. 2011). As elementary school has been identified in several research studies as the crucial period in which interest in science can be influenced (NRC 2007; Strobel et al. 2011), a multi-sensory communication experience can deepen the learning within science and engineering education across levels. Students are shown to often disengage from science, as learners they are forced to take on more passive roles within the traditionally taught rote/lecture pedagogy (Ainsworth et al. 2011). The new design pedagogy needs to reflect real practice, where scientists make observations to learn about the world, and observations seen as a central component within scientific activity of all siloed disciplines, leading to every scientific decision and discovery (Eberbach and Crowley 2009).

David Orr (1994) provides a natural base for learning and teaching good design within science education, creating *biologic-* an ecologically competent community.

> Ecological design competence means maximizing resource and energy efficiency, taking advantage of free services of nature, recycling wastes, making ecologically smarter things, and education ecologically smarter people. It means incorporating intelligence about how nature works, what David Wann (1990) called "biologic", into the way we think, design, build, and live. Design applies to the making of nearly everything that directly or indirectly requires energy and materials or governs their use, including farms, houses, communities, neighborhoods, cities, transportation systems, technologies, economies, and energy policies. When human artifacts and systems are well designed, they are in harmony with the larger patterns in which they are embedded. When poorly designed they undermine those larger patterns, creating pollution, higher costs, and social stress in the name of a spurious and short-run economizing. Bad design is not simply and engineering problem, although better engineering would often help. Its roots go deeper.... Good design everywhere has certain common characteristics including the following: right scale, simplicity, efficient use of resources, a close fit between means and ends, durability, redundance, and resilience. Good design also solves more than one problem at a time... When good design becomes part of the social fabric at all levels, unanticipated positive side effects (synergies) multiply. When people fail to design carefully lovingly, and competently, unwanted side effects and disasters multiply... Good design uses nature as a standard and so requires ecological intelligence, by which I mean a broad and intimate familiarity with how nature works. (Orr 1994, pp. 104–106)

Here our goal is to transform science teaching by engaging educators experientially in local, project-based research projects with the support of the graphic design processes. We have begun by creating a common language that can be used across disciplines, by taking a fresh look at the role of scientific observation within a multisensory graphic design lens. For example, the scientific observation process includes data collection that can benefit from the observational strategies used in graphic design, including rich visual (multi-modal) research and analysis through diagramming a combination of images and text in describing a particular environmental issue (Fig. 9.6). We also focus on integrating the graphic design core concepts into the wonder and communicate phases of the scientific process, as the graphic design process can support deepening the exploration and access in understanding the concept at both of these stages. Within the engineering design process, this support can similarly extend ideas within the define, research/brainstorm, and communicate steps (see Example 9.2; in grey box).

Translation into Classroom Practice

Within the literature, there are few instances where support is provided for learning and practicing the visual pedagogy within the classroom, for use as a learning strategy (Van Meter and Garner 2005). Often, educators only have the mentored example of taking designs off the web or sharing existing resources that are "not quite right" to communicate an idea, both out of convenience and not knowing an

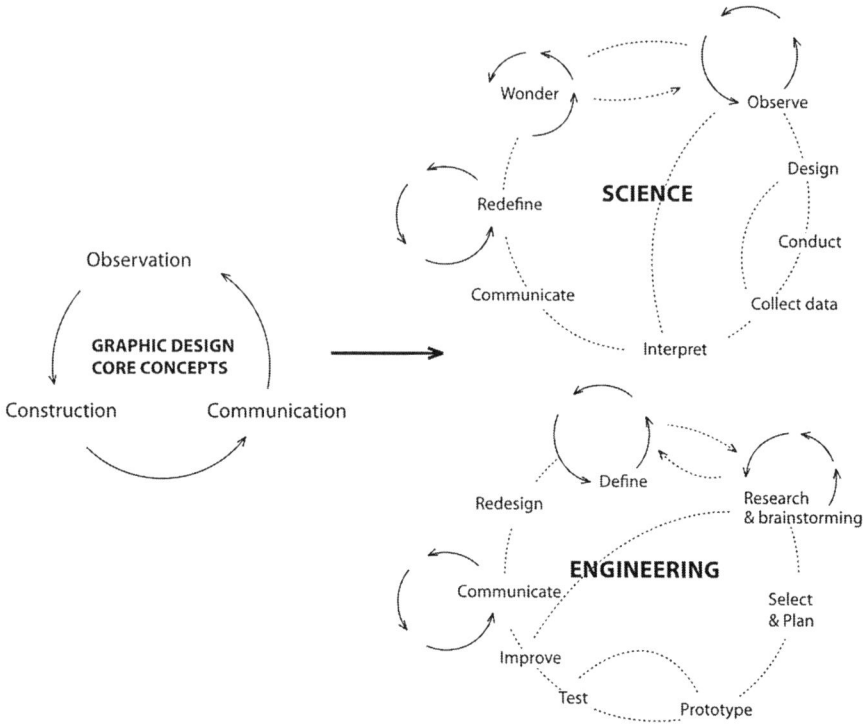

Fig. 9.6 Focus areas of integration within our science in education program

alternative. These visual interpretations may be detached from the educators' intent or even their pedagogical stance. Learning graphic design techniques has great implications for educators for communicating complex scientific concepts, designing visual communication that can take the form of info graphics, digital presentation (i.e. Powerpoint), or video. One example of success is Salman Khan, of Khan Academy, who has found a way to reinforce his science education pedagogy with his own unique visual language strategy. Here we will share teacher-generated examples of multiple means of communicating information, knowledge, and understanding within the classroom.

For our first example, most teachers are familiar with the science fair poster, seen as a creative science project. These posters usually contain three aspects of language: text (titles, body text, quotes, the visual aspects of text such as the font and its size and color) images (photographs, illustrations, drawings) and shapes (symbols, diagrammatic language). Here we share how a traditional poster can transform into a more multi-sensory experience for both the learner and client (parent and teacher) populations (Fig. 9.7). Text on posters cannot be used alone simply because images and shapes support student understanding and demonstration in ways that text alone is unable to communicate, there is an interdependent integrated language is visual language *(or multi-modal)* communication. Here a science question or

9 Language of Design Within Science and Engineering

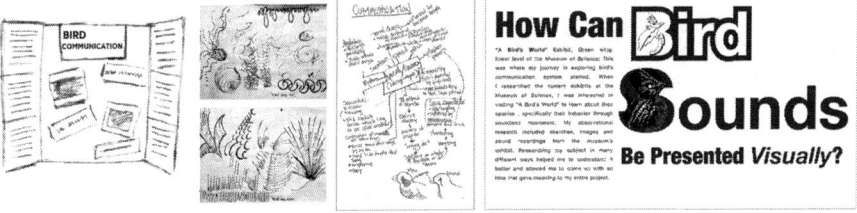

Fig. 9.7 Going deeper from the traditional science poster to designing a multi-modal language system with a common language of communication

engineering problem (SEP.1, NRC 2012) can focus the poster content, and the process of working through the observation and brainstorming phases can be supported through the multi-sensory graphic design process of creating a visual language that of supports the learning and understanding of the work.

Again this can be directly connected to disciplinary core idea of engineering, technology, and applications of science (ETS) and within the science and engineering practices (SEP). Within the ETS, this touches on two areas within the engineering design process, of delimiting the issue (ETS1.A) and designing solutions (ETS1.C). In delimiting the idea or problem, the designer has to consider the targeted audience (NRC 2012). Then within the development of the solution, the designer has to reconsider the audience and the effectiveness of communicating the information through a prototype (initial design), then testing the design through user feedback. In addition, within the SEP.1–8 from the engineering lens (NRC 2012), all of the practices can be linked to this process.

Our second example bridges what we have learned within the Language of Design course translated to the transdiciplinary classroom application and assessment of student learning. Building on the assessment created for elementary students by Lisa Donovan and Kristina Lamour Sansone *(co-author)* at Lesley University, educators can co-develop an assessment tool that honors multi-modal language as the primary language for STEM practices. This visible assessment is flexible in monitoring student progress, and captures multi-modal understandings via text (textual and visual content), drawing, photography, and video. In the example (Fig. 9.8), an educator tracks the progress of student learning within their science notebook prior to and during a unit on ornithology, noting that student observations become more multi-model as they are exposed to the design research experiences. The student work after the experience becomes more inclusive of design elements, with the student taking on more ownership of the learning and communication process.

Lastly, we continue to develop the common language across our disciplines, and discuss similar habits of mind within our siloed disciplines that can provide solid legs for teaching and learning. We have begun to tease out the unique attributes that provide support for multi-modal learning and representations *(such as science fair poster boards)*. Science education and graphic design faculty are continuing to work collaboratively to develop the common language, leaning supports, and assessment

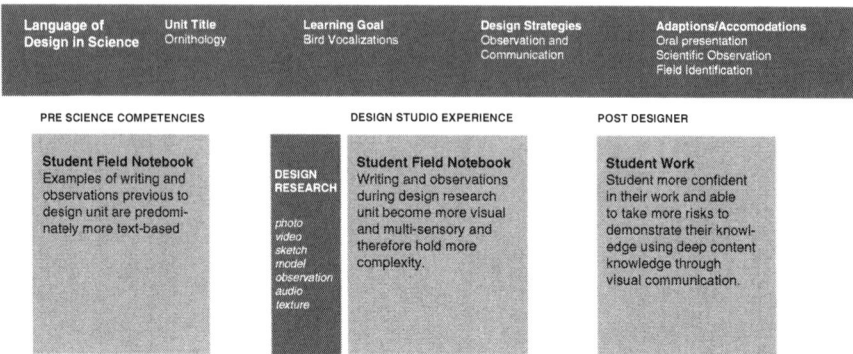

Fig. 9.8 Visible assessment: flexible student assessments that capture multi-modal understandings via text (textual and visual content), drawing, photography, and video. Building on the multi-modal assessment of Lisa Donovan and Kristina Lamour Sansone, a model that uses the visual aspects seen in the student work as part of the visible evidence

models that challenge the current practices, like the science fair dissemination or e-Portfolio work, and we will continue to visit each others classrooms to support this growth and understanding.

Acknowledgments Thank you to Yoselin Rodriguez who was a first semester freshman when this work was produced. Since then, she has been accepted into the Bachelor of Fine Arts/Master of Education Dual Degree in Visual Art.

Example 9.1: *Life Science Course Supported by the Engineering Design Process*- Green Choices Case Study

Green Spaces are all around you. Here is a scenario to think of when designing your plant research experiment. We all have green spaces around us, however are they being utilized in a way that would benefit us, our students, and other local biodiversity. When thinking of your biology exploration, think of how the green spaces can be addressed. We will then begin to create the links between content and practices within the life science and engineering core ideas, as presented in the new framework of the National Research Council (2012) to begin to visualize our research in a cross-disciplinary context. First we will learn about the case study, and examine possible solutions as a class. Then we will look at our individual plant research projects to see how the experiment itself can provide some supportive data for some of the possible solutions.

In Boston, schools are often surrounded by potential green areas *(like an open lot, roof, or parking lot)*, which can be utilized for scientific educational

(continued)

Example 9.1 (continued)

purposes. The main modes of experimentation in and around the school campuses are classroom or even lab based, where they are often not equip to handle long term and/or "natural" science projects, simply due to the shear volume of individuals needing the space to work or the facilities itself. By looking at the local habitats, some schools may be able to provide a "natural lab" where students can explore outside the walls of the classroom.

Here is a charter school where the campus lab only has an indoor facility available for students to work, not allowing students to maintain experiments beyond a week, due to the number of students in the school. So two teachers started to look beyond their school walls, where they focused on the possibility of an abandon lot across the street from the school and a fenced in area within the school parking lot. They have received a small fund to employ 8 students over the summer to work at the school on this project, however need help in figuring out what the areas can be used for and how to get students (*and teachers*) motivated. They have turned to you to assist them.

Here are the session components at a glance: **Engineering Design Context/** *Scientific Method Support*

Step 1: Define/*Wonder* Begin to define the school need and consider the plant experiment through the lens of the case study.

Step 2: Research/*Observe* Construct a background framework of the current system by defining different components; including possible constraints, assumptions needed for success, the stakeholders involved (including target audience and other important groups), and the stakeholder needs within the problem. Discuss the framework of the current system with the group, along with possible supports of plant research *(like sun vs. shade or soil testing experiment to choose best plant options for the open lot)*.

Step 3: Brainstorm/*Design to Collect* Further develop your solution, first by looking at what was discussed with the group; take examples that fit well from the discussion, and add/modify components to fit your framework. Then define possible solution paths (based on prior art, brainstorming, etc.), compare alternative solution possibilities, and identify information (data) needed from science experiment to support your proposal.

(continued)

Step 4: Select and Plan/*Interpret to Communicate* Choose a best scenario based on identifying the strengths, weaknesses, and assumptions associated with each conceptual solution by using a Pugh Matrix. Identify the top three solutions and create a written narrative to describe your solution, weaving in the results from the plant experiment where appropriate.

Step 5: Create Prototype/*Redefine* Once a solution is selected, carry out a pilot test of what that may look like. Share this with the group and discuss.

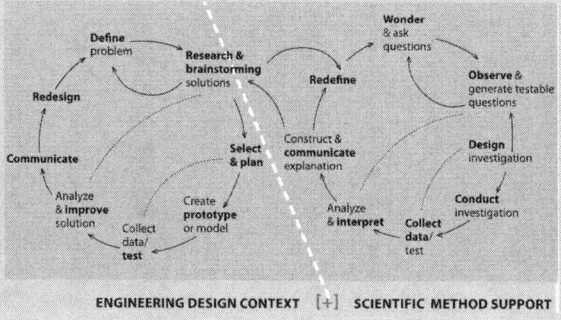

Example 9.2: *Engineering STEM Solutions Course Supported by the Graphic Design Process-* Eco Design Choices Case

Natural Spaces are all around you. We all have natural ecosystems within the spaces around us, however are they being utilized in a way that would benefit us, our students, and other local biodiversity. When thinking of local engineering classroom extensions, here are two very different scenarios to think of how the natural ecosystems within the spaces around us can be re-purposed into a **Biophilic Design** to promote learning. We will then begin to create the links between content and practices within science and engineering, as presented in the new framework of the National Research Council (2011) to begin to visualize our research in an interdisciplinary context. First we will learn about the case study, and examine possible solutions. Then we will look at our research through the graphic design process to see how this can provide supportive data for the possible solutions.

(continued)

Example 9.2 (continued)

Schools often are surrounded by potential research areas *(like a playground or parking lot)*, which can support scientific and engineering practices if you take the time to define the experience. By looking at the local habitats, some schools may be able to provide a "natural lab" where students can explore outside the walls of the classroom, and potentially the walls of the school building itself. Here is a school where the campus lab only has an indoor facility available for students to work on short-term projects, and an outside container garden that has been abandoned for the last 2 years. Two teachers at the school started to look beyond their school walls, where they focused on the possibility of a local playground across the street from the school and a garden area within the school parking lot. They have received a small grant to develop a science and engineering curriculum based on *biophilic* design ($2000), and have the summer to develop the key activities for the school to incorporate specific grade level projects (3–6th grade). The teachers now need to decide what the areas should be used for specific activities, how to get students *(and teachers)* motivated to use these areas, and how to connect grade level projects to the overarching theme of *biolphilic* design. The school administration is very interested in this project, and has additional funds to incorporate other areas of the school ($3000), depending on how this project develops. They have turned to you to assist them.

Here are the session components at a glance:

Step 1: Define/*Observe Within a Graphic Design Lens (GDL)* Begin to consider the case study, defining the problem faced, taking the time to work through the graphic design process within this step. Some may work through the entire process here or remain in the observation stage; this is dependent on how the GDL helps in your process.

(continued)

Example 9.2 (continued)

Step 2: Research/*Construction Within a GDL* Construct a background framework of the current system by defining different components; including possible constraints, assumptions needed for success, the stakeholders involved (including target audience and other important groups), and the stakeholder needs within the problem. Discuss the framework of the current system with the group.

Step 3: Brainstorm/*Communication Within a GDL* Further develop your solution, first by looking at what was discussed with the group *(in both their design and feedback to your proposal)*; and add/modify components to fit your framework. Then define possible solution paths (based on prior art, brainstorming, etc.), compare alternative solution possibilities, and identify information needed to support your proposal.

Step 4: Select and Plan Choose a best scenario based on identifying the strengths, weaknesses, and assumptions associated with each conceptual solution by using a Pugh Matrix. Identify the top three solutions and create a written narrative to describe your solution.

Step 5: Prototype to Improve Create a prototype of the solution selected. Share this with the group and discuss. Then you will test and improve your model, and weave in your results from your experiments where appropriate.

Step 6: Communicate/GDL Present your proposal to the client.

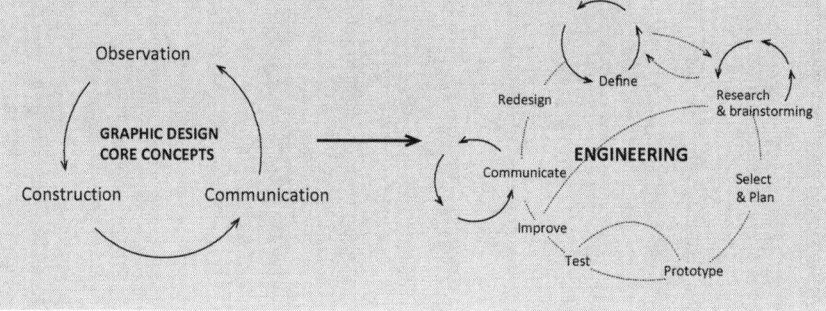

References

Ainsworth, S., Prain, V., & Tytler, R. (2011). Drawing to learn in science. *Science, 333*, 1096–1097.

Dahaene, S. (2009). *Reading in the brain: The science and evolution of a human invention.* New York: Viking.

DeHann, R. (2011). Teaching creative science thinking. *Science, 334*, 1499.

Eberbach, C., & Crowley, K. (2009). From everyday to scientific observation: How children learn to observe the biologist's world. *Review of Educational Research, 79*(1), 39–68.

Hackling, M., & Prain, V. (2005). *Primary connections: Stage2trial.* Canberra: Australian Academy of Science.

Kosslyn, S. (2007). *Clear and to the point: 8 psychological principles for compelling Powerpoint presentations.* Oxford: Oxford University Press.

Kroodsma, D. E., & Miller, E. H. (1996). *Ecology and evolution of scoustic communication in birds.* Ithaca: Cornell University Press.

McWilliam, E., Poronnik, P., & Taylor, P. (2008). Re-designing science pedagogy: Reversing the flight from science. *Journal of Science Education Technology, 17*, 226–235.

Musante, S. (2011). Teaching biology for a sustainable future. *BioScience, 61*(10), 751.

National Academy of Engineering and National Research Council of the National Academies (NAE & NRC). (2009). *Engineering in K-12 education: Understanding the status and improving the prospects.* Washington, DC: National Academies Press.

National Center for Best Practices. (2011). *Building a science, technology, engineering and math education Agenda: An update of state actions.* Washington, DC: Black Point Policy Solutions, LLC.

National Research Council. (2007). *Taking science to school: Learning and teaching science in grades K-8.* Committee on science learning, Kindergarten through eighth grade. In Richard A. Duschl, Heidi A. Schweingruber, & Andrew W. Shouse (Eds.), Board on Science Education, Center for Education. Division of Behavioral and Social Sciences and Education. Washington, DC: The National Academies Press.

National Research Council. (2009). *A new biology for the 21st century.* Washington, DC: The National Academies Press.

National Research Council. (2011). *A framework for K-12 science education: Practices, crosscutting concepts, and core ideas.* Washington, DC: Committee on a Conceptual Framework for New K-12 Science Education Standards. Board on Science Education, Division of Behavioral and Social Sciences and Education. The National Academies Press.

National Research Council. (2012). *A framework for K-12 science education: Practices, crosscutting concepts, and core ideas.* Committee on a conceptual framework for new K-12 science education standards. Board on Science Education, Division of Behavioral and Social Sciences and Education. Washington, DC: The National Academies Press.

NGSS Lead States. (2013). *Next generation science standards: For states, by states.* Washington, DC: The National Academies Press.

North American Association for Environmental Education (NAAEE). (2011). *Assessing environmental literacy: A proposed framework for the Programme for International Student Assessment (PISA) 2015* (p. 44). Washington, DC: NAAEE. http://www.naaee.net/.

Orr, D. W. (1994). *Earth in mind. On education, environment, and the human prospect.* Washington, DC: Island Press.

Strobel, J., Weber, N., Dyehouse, M., & Gajdzik, E. (2011). Recommendations to realign the national STEM education agenda, *ASQ Higher Education Brief, 4*(1). http://asq.org/edu/2011/02/engineering/recommendations-to-realign-the-national-stem-education-agenda.pdf?WT.dcsvid=Nzg3NTMxNDExS0&WT.mc_id

Van Meter, P., & Garner, J. (2005). The promise and practice of learner-generated drawing: Literature review and synthesis. *Educational Psychology Review, 17*(4), 285–325.

Chapter 10
Teaching with Design Thinking: Developing New Vision and Approaches to Twenty-First Century Learning

Shelley Goldman and Molly B. Zielezinski

> *Give the pupils something to do, not something to learn; and the doing is of such a nature as to demand thinking; learning naturally results.*
>
> –John Dewey

What will the world be like in 2026? Predications are not easy, but it is easy to count on change as a huge factor. Economic, social, natural, and political forces are in flux and will continue to defy our traditional models and processes for thinking and acting. By 2026, another eight technology innovation cycles will have occurred. Jobs will have shifted even more towards science, engineering and technology sectors. The 50 million K-12 students in public schools will have moved through the education system on their way to further education, work, and adulthood. The year 2026 is around the corner, and it is imperative for learners to be prepared for a continually evolving and changing world. Can schools respond as needed? The skills for adapting and problem solving into the future certainly go beyond the skills and know-how that currently dominate school programs and curriculum. Calls for movement beyond what is currently taught in schools have persisted for years, and its recognized that students are likely to need competencies such as communication and collaboration, research and information fluency, critical thinking, creative problem solving, decision-making, digital citizenship, and technology operations and concepts (Pellegrino and Hilton 2012). It is imperative to integrate these new skills and know-how into the K-12 curriculum. Teachers are the front line professionals of twenty-first century education and are key to how students will be prepared. Helping teachers integrate current and new teaching practices is critically important. Our direction has been to understand how the standards can be seen as a blueprint for a twenty-first century education and how design thinking, which embodies many of the twenty-first century competencies, can be integrated into the K-12 schools. This

---

S. Goldman (✉) • M.B. Zielezinski
Stanford University, Stanford CA, USA
e-mail: sgoldman@stanford.edu

chapter takes a detailed look into how we imagine design thinking can work in K-12 and how we bring it into focus with teachers. We take three directions. First, we describe design thinking, discuss its features and its potential, including how it corresponds with and supports the current standards, and why it is appealing to educators. Second, we describe two cases from the work we have been doing with teachers to review how they make design thinking a reality for their students and a resource for learning. Finally, we discuss the characteristics of professional development that we have found successful for helping teachers consider design thinking pedagogies.

Design Thinking and How It Works in K-12?

The particular version of design thinking that we are implementing is an approach to teaching and learning that fosters students' abilities to find answers to complex problems that have multiple viable solutions. It develops students' skills, dispositions and mindsets, so they can become active participants in a changing world with many problems to solve. It also has a focus on developing creative competence in teachers and students--an ability to tap into principles and strategies that help people approach and solve problems throughout life (Kelly and Kelly 2013).[1]

We take this conception of design thinking and show teachers its potential and malleability in the K-12 learning context. Design Thinking is a human-centered enterprise, and the process is defined by deep and radical collaborations, rapid prototyping, feedback and revision. Design thinking can take on "wicked problems" that may be ill defined or ill structured (Rittel and Webber 1973), and may not be conducive to conventional or incremental methods for problem solving. Tried and true solutions might be absent, and in some cases, the resources for problem solving might seem insufficient (Cross 2006). As an approach for teaching and learning, design thinking embraces active problem solving in the world and aims to create change (Dewey 1916). It is deeply reciprocal and nets outcomes for both the design recipient and the design thinker.

Design thinking is similar to project–based and learning, but it is useful to distinguish between the two. Both project-based teaching and learning and design thinking engage students in sustained, in-depth investigations in topics of real-life importance. They embody twenty-first century teaching and learning competencies such as critical thinking, collaboration and communication, and use of technologies. Design thinking differs on several fronts. It is always driving towards an innovative solution rather than predetermined or pre-understood outcomes. The version of design thinking that we ascribe to at Stanford always takes a human-centered approach to problem solving and change, so in-depth research and learning is put to

---

[1] The design thinking approach we use is adapted from the one that was developed at IDEO by David Kelley and Tom Kelley and taught at Stanford University.

use in relation to a person's needs. We believe that this empathy factor helps to establish relevance, supports engagement, and offers an answer to the age-old question, "Why are we doing this?" A student who is doing design thinking never solves a problem as a mere intellectual exercise or by designing for his or her own needs; a problem is always being solved for the actual needs of another as they are observed and their needs are unpacked by the designer. The user-centered aspects can be engaging for students who might not be intrinsically disposed to complex problem solving. Finally, design thinking promotes commitments to inter-disciplinary collaborations and teamwork. The process offers outlets for all types of learners to participate successfully and scaffolds involvement regardless of language status, learning preferences, areas of expertise, or personality. The process and outcomes are not about individual achievement; they are about the synergy of people with diverse ideas, approaches and talents. Design thinking benefits the problem solvers by helping them develop new mindsets, which are deeply engrained ways of thinking, orienting to problems, and acting on them (Goldman et al. 2012). Becoming a design thinker is a process that can be defined by moments or experiences of insight or shifts in a person's understandings and dispositions. We like to help learner's accomplish these "*mindshifts*" (e.g., being human-centered and empathetic in their approaches to problem-solving, working in deeply collaborative ways, and recognizing that failure can be a powerful part of the learning process).

Our particular approach integrates and aligns the conceptual and process-related underpinnings of STEM learning and design thinking such as collaboration, deep critical thinking, active problem solving, and a bias towards action. Teachers and students engage in hands-on design challenges that focus on developing empathy, promoting a bias toward action, encouraging ideation, and developing metacognitive awareness (Carroll et al. 2010). Design thinking fosters active and iterative problem solving and solution generation, making it relevant to problem-solving projects in STEM subjects while adding an inventive, innovation-imperative that is highly consistent with the development of twenty-first century competencies. These include innovation, creativity, critical thinking, problem solving, communication, and collaboration skills, which all seen as the basis of a twenty-first century education (Partnership for 21st Century Skills 2008). Design thinking facilitates the learning of skills such as working in groups, following a process, defining problems, and creating solutions. Vande Zande (2007) characterizes design thinking as a means of creative problem solving that relates thought and action directly and dynamically.

There are no easy recipes for how to teach with design thinking and implement it in K-12 classrooms. With its process, skills, and mindsets, there is much to learn and accomplish to make it a reality, integrate it into the subject areas, and to instantiate it as a classroom staple. We explore what is possible with children, teachers, teacher leaders, parents, and educators in supplemental settings such as after-school, summer school, and camp settings. We are specifically interested in how design thinking can be a resource for twenty-first century learning *and* its accompanying challenges. While it is relatively cavalier to say that teaching must change to meet the demands of the future, we are aware that this is an extremely difficult goal. Teaching is an incredibly complicated and diverse set of epistemologies, experiences, skills, practices, and mindsets that are influenced by the many factors and pressures affecting the profession and in-classroom practices (Berry et al. 2011). Teaching practice takes into account individual and community practices and resources, pressures, and imperatives. Change is complex, and innovation in terms of twenty-first century competencies is sometimes difficult to achieve (Chen 2010; Goldman and Lucas 2012; Hess et al. 2009), although predictions imagine these changes are possible (Berry et al. 2011).

We are optimistic about making change on the ground with teachers. Over the past 6 years, we have been working with teachers in order to introduce them to teaching and learning with design thinking, showing them how it connects to standards, helping them start implementation in their classrooms and schools, and trying to understand both their accomplishments and frustrations. We have done the bulk of this work with teachers through *d.loft STEM Learning*, a project devoted to bringing design thinking to interdisciplinary STEM topics.[2] The inspiration for *d.loft STEM Learning* is the "Design for the Other 90 %" movement (Smith 2007), which consists of engineers, designers, scientists, technologists, architects, and mathematicians engaged in designing low-cost innovative solutions for large portion of the world's population who do not have access to basic services and products. We emulate that work by introducing design challenges that engage participants in relevant STEM topics such as access to, and conservation of water, energy, shelter, and food.[3]

Our process with teachers is to immerse them in a design thinking challenge that engages them in creating solutions for interdisciplinary STEM challenges. Usually, a workshop is a 2-day experience where a topic such as access to or conservation of clean water drives the design thinking challenge. The teachers are put into a "team" that is introduced to a "client" or "user", and it is their job to design for that person's water-related needs. We step them through the design thinking process (see Fig. 10.1), from understanding the problem space, to developing knowledge about their user, to learning how to develop empathy and gain insights, to creating a needs statement for their user, to brainstorming, prototyping and gaining feedback from the user about their solutions.

Through the process, the teachers experience new ways to solve problems and learn, reflect on new ways to teach, and even experience design thinking relevant *mindshifts*. The teachers are often pleasantly surprised that their team solution is so creative and appropriate for their client; they are also impressed with the diverse solutions presented by other teams. Some have experienced *mindshifts*, and see the value of them such as being empathy driven, rethinking failure or gaining insights

---

[2] Read more about d.loft STEM at dloft.stanford.edu. D.loft STEM is an NSF ITEST project, number 1029929. Any opinions or research reported on is the authors', and are not the opinion of the NSF.

[3] We also produce curriculum materials on these topics. They are available at: http://tinyurl.com/designthinkingcurriculum

10 Teaching with Design Thinking: Developing New Vision and Approaches...

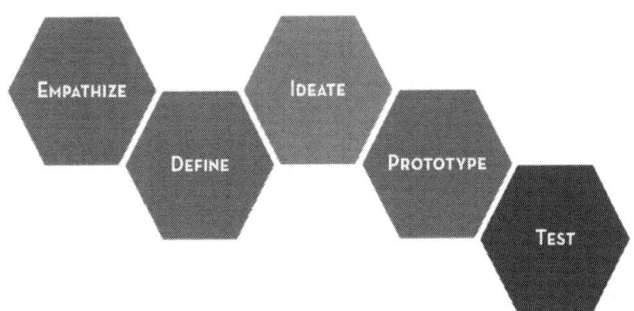

Fig. 10.1 Stages for learning design thinking
We teach six stages of the Design Thinking Process as represented in this graphic. Starting from an open-ended problem space, students go through the Empathy process in order to Define a concrete design problem to be solved. In the Ideate, Prototype and Test stages they generate, refine, and communicate possible solutions in an iterative fashion

	Common School Value	Value from *Mindshift*
Assignments	Scripted problems	Authentic challenges
Problem-Solving	Solution-driven	Human-centered and empathy-driven
Valued Outcome	Correct Answers	Innovative Solutions
Evaluation	Polished final products used to assign grades at the end of an assignment	Low-resolution prototypes used to elicit feedback throughout the project
Nature of Work	Independent work or team work with individual grades	Deep collaboration in service of innovation
Feedback	From your teacher, feedback explains your grade	From specific user, feedback drives iteration
View of failure	Failure, F, a grade that is not passing, something you are ashamed of and wish to forget or hide	Failing forward- failure offers key information for your next steps, to be learned from and be proud of, breeds persistence
Audience	Teacher, possibly other students	People, community members, specific users
Goals	Learn specific content, hope for later application	Learn content, make contributions to meet people's needs, develop creative confidence and efficacy

Fig. 10.2 Design thinking *mindshifts*

about why prototyping solutions can be powerful (see Fig. 10.2). Learning design thinking can be a very exhilarating process, and we capitalize on that positive energy to reflect with teachers about how design thinking can be generative or integrated in school subjects and disciplines. The last part of the professional development experience is to give teachers to time to sit together and reflect, then do some active planning for how they might bring design thinking back to their students.

Through this process, we provide educators the opportunity to engage in an experiential and hands-on process for creative problem solving that both models and is inclusive of twenty-first century skills. We have them experience the tools they need for taking some aspects of design thinking back to their classrooms and schools. Then, we follow up with teachers in several ways. One form of follow up is to invite teachers to the next level of workshops focused on designing lessons so they can further develop design thinking skills and learn to coach or train other teachers. Another involves periodically checking in with the teachers to offer them support and coaching regarding the implementation of design thinking in their classrooms, a method that helps us understand the successes and challenges associated with implementation, mitigate the challenges, and promote further successes. We understand that even expert design thinkers are always in learning mode, and we do not expect that it is easy for teachers to come from one professional development experience and then be willing or ready to replicate it in their classrooms. We do encourage teachers to bring any aspect or tool of design thinking they are comfortable with into their classrooms or schools. This has produced a diverse set of post-workshop experiences. We have had teachers who attended a workshop in a school team implement a design thinking project for their entire 9th grade. Several other teachers at the same workshop used brainstorming approaches back in their classrooms. A few others began using empathy mapping with their students, which is a particularly powerful tool we teach for the development of perspective taking. We have also had teachers who tell us they consider design thinking and apply it to their own lives, and feel comfortable with this as a starting point before applying it within their classrooms.

Design Thinking and the Standards

We work with K-12 teachers, educators working in supplemental settings such as after-school programs and museums, and their administrators. We have worked with over 300 educators to date. We have found that an effective way to engage teachers in the potential of design thinking is to apply it to the challenges they face back in their classrooms. We connect design thinking practices to curriculum challenges in the context of the current accountability demands, including the implementation of the Common Core Standards (CCSS) and the Next Generation Science Standards (NGSS). We believe that each document contains a critical mass of standards that are well aligned to the design thinking process and mindsets and as such, teachers have a warrant to apply these methods in their classrooms in service of the standards. This connection for teachers is crucial to their consideration of design thinking in their work with students.

The Common Core Standards for Literacy across the Content Areas (CCSS)

How do the standards support design thinking? The *Common Core Standards for Literacy Across the Content Areas* (CCSS) privilege specific academic skills such as multiple perspective taking, the synthesis of information from multiple sources, and in the disciplines, the application of understanding through argumentation & justification. As you will see in the case of work we do with educators in Salt Lake City, Utah, these practices are integral within design thinking.

Additionally, the CCSS "offer a portrait of students who meet the standards set out in this document" (National Governors Association Center for Best Practices & Council of Chief State School Officers 2010: p. 7). In this portrait, writers of the standards present seven capacities of students demonstrating college and career readiness in speaking, listening, reading, and writing across the content areas. These capacities are elaborated more specifically within the standards themselves. We believe that design thinking practices and mindsets have the potential to directly support the development of four of these seven capacities. This is not just evident in the presentation of the capacities themselves but also in terms of the individual standards that are aligned to each. Figure 10.3 names the four capacities of college and career readiness that we feel are explicitly aligned to design thinking, examples of standards that are representative of the capacities, and examples of activities within design thinking that can be used to build students' capacities.

In our professional development sessions, we teach the design thinking process as a hands-on fast moving set of steps. We feel that it is helpful for teachers to complete an embodied experience of the process in order to gain insights about how these tools can be applied to their practice. The teachers engage in the design thinking process primarily as learners, working collaboratively with others to solve real-world problems using tools specific to design thinking, and as such, have the experience of learning aligned to the core. We help teachers identify the ways their design thinking experience is aligned to the standards. By providing them with scaffolded discussion connecting their experience of design thinking to current standards, we avoid preaching a pre-specified fit that feels misaligned, and instead, afford the opportunity to visualize and plan for how this pedagogy could both fit into their current practice and meet the standards. Over several years of workshops, we have come to understand that the capacities and standards detailed in Fig. 10.3 are those most commonly identified by educators as core-aligned.

With regards to the modalities delineated in the core standards—that speaking, listening, reading, and writing are the domains for student engagement—our introductory workshops typically involve prioritizing tools for maximizing engagement in speaking and listening while educators learn the design thinking process. Then, when teachers develop an application of design thinking to their classrooms and instructional environments, we present examples of design thinking curricula that directly engage students in all four domains for engagement. Additionally, we provide one-on-one coaching, access to teacher developed resources, and connections

Capacity for College and Career Readiness	Sample of Standards Aligned to Capacity	Relevant Stage, Tool or Mindset used in Design Thinking
COMPREHEND & CRITIQUE: "Students are engaged and open-minded—but discerning—readers and listeners. They work diligently to understand precisely what an author or speaker is saying, but they also question an author's or speaker's assumptions and premises and assess the veracity of claims and the soundness of reasoning."(pg. 7)	CCSS.ELA-Literacy.SL.6.1c Pose and respond to specific questions with elaboration and detail by making comments that contribute to the topic, text, or issue under discussion. CCSS.ELA-Literacy.SL.6.4 Present claims and findings, sequencing ideas logically and using pertinent descriptions, facts, and details to accentuate main ideas or themes. CCSS.ELA-Literacy.WHST.6-8.2d Use precise language and domain-specific vocabulary to inform about or explain the topic.	EMPATHY: ⇒ Interviewing ⇒ Note-taking during field observations ⇒ Note-taking during interviews ⇒ Collaborative sharing of notes from observations interviews ⇒ Do and say portion of the empathy map ⇒ Capturing user feedback from multiple sources. PROTOTYPE, TEST: ⇒ Developing and presenting solutions.
VALUE EVIDENCE: "Students cite specific evidence when offering an oral or written interpretation of a text. They use relevant evidence when supporting their own points in writing and speaking, making their reasoning clear to the reader or listener, and they constructively evaluate others' use of evidence." (pg. 7)	CCSS.ELA-Literacy.SL.6.3 Delineate a speaker's argument and specific claims, distinguishing claims that are supported by reasons and evidence from claims that are not. CCSS.ELA-Literacy.WHST.6-8.2b Develop the topic with relevant, well-chosen facts, definitions, concrete details, quotations, or other information and examples. CCSS.ELA-Literacy.W.6.1b Support claim(s) with clear reasons and relevant evidence, using credible sources and demonstrating an understanding of the topic or text.	EMPATHY: ⇒ Using "do" and "say" quadrants of the empathy map to make inferences recorded in the think and feel quadrants. DEFINE: ⇒ Writing and revising point of view statements. TEST & ITERATE: ⇒ Collecting user feedback from multiple sources ⇒ Iterate on prototype using information from user testing.
ESTABLISH KNOWLEDGE BASE: "Students establish a base of knowledge across a wide range of subject matter by engaging with works of quality and substance. They become proficient in new areas through research and study. They read purposefully and listen attentively to gain both general knowledge and discipline-specific expertise. They refine and share their knowledge through writing and speaking." (pg. 7)	CCSS.ELA-Literacy.CCRA.W.7 Conduct short as well as more sustained research projects based on focused questions, demonstrating understanding of the subject under investigation. CCSS.ELA-Literacy.CCRA.W.8 Gather relevant information from multiple print and digital sources, assess the credibility and accuracy of each source, and integrate the information while avoiding plagiarism. CCSS.ELA-Literacy.CCRA.W.9 Draw evidence from literary or informational texts to support analysis, reflection, and research.	EMPATHY: ⇒ Interviewing ⇒ Note-taking during field observations ⇒ Note-taking during interviews ⇒ Using fiction and non-fiction text as a data source to inform the problem space DEFINE: ⇒ Writing and revising point of view statements. IDEATION: ⇒ Collaborative brainstorming based on point of view statement PROTOTYPE: ⇒ Building low-resolution prototypes to meet a users need within a problem space TEST & ITERATE: ⇒ Presenting prototypes to users ⇒ Gathering feedback from users ⇒ Making changes based on feedback

Fig. 10.3 Design thinking in the common core state standards for english language arts and literacy in history/social studies, science, and technical subjects. First two columns are quoted text from National Governors Association Center for Best Practices & Council of Chief State School Officers (2010b)

UNDERSTAND OTHER PERSEPCTIVES & CULTURES: "Students appreciate that the twenty-first-century classroom and workplace are settings in which people from often widely divergent cultures and who represent diverse experiences and perspectives must learn and work together. Students actively seek to understand other perspectives and cultures through reading and listening, and they are able to communicate effectively with people of varied backgrounds. They evaluate other points of view critically and constructively. Through reading great classic and contemporary works of literature representative of a variety of periods, cultures, and worldviews, students can vicariously inhabit worlds and have experiences much different than their own" (pg. 7)	CCSS.ELA-Literacy.SL.6.1d Review the key ideas expressed and demonstrate understanding of multiple perspectives through reflection and paraphrasing. CCSS.ELA-Literacy.SL.6.1 Engage effectively in a range of collaborative discussions (one-on-one, in groups, and teacher-led) with diverse partners on grade 6 topics, texts, and issues, building on others' ideas and expressing their own clearly.	Perspective taking is accomplished not only via addressing this capacity but also as an extension of the each of the capacities & activities presented above. MINDSETS ⇒ Perspective taking through user cantered design. ⇒ Perspective taking through radical collaboration. ⇒ Perspective taking through interviewing and empathy mapping. ⇒ Perspective taking through iteration based on feedback

Fig. 10.3 (continued)

within a network of educators who are applying design thinking in service of students' mastery of the core standards.

Next Generation Science Standards (NGSS)

Design Thinking pedagogy also attends to and aligns with the Next Generation Science Standards (NGSS). NGSS were developed as a coherent companion to the CCSS but the utility of design thinking to these standards is not limited to it's shared relevance to the CCSS. Our work in design thinking has concrete relevance to many of the disciplinary ideas, crosscutting concepts, and science and engineering practices that are broken down in the NGSS. For example, in an international water challenge, design teams read a user profile detailing the struggles of an intergenerational family of farmers from Hyderabad, India. This firsthand account describes the need for ever-deeper wells as water continued to grow scarcer because of government drilling and water overuse by humans in a climate not suitable for farming. This introduces students to the disciplinary core idea in ESS3.C[4] of the NGSS, "typically as human populations and per capita consumption of natural resources increase, so do the negative impacts on the earth unless the activities and technologies involved are an engineered otherwise." In this activity, students first define the problem from the perspective of the user, a task that necessitates discussion of specific behaviors and their effects over time on the ecosystem (thus addressing crosscutting concepts cause-and-effect and systems). Next, they engage in the science

---
[4] http://www.nextgenscience.org/ms-ess3-3-earth-and-human-activity

and engineering practice "apply scientific principles to design an object, tool, process, or system" as they brainstorm and prototype solutions to meet the users needs while also taking into account the environmental constraints.

In another design task, the water filtration exploration, learners interrogate disciplinary core idea ETS1. B. This standard is explicated in performance expectation MS-LS2-5,[5] which indicates that in covering this standard, learners "Evaluate competing design solutions for maintaining biodiversity and ecosystem services". Water purification, an ecosystem service, is explored as design teams plan, prototype, and test filtration devices that use different combinations of natural materials. As the students compare and evaluate designs within and across teams, they take up a key science and engineering practice: engaging in argument from evidence. Design teams work together to identify the best possible design drawing on evidence recorded during the challenge. This relates to the crosscutting concept of stability and change as designers have the opportunity to observe firsthand how small differences among purification systems can result in large changes in outcome. Through these and other activities, the water curriculum provides teachers with a proof of concept regarding the application of design thinking to support knowledge development around disciplinary core ideas and crosscutting concepts in the NGSS.

While uptake of the disciplinary core ideas and crosscutting concepts is absolutely integral to our work, the strongest alignment between design thinking and the NGSS is definitely in the science and engineering practices listed for each standard. This section of the standards explains broadly what a teacher should do, but there is no elaboration on the methods or pedagogical tools needed for operationalizing these steps. By contrast, our design thinking professional development provides teachers with specific tools, processes, and strategies for building students' capacities to engage in these practices. Figure 10.4 shows several Science and Engineering Practices that should be used in instruction of the Motion and Stability standards for middle school students. The first, asking questions and defining problems, indicates that teachers should have students ask questions that can be answered in local contexts and, if appropriate, follow up with observations and hypotheses. This leaves a reader wondering exactly how this is done. You cannot just tell a middle school student to go out and ask answerable questions. The process and mindsets utilized in design thinking provide teachers with a set of tools, strategies, and coaching techniques for this and other engineering practices. These practices are addressed in the international water challenge and filtration exploration described above. While these activities are from our curriculum for students, teachers are first introduced to the tools during workshops where the curriculum is not the focus. As an example, we take the NGSS practice "asking questions and defining problems". At the start of a 2-day workshop, teachers are given a grand challenge and, as learners, they are introduced to structured protocols for observation and graphic organizers for synthesizing data and evidence. Facilitators coach teams through a multi-step process

---

[5] http://www.nextgenscience.org/msls-ire-interdependent-relationships-ecosystems

> **Science and Engineering Practices**
>
> **Asking Questions and Defining Problems**
> Asking questions and defining problems in grades 6–8 builds from grades K–5 experiences and progresses to specifying relationships between variables, and clarifying arguments and models.
> - Ask questions that can be investigated within the scope of the classroom, outdoor environment, and museums and other public facilities with available resources and, when appropriate, frame a hypothesis based on observations and scientific principles. (MS-PS2-3)
>
> **Planning and Carrying Out Investigations**
> Planning and carrying out investigations to answer questions or test solutions to problems in 6–8 builds on K–5 experiences and progresses to include investigations that use <u>multiple variables</u> and provide evidence to support explanations or design solutions.
> - Plan an investigation individually and collaboratively, and in the design: identify independent and dependent variables and controls, what tools are needed to do the gathering, how measurements will be recorded, and how many data are needed to support a claim. (MS-PS2-2)
> - Conduct an investigation and evaluate the experimental design to produce data to serve as the basis for evidence that can meet the goals of the investigation. (MS-PS2-5)

Fig. 10.4 Example of science and engineering practices (Excerpted from: Achieve, Inc., Disciplinary Core Ideas (2013))

for developing context-specific problem statements. Through firsthand experience and reflection, teachers develop a clear understanding of one way for students to ask answerable questions and define problems. Since the engineering practices in the NGSS and the design thinking process are each informed to some degree by the engineering design process, the relevance of our tools is not restricted to asking questions. Various techniques are introduced for each phase of a design challenge, and through this variety in method, teachers develop a robust pedagogical toolkit (See Fig. 10.5 for additional information about the link between design thinking and the NGSS Science and Engineering Practices).

We do not contend that our tools are the only ones that can be used to accomplish the engineering practices, only that they are suitable, contain sufficient detail to be actionable, and can be used flexibly (as a collective set or one at a time as needed). By sharing some examples, we are drawing attention to the applicability of design thinking to the new standards. Furthermore, we are making the claim that the inclusion of science and engineering practices in the NGSS provides a warrant for the use design thinking and alignment details as evidence that design thinking is a highly relevant process to teaching of both the CCSS and NGSS. Teachers, often exhilarated by the tools we offer, should see the new standards as opening the door for applying design thinking in service of the new standards.

Science & Engineering Practices in NGSS	Relevant Stage and Technique/Tool used in Design Thinking
1. Asking questions (science) & defining problems (engineering)	DEFINE: ⇒ Characterizing the user ⇒ Characterizing the needs of a user ⇒ Writing and revising point of view statements
2. Developing & using models	PROTOTYPE: ⇒ Building low-resolution prototypes to meet a users need
3. Planning & carrying out investigations	This occurs when a project covers each stage of the design thinking process and includes some or all components associated with EMPATHY, DEFINE, PROTOTYPE, TEST, ITERATE
4. Analyzing & interpreting data	EMPATHY: ⇒ Triangulating evidence in "do" and "say" quadrants of empathy map to make inferences recorded in the "think" and "feel" quadrants. DEFINE: ⇒ Making deep, user specific inferences explaining tendencies of a particular user by triangulating data collected during empathy phase IDEATION: ⇒ Organizing potential solutions into categories ⇒ Rank ordering potential solutions based on specific criteria TEST & ITERATE: ⇒ Synthesizing feedback gathered during testing ⇒ Engaging in collaborative decision making about iterations to models and prototypes based on this feedback
5. Using mathematics	N/A
6. Constructing explanations (science) & designing solutions (engineering)	PROTOTYPE: ⇒ Planning for low-resolution prototype ⇒ Building low-resolution prototypes to meet a users need within a problem space ⇒ Increasing resolution of prototype after numerous feedback driven iterations
7. Engaging in argument from evidence	EMPATHY: ⇒ Justifying inferences recorded in the "think" and "feel" quadrants using direct observations recorded in the "say" and "do" column IDEATION: ⇒ Using point of view statement and evidence from empathy map to identify, discuss and select most relevant solution(s) to prototype
8. Obtaining, evaluating, & communicating information	EMPATHY: ⇒ Interviewing & field observations ⇒ Note-taking during field observations and/or interviews ⇒ Collaborative sharing of notes from observations interviews IDEATION: ⇒ Collaborative brainstorming of hundreds of possible solutions for a problem statement PROTOTYPE: ⇒ Dialogue supporting collaborative development of prototypes TEST: ⇒ Presenting prototypes to users, clients, and other design teams ⇒ Collecting user feedback from multiple sources

Fig. 10.5 Alignment between design thinking and NGSS science and engineering practices

Two Case Studies

We present two case studies of work we have done with teachers. They are meant to be snapshots of the professional development experiences we have been creating and where teachers have taken them. The workshops are an active and engaged process for the teachers. They introduce design thinking process, techniques and mindsets through interdisciplinary STEM topics as we are aware of how important it can be for professional development to do so in relation to content (Garet et al. 2001). As discussed, we provide opportunities to understand how design thinking aligns with the standards. In each instance, our work draws on strategic partnerships so that the workshops we offer teachers do not end up being stand-alone in nature. We also include opportunities for school or team planning and conversations with other teacher coaches and mentors (Lieberman 1996), and then we work with our partners on follow-up, in the hopes that workshop essentials can develop within the educators' practice.

The two cases represent different partnerships and models of how professional development with design thinking can be implemented. They are not meant to be the only ways we can imagine bringing teachers to design thinking integration. In fact, we have tried other models, including stand-alone workshops with entire school faculties and "apprenticeship models" where we teach design thinking side-by-side with individual teachers. We concentrate on two cases because they exemplify efforts for which we have seen positive results. Even so, they surface issues about the challenges that lie ahead with design thinking as pedagogy in K-12 classrooms. The first case tells a story of how design thinking can influence pre-service teacher education while also impacting veteran teachers and students. In the second case, we discuss the work that we have done in Salt Lake City, Utah with in-service teachers, and profile how one teacher who attended our workshops has fared with design thinking. The two cases illustrate how and why we work with teachers, how they can innovate with design thinking based on the situations and conditions in their classrooms, schools, and school communities, and how to see some of the issues teachers face. These cases inform the discussion that follows them.

Case 1: The Introduction to Teacher Education

The 8th grade classroom is buzzing with students huddled in groups at tables and desks. They are intent on solving a design challenge. They are making "boats" and seeing how much weight in pennies their boats can carry without sinking as they float them in a huge tub of water. Their boat material is aluminium foil. The first round brings results shouted from around the room: 68! 92! 47! Their teacher tells them to try again, redesigning their boats on the basis of observations from the first round. Pre-service teachers from the Stanford Teacher Education Program (STEP) are mixed in the groups. They are observing and coaching the students, helping them reflect on what happened in the first round as they plan for their next, hopefully, better performing boat. Five minutes later the groups are chanting their counts as they place pennies one-by-one in the new boats. This time some students are more careful to place pennies in one at a time. One group is strategic about where in their boat they place each next penny. They are talking around the topic of surface area as they place coins. Some students are now confident that their boats will be more successful. "Ms. G, we already have 100 and our boat is holding." A second group has 115 pennies. Another couple of groups are on their third boat. Finally, a group puts the 158th penny on and the last boat sinks. Time has run out. Cheers go up, and then the students start to figure out what made their second round designs more successful. The students come up with ideas based on the experience. They discuss density and surface area as it pertains to water. They also discuss the prototyping process and what was learned from each cycle of design and iteration.

After class that day, the master teacher met with the STEP students to reflect on the day's activities and plan for the next few days. The novice teachers were curious. They were excited about the eighth graders enthusiasm and know-how. When asked about how they were feeling about the class, one young teacher-to-be responded that she had never seen anything like this: "The students have not picked up one textbook, yet they are learning so much from the activities and each other. I don't know how this is happening." We talked briefly about the content being considered by the students, then moved quickly to a discussion of the range of teaching methods that are possible in classrooms. As we talked, it was revealed that most of these novice teachers had been taught with traditional methods when they were in school. What they were part of now was strange to them, but they were curious and felt engaged. A few related how students were excited about the class activities and thought they were generating reasonable solutions and ideas about them in the activities. Over the next 4 weeks and 80 h of summer school, they were exposed to how design thinking presented new possibilities for how activity in the science classroom could be structured to increase engagement, involvement and active learning.

This practicum experience for pre-service teachers provided one route to introducing design thinking in the classroom. Even with prospective teachers there is a need to experience new and varied ways of teaching STEM topics, and before this class, design thinking was not on their radar. The design thinking summer school classrooms were a relatively low-stakes way to give the pre-service teachers an immersion view of new practices that have potential for shaping their professional vision. We were hoping that it would provide a foundational experience.

The teacher education program at our university works in partnership with a local school district. It is committed to having theory and practice meet in the classroom, and the summer practicum is one of the first sites for new pre-service teachers to begin understanding the complexity of teaching. With immersion of all pre-service teachers in one school's summer school classrooms, the program takes aim: "The links between theory and practice, university and school, experience and standards, are the links of learning" (STEP website 2013).

The *d.loft STEM Learning* team developed the design thinking-based curriculum units that were used as the summer school science curriculum. The summer school serves rising 5th through 7th graders in the district, and the science classrooms serves up to 250 students in any given year. The summer school has an extremely diverse population that mirrors the district diversity, which is: 19.7 % white, 2.5 % Black or African American, 42.0 % Hispanic or Latino, 23.5 % Asian, 7.6 % Filipino, 4.7 % other races or mixed race. Within the district, 36.5 % of students are English Language Learners and 47.6 % of students receive free and reduced lunch. The science faculty for the school consists of four veteran teachers who have supervisory experience and capability. Added into the yearly mix are 15–20 newly enrolled pre-service teacher education students with interests in science teaching.

While we were able to observe in summer school and talk with all of the teachers, it was difficult to find out how the in-service teachers incorporated design thinking into their work in classrooms. Anecdotally, we learned that several did complete lessons or planned their required teaching unit with design thinking. We are finding ways to do more to reinforce the early learning of the pre-service teachers and we are working with the STEP program to figure out how to better supplement the summer experience, given the intense requirements and fast pace of this 1-year masters level program.

While it has been difficult to track how the pre-service teachers were affected by the summer school experience, we have been able to see the effects that the design thinking approach has on the summer school master teachers. In our first year, the master teachers were also new to design thinking, yet had taken the positions knowing they would have the chance to teach it integrated with science. The master teachers attend a 1-day workshop where we experienced a design thinking challenge, discussed it in concept and practice, and then completed a read through of the curriculum unit that our project team developed. Our team answered questions about specific activities and more general ideas in design thinking. The teachers then set up their classrooms and prepared to help the pre-service teachers to fit in as observers and helpers. Their model is very similar to an apprenticeship model in approach, with master teacher orchestrating the classroom and novices observing and helping.

And what happens to the master teacher? Over the 3 years that we have held this partnership we have worked with nine master teachers and 45–55 pre-service teachers. We hope that the work benefits everyone, from student, to teacher education student, to master teacher. The teacher education students get to experience first-hand new, twenty-first century teaching and learning, and to see how powerful it can be for students. It starts to help them learn new possibilities for teaching that go far beyond the ways they were taught. Even though we think that those entering the profession are digital natives or products of standards-based teaching, we learn that their schooling experiences were predominately traditional in style. Master teachers tell us they were there to enhance their abilities to add new teaching ideas and practices to their repertoires as they supervise novice teachers. There are learning goals and new horizons being sought for all involved.

One master teacher, Claudia, became very enthusiastic, and following the summer, took design thinking back to her school in a nearby district. She entered into collaboration with another teacher, and together they established a new after-school program that was STEM focused. Twenty-five students were chosen to participate in design thinking, leadership and teamwork activities. The teachers loosely based their program on the *d.loft STEM Learning* water curriculum that Claudia had used in the summer school. They worked on a global warming challenge that involved designing ways to conserve water and energy at their school.

Students utilized data collected by the district to improve the amount of energy the school saves by focusing on the shutdown of electronic devices before extended weekends and vacations. Once the children came up with designs, they educated others in the school about their program by visiting every classroom. With implementation, the school actually improved its energy conservation significantly. The program won high marks from all involved, and the *San Jose Mercury News* featured an article about the program (Wilson 2013). In the article Claudia was quoted: "It's always about targeting those other ways of thinking in kids that can help them learn something more"…"The whole concept revolves around energy conservation, which they can bring home and expand it, replicate it and use those skills in real life. That's every teacher's goal."

Claudia was not alone. Another mentor teacher returned for a second year, and was joined by three new mentor teachers. The mentor teachers who taught in public schools were looking forward to using design thinking back in their classrooms. Like Claudia, we are hoping that the experience of being a mentor teacher in partnership with a teacher education program helps build the capacity for leadership with other colleagues. We consider this a strategic and useful way to introduce and spread design thinking practices.

Case 2: The Utah Experience–Supporting the Utah Core Standards Implementation

Melinda, a 6th grade teacher, sits in a chair facing 20 other teachers in the middle of a school multi-purpose room. She's been asked to tell them her story of how she came to teach with design thinking. She explains how she attended a workshop and then started using a few design thinking-based lessons in her classroom. In speaking about what really invested her in design thinking, she recalls they ways her students responded to an after-school design thinking class she began offering 2 days a week. Through that class she:

> …got to look a lot at the different parts of the process and skills, and the kids loved it. They kept saying, "Oh could we do this more in the classroom, more in the classroom because we are so engaged. This feels real to us, this is real." It made me think, well, when they get into the workplace, this is real, this is what it will really be like. So I started putting more [design thinking] in my classroom and writing more and more lessons and unit plans that dealt around the whole thing.

While the appreciation of the students for how they were learning was impressive, Melinda was also encouraged to go further because of how she saw design thinking aligning with the standards and how she might engage them in the classroom. From the first workshop she attended, she saw the connections between design thinking and the standards:

> I'd been working with the ELA core standards for a bit. The whole way through every new step that we did, I'm like, oh my heavens, there's inferencing, there's taking multiple perspectives, there's providing evidence. So I could see the core standards were just built into the whole process, but at a deep, using level. So I'm like, Oh, this is how I'm going to

deepen my instruction in my classroom for the depth of knowledge. I'm going to put these pieces in.

Melinda took the initial connections she saw between design thinking and the standards and began to work at pairing them systematically in her classroom. She brought individual aspects of design thinking into play such as observation, brainstorming and empathy mapping. During a follow-up workshop focused on curriculum construction, Melinda worked with our team to construct a design thinking workshop for the school-community council; this was a strategic effort to demonstrate to parents the benefits of design thinking for their children and the school community. The resulting success with the community council got the attention of her fellow teachers and administrators, prompting their attendance to the next available design thinking workshop offered by our team. Melinda left this workshop having planned a design challenge on Ancient Egypt, a required set of standards in her 6th grade curriculum. In this challenge, her students designed an Egyptian museum that was visited by the entire student body, teachers and parents.

The teachers at the workshop asked Melinda questions about how she made these connections with design thinking and how she gained the support of her principal. She admitted that her principal had been skeptical when she first was suggesting after-school classes, parent workshops, and family fun nights. His interests changed when parents began approaching him with excited comments about what their children were doing. When the students' test scores, that had been flat for the prior few years, showed significant improvement, the principal was officially on board.

Melinda explained that she was an early adopter of the new core standards because she saw them as a more rigorous approach. She started planning lessons that would help her students achieve the new standards, and her students told her they then had an easy time with the high-stakes tests (that still measured the older standards). When design thinking came along, it gave her a way to deepen the implementation of these standards by affording relevance and engagement with the new skills and competencies required. As a result, her students' test scores rose. She emphasized that what she did was not about test scores; it was about approaching the comprehensiveness and deep conceptualizing of the core standards in a meaningful way.

The museum challenge and the improved test results sparked the interest of the other 6th grade teachers, and they came on board and helped plan three standards-based design challenges that would take place in the upcoming school year.

Another example of how Melinda and her grade team plan shows how they accomplish the implementation of design thinking with standards in an interdisciplinary course of study that is both STEM and language arts based. Melinda noticed that every year, her students did not score well on the state standards related to the phases of the moon and seasons. She and the others thought about how they could incorporate some "deeper" learning since they were not performing as well as on other concepts. In the design challenge the plan was for the students to take up one of the two topic standards, ask some essential questions, and conduct background research. Then in design teams, the students would interview second graders, who also have those topics in their core. The interviews would lead the older students to plan and write informational texts about the moon and the seasons. Melinda feels that if the sixth graders can interview, do the research, create the narrative stories and informational texts, and read them with the younger students, they will learn the science.

In Melinda's words, "it just keeps snowballing." She claims she is not an extraordinary teacher

> And like I say, I don't consider myself a fabulous teacher or anything. I just take the things that I have—the cognitive rigor matrix, the design thinking. "Where are my students low? What can I do to impact that area and help them out? So design thinking has really hit a lot of those areas.

Many would think that Melinda is very humble, and that she is actually exceptional teacher. Melinda is certainly an early adopter and a teacher who like to get done what she is responsive to her students' needs. She uses the tools she has available, in this case, the Utah Core Standards, the cognitive rigor matrix,[6] available data on her students' progress, and the support she can get from others to implement a practice of design thinking pedagogy. Even though she is modest, she may be the definition of the kind of teacher we need to really prepare kids as twenty-first century learners. When she tells other teachers how she does what she does to help her students learn, it makes sense to them. They ask her questions about how they might get started, and double-check that she said this helped raise her students' test scores.

The work that Melinda does in her classroom and what she does by helping to work with other teachers and after-school educators is extremely important. The amazing part of her story is how far she stretches to help others with design thinking. In the 2 years since she was first introduced to design thinking, she has run a year-long after school program and a week-long camp at the Utah Museum of Natural History, and facilitated design thinking workshops for after school educators. She goes "on the road" to do workshops. She is creating and on-line professional development course for other 6th grade teachers that profiles design thinking and standards integration. She is early adopter who is energized, and loves to spread the word. She does this advocacy work because she thinks it is so important to find ways to help children become accountable and successful learners. She wants the students to be ready for whatever comes next in their lives, and for right now, design thinking is one of the big ways she is helping them achieve that kind of learning. She wants to help other educators get on board because the needs are great.

Melinda does not standalone. In the time we have been doing design thinking workshops in Utah, 150 teachers have attended our introductory workshop, and 25–30 have returned one or more times to learn more design thinking and learn to coach others. The teachers who have returned have made strides in incorporating design thinking into their schools and classrooms. One group of six to eight from a STEM magnet school have done several design thinking challenges in their school, from having the entire 9th grade complete a school-wide challenge, to incorporating

---

[6] See Hess et al. (2009).

lessons based on design thinking methods such as developing empathy and creating empathy maps into classroom subjects, and creating an design thinking themed elective period. There have been partnerships among the science and language arts teachers at the school around design challenges. Three or four of the teachers have returned to workshops to coach others and talk with them about the nuts and bolts. Other teachers have developed an online course for teachers who teach British Literature, so they can learn to create design challenges inside of their language arts classes.

While we can't be sure of what every teacher takes away from the design thinking experience, we have been able to learn why the teachers who have returned are doing so. The teachers seek out the workshops and the follow up implementation experiences because they are hoping to better meet the learning and life needs of their students. They are hoping to find ways to have their students develop an interest in varied ways of learning, to love learning, and to be prepared for what happens outside of school. They are hoping that students can make more progress than they have to date. They are mission driven, and aware that the record in Utah needs improvements.

Public education in Utah provides a context of need. There are almost 600,000 students in Utah with 77 % white, 15 % Hispanic, and the other 8 % divided between American Indian, Asian, African-American, Pacific Islander, and multiple races. Fiscal year 2011 shows Utah spend the least amount of money per pupil of any state in the country at only $6212.00 (Governing the States and Localities 2013). The average pupil-teacher ratio in Utah is 22.0 as compared the national average 15.4 students. The high school graduation rate is 79 % with 59 % of those seniors who drop out being English Learners (National Center on Educational Statistics 2013). Almost 60 % of high school graduates enter college, yet the University of Utah reports that the graduation rate is lower, especially among women who lag behind the US average (University of Utah 2007). Salt Lake City is a US designated refugee settlement city, and also has many students who are new to the country and US culture. Utah educators at all levels of the state and partner organizations have made a commitment to creating innovative approaches to marshaling resources to benefit the children and youth of Utah from pre-kindergarten to post-secondary education. A huge issue is how to develop opportunities for all Utah students to be career and college ready (Prosperity 2020 Initiative (2015)).

It is against this backdrop, and in support of the new standards that our group has partnered with the Utah State Board of Education to offer a design thinking approach. We have tried to leverage our partnership so that it can reach beyond the typical boundaries that exist around the academic subjects by reaching out to teachers, school leaders, supplemental educators, and those working across the subject areas.

Melinda and the other Utah teachers have been part of activities to help see design thinking as a viable form of pedagogy that connects to content, the standards, and the outcomes they would like for students. They are supported through the professional development workshops they attend, the time, resources and support that they are given through the State Office of Education and in their districts and schools for individual and team planning and implementation work. Several teachers have been working on the standards implementation and assessments have attended the workshops, reinforcing the connection between the standards and activities such as design thinking. With the support given by our team, state office personnel, the welcoming staff of the Utah Museum of Natural History and the Museum of Natural Curiosity in terms of facilities and access to scientists and designers who can help in design challenges (as experts and users concerning environmental topics such as those we introduced), the professional development of after-school educators, we have set in to motion the goal of making design thinking one of the viable pedagogies available to Utah teachers.

Discussion and Conclusion

As the cases illustrate, trying to make design thinking a choice in teachers' toolkits is a lofty goal, and there are many pathways that are possible. We take several in our professional development work. The cases share some features: (1) they each take an approach to immerse teachers in design thinking as learners; (2) they introduce teachers to interdisciplinary teaching and learning, providing opportunities for discussion, reflection and planning; and (3), they leverage partnerships with organizations that have the capacity to help the teachers carry forward and amplify the work. We discuss each feature with regards to how it contributes to the potential for strengthening teacher practice through the uptake of design thinking.

Immerse Teachers in Design Thinking as Learners

Educators who engage in our workshops experience authentic twenty-first century instruction as learners. Our model of professional development honours the fact that teachers, like their students, are independent thinkers and learners who develop mastery based on authentic experiences, collegial collaborations, and opportunities to reflect. When asked what they liked about our professional development in a post workshop evaluation, we received many comments along these lines:

> [I liked] how well playing the role of student helped me to understand ways to teach the material to my students and made the workshop more fun!
> –*French Teacher, 10 years experience*
>
> I loved how it was facilitated through movement laden non traditional techniques. We not only reimagined education but [also] the classroom, a sense of time, & what it means to work as a group.
> –*School Administrator, 5 years experience*

These comments are echoed throughout the evaluations and strengthen the notion that teachers benefit from experiencing new educational practices as learners prior to being asked to adjust their teaching. In the workshops, the teachers learn design thinking through completing a design thinking challenge. In teams, they were introduced to a problem space (such as designing an energy solution). They discussed interviewing, then prepared questions and interviewed an energy user. They processed the interview information by being guided to create an empathy map that helps them draw insights about their user, and then to more specifically define their user's energy needs. They then brainstormed possible solutions and chose one solution to prototype. Once they constructed a prototype, they tested it with their user and had a chance to revise it. At every step in the process, they learned how to take the steps, and saw how those steps could be taught in a classroom. There were times for questions and answers concerning the process and how to teach it. Once the design challenge is completed, we extend the authenticity of the professional development by asking teachers to imagine how they could use the design thinking tools to meet the standards, thus further honouring them as learners.

Finally, rather than providing them a detailed implementation guide, full of constraints, we let groups of teachers work together to develop personal plans for implementation. One technique we used was to ask teachers to examine a lesson they will teach in the next week by seeing how it aligns within the four levels of the Cognitive Rigor Matrix (Hess et al. 2009). Once teachers mark the level of their lesson by skill, we suggest to them that they try to use some design thinking processes to move the activity to a more complex level of work for the students. An example would be: In level 1, students "Recall, recognize, or locate basic facts, details of events, or ideas explicit in text." A teacher might have been planning to have students describe a character in a story. Instead, the teacher revises that plan to create an activity where students use an empathy map. The empathy process would drive students beyond simple recall of facts about a character in a story to generating inferences about the character based on their interpretations of what the character said, did, and even felt. This switch to an empathy activity would take the lesson from being a level 1 activity to a level 2. Our aim in having teachers alter an upcoming lesson and vet it with their colleagues helps them to make use of what they learned about design thinking and some of its tools back in their classrooms. Throughout the workshop, teachers are learning about the design thinking process, how it applies to the standards, and how to apply it in a small way in their classrooms.

Provide Teachers Interdisciplinary Teaching and Learning Experiences

We bring teachers together from a variety of disciplines and experience levels who teach at schools with a range of nationalities, socioeconomic statuses, and language statuses. We place teachers who work closely together on separate design teams, because we want teachers to check their everyday baggage at the door. At first, some teachers groan about being separated from each other and question the relevance of learning outside their school site teams (teams we let them return to for the reflection and lesson implementation portions of the workshop). Despite the initial complaints, we are thanked for this opportunity at every workshop. Here are just a few examples:

> [I liked the] cooperative opportunities, collegial atmosphere, passion of instructors, relevance of problem, "next day" applicability.
> –Social Studies Director, 12 years experience
>
> [I liked the] collaboration, feedback process, [it provided] encouragement that pushes to keep [us] doing more.
> –Science Coach, 13 years experience

While the comments do not capture the complexities of cross-disciplinary collaboration, they do demonstrate that what is first thought to be an uncomfortable request is beneficial to the outcome of the professional development. Cross-disciplinary teams afford their participants the opportunities to "try on" different approaches and disciplinary views. Teaching is often an isolated profession with islands of innovation separated by oceans of mandates. Allowing teachers new collaborators offers exposure to the way others are parsing the mandates as well as demonstrates design thinking's idea that radical and unusual collaborations lead to innovation (Goldman et al. 2013).

Furthermore, twenty-first century problems, for example, the aftermath of the earthquake in Haiti, do not occur in specific domains such as language arts or in Algebra 1. Real world problems cross boundaries, however messy that may feel. For this reason, we integrate STEM topics such as access to and conservation of water, energy, and shelter to illustrate how various disciplines can make contributions to the topic and solutions for users. We try to show how teachers from vastly different subjects such as science, math, social studies and language arts can all find ways into the materials and activities of the design challenges. Experiencing successful problem solving on interdisciplinary teams gives teachers an experience they might start to model with their students. This involves synthesizing input from multiple areas of expertise to develop a working solution to a real (and complicated) problem. Design thinking scaffolds multiple vehicles for valid participation as well as tools for taking the perspectives of others. Not only does this open the possibility of more learning for the participants, it leads to more nuanced, multifaceted solutions that are better equipped to stand up to the complexities of the real world.

Leverage Partnerships

We seek partners who share commitments to teachers, some ideas about best teaching and learning practices, support of the standards, and helping students move though schools towards happy and productive futures. The strategic partnerships we have been able to form exponentially magnify our ability to bring deep experiences with and about design thinking to K-12 teachers. Two of our partners were highlighted in the cases: the partnership with a teacher education program, and the partnership with colleagues at the State Office of Education.

Our partnership with the Stanford Teacher Education Program helped us place pre-service teachers in apprenticeship roles with teachers who are teaching design thinking. The fact that pre-service teachers have their very first, 20-day, intensive practicum with a design thinking pedagogy is foundational and it is a statement about the nature of twenty-first century classrooms (Stanford Teacher Education Program 2014). It is both symbolic and practical in nature. It predates pre-service teachers entering their student teaching experiences where the full pressure of the existing system is pressing into the new teacher's classroom realities and psyches. We are delighted that we have the chance to make impact at such a formative time for new teachers.

The partnerships in Utah have helped us gain access to sustained work with in-service teachers. The State Office of Education organizes a huge number of activities in support of the Utah Core Standards, from content-based workshops, to e-text and book development, development of state assessments, to workshops on design thinking. Our colleagues there are committed to developing capacity in teachers who are implementing the standards. Their work with us is designed to help teachers realize that huge changes in practices are necessary for meeting the standards, and that business as usual in the classrooms will not meet the goals. Our advocate at the State Office sponsored our PD workshops and invited schools and teachers. She has provided support for follow-up and planning sessions for teachers, and reached out to school principals, representatives of city, county, and state-wide education initiatives to spread the word about design thinking.

She secured venues and brokered relationships with other Utah partners such as The Natural History Museum of Utah and the Museum of Natural Curiosity as partners. Both museums opened their doors to workshops, helped us create challenges that drew on their expertise and exhibits, and had their staffs participate in the design workshops as learners and experts. The combined efforts of various partners provided momentum and resources for follow-up, helping teachers to develop further experiences with design thinking, and developing teachers into design thinking mentors and coaches.

We cannot underestimate the impact of these partnerships on the success of the work we have done, and we see them as essential to seek out and develop.

Overcoming Obstacles

We have observed that teachers who facilitate design thinking in their classrooms are generally pleased. We profiled two teachers who were especially successful at implementation. Yet we recognize that there were frustrations that surfaced in each of the cases and that each teacher took a different route in instituting design thinking into her professional practice. We realize that it is important to have many pathways to adoption. Not all teachers will implement whole design challenges after attending the professional development. We advocate that teachers build their competency with design thinking in the classroom over time. They might start with a small challenge, or by implementing a part of the process such as brainstorming or empathy mapping. They need to see themselves as designers, their students as "users," and build on what they are doing based on feedback. They may need to see bits of design thinking return as "results" such as content-engaged and accomplished students, complementary parents, or supportive administrators. They may need to enlist their colleagues, and have time to plan for implementing new strategies.

Our biggest advice to teachers is that they try out part or all of the design thinking process and witness the impacts in their students. Sometimes the impacts seem tiny such as when a student participates with new enthusiasm. Other times, the impact can be unexpected such as when an evaluation was conducted on Melinda's after-school design thinking program and students reported better attention in school classes once they participated in the design thinking course. Teachers may be required to take a leap to develop confidence that teaching towards innovation, rather than back-to-basics, may be what their students need. Our work is primarily about helping educators to embrace that change in mindset.

Conclusion

It is still too early to know what the 300+ teachers who we have introduced to design thinking will accomplish. Melinda and Claudia have jumped in enthusiastically, and some of the other teachers who have attended professional development have dabbled in design thinking, implementing parts of the process when and where they see the fit. And we know of a few who have done little in their classrooms. Those teachers cite various reasons: they need to stay on basics, they cannot get support from the administration, there is no time, and design thinking seems like a huge reorganization for them. The new standards have recently begun to be implemented, and the first high-stakes testing began in 2014. For many of the teachers, design thinking is an attractive theoretical possibility rather than a concrete strategy for helping students to accomplish standards-based learning. Once the new standards and assessments are in place, we expect some additional shifts to take place as teachers develop strategies that work for their students. We are seeking new ways to address teachers' needs as they evolve.

If schools are to prepare students for the world they will face in 2026, a significant change in teacher practice is necessary. Business as usual will leave students ill prepared for life and work in twenty-first century. The wave of new standards is introducing new possibilities. We believe the introduction of design thinking into K-12 education has the potential to support student development as engaged, adaptive, deep learners, creative individuals, and productive citizens. We utilize teacher professional development as one means towards these ends. We have learned valuable lessons about professional development generally and specifically through bringing design thinking to teachers. In providing professional development, we have gained traction by forming partnerships with relevant community organizations and leveraging them to create a space for teachers to be learners, engaging in hands-on work with non-traditional interdisciplinary teams. While facilitating this process we have seen first-hand the power and relevance of design thinking for addressing new standards, affording concrete strategies for the development of twenty-first century competencies, and increasing teachers' creative confidence. By supporting educators through user-centred design, we give them the time, space, and experience needed to begin thinking differently about their practice. While this is not the only way to stir the winds of change, our work has illuminated the process and mindsets of design thinking to be powerful tools, suitable and effective, flexible and robust, ready for use today in support of a better tomorrow.[7]

Acknowledgements We thank the many teachers and educators who have partnered and learned with us, especially Christelle Estrada at the Utah State Office of Education, the Utah Museum of Natural History, The Museum of Natural Curiosity, and the Stanford Teacher Education Program. We also owe a special thanks to our d.loft team members who have worked with teachers: Stephanie Bacas-Daunert, Maureen Carroll, Tanner Vea, Ugochi Acholonu, Zaza Kabayadondo, Aaron Loh, David Kwek and Eng Seng Ng. Without this collective effort, design thinking would not be in the hands of K-12 teachers. This material is based upon work supported by the National Science Foundation under Grant #1029929. Any opinions, findings, and conclusions or recommendations expressed in this material are those of the author(s) and do not necessarily reflect the views of the National Science Foundation.

References

Achieve, Inc., on Behalf of the Twenty-Six States and Partners that Collaborated on the NGSS. (2013). Next generation science standards. Achieve, Inc. on behalf of the twenty-six states and partners that collaborated on the NGSS. Washington, DC: The National Academies Press

Berry, B., & The Teacher Solutions 2030 Team. (2011). *Teaching 2030*. New York: Teachers College Press.

Carroll, M., Goldman, S., Britos, L., Koh, J., Royalty, A., & Hornstein, M. (2010). Destination, imagination, and the fires within: Design thinking in a middle school classroom. *International Journal of Art & Design Education, 29*(1), 37–53.

Chen, M. (2010). *Education nation: Seven leading edges of innovation in our schools*. San Francisco: Jossey-Bass.

Cross, N. (2006). *Designerly ways of knowing*. London: Springer.

Dewey, J. (1916). *Democracy and education*. New York: Macmillan.

Garet, M. S., Porter, A. C., Desimone, L., Birman, B. F., Suk Yoon, K. (2001). *American Educational Research Journal, 38*(4) (Winter), 915–945. Stable URL: http://www.jstor.org/stable/3202507

---

[7] We offer curriculum units that have been created, tested and revised based on their use in a range of classroom and after-school situations. In some ways, the curriculum challenges are our tried and true resources that we bring forward. We also develop and share formats for professional development that can be put into practice by others once they have been introduced to design thinking. Visit http://tinyurl.com/designthinkingcurriculum for more information.

Goldman, S., & Lucas, R. (2012, March). Issues in the transformation of teaching with technology. In Society for Information Technology & Teacher Education International Conference (Vol. 2012, No. 1, pp. 1792–1800).

Goldman, S., Carroll, M. P., Kabayadondo, Z., Britos Cavagnaro, L., Royalty, A., Roth, B., Kwek, S. W., & Kim, J. (2012). Assessing d.learning: Capturing the journey of becoming a design thinker. In C. Meinel, L. Leifer, & H. Plattner (Eds.), *Design thinking research: Measuring performance in context* (pp. 13–33). Springer, London.

Goldman, S., Kabayadondo, Z., Royalty, A., Carroll, M., & Roth, B. (2013). Student teams in search of design thinking. In C. Meinel, L. Leifer, & H. Plattner (Eds.), *Directions in design thinking research* (Vol. 3). Springer International Publishing.

Governing the States and Localities (2013). Education Spending Per Student by State. Accessed at: http://www.governing.com/gov-data/education-data/state-education-spending-per-pupil-data.html.

Hess, K. K., Carlock, D., Jones, B., & Walkup, J. R. (2009). *What exactly do "fewer, clearer, and higher standards" really look like in the classroom? Using a cognitive rigor matrix to analyze curriculum, plan lessons, and implement assessments*. Hess' local assessment toolkit: Exploring cognitive rigor. Available [online] http://www.nciea.org/cgi-bin/pubspage.cgi

Kelly, D., & Kelly, T. (2013). *Creative confidence: Unleashing the creative potential within us all*. New York: Crown Business.

Lieberman, A. (Ed.) (1996). Practices that support teacher development: Transforming conceptions of professional learning. In M. W. McLaughlin & I. Oberman (Eds.), *Teacher learning: New policies, new practices* (pp. 185–201). New York: Teachers College Press.

National Center on Educational Statistics. (2013). *Public school graduates and dropouts from the Common Core of Data: School Year 2009–10*. NCES 2013-309rev. US Department of Education.

National Governors Association Center for Best Practices & Council of Chief State School Officers. (2010a). *Common core state standards*. Washington, DC: National Governors Association Center for Best Practices & Council of Chief State School Officers.

National Governors Association Center for Best Practices & Council of Chief State School Officers. (2010b). *Common core state standards for English language arts and literacy in history/social studies, science, and technical subjects*. Washington, DC: Authors.

National Governors Association Center for Best Practices & Council of Chief State School Officers. (2010c). *Common core state standards for mathematics*. Washington, DC: Authors.

Partnership for 21st Century Skills. (2008). *21st century skills, education & competitiveness: A resource and policy guide*. Tucson: Partnership for 21st Century Skills.

Pellegrino, J. W., & Hilton, M. L. (Eds.). (2012). *Education for life and work: Developing transferable knowledge and skills in the 21st century*. Washington, DC: National Academy Press.

Prosperity 2020 Initiative (2015). http://prosperity2020.com/the-vision/.

Rittel, H., & Webber, M. (1973). Dilemmas in a general theory of planning. *Policy Sciences, 4*, 155–169. [Reprinted in Cross, N. (Ed.). *Developments in design methodology*, (pp. 135–144). J. Wiley & Sons].

Smith, C. (2007). Design for the other 90%, Cooper-Hewitt, National Design Museum. New York: Smithsonian Organization.

Stanford Teacher Education Program (2014). https://gse-step.stanford.edu/academics.

University of Utah. (2007). New statewide initiative helps underserved high school students get to college [Press release]. Accessed at: http://unews.utah.edu/p/?r=083007-1#Media_Contacts

Vande Zande, R. (2007). Design education as community outreach and interdisciplinary study. *Journal for Learning through the Arts: A Research Journal on Arts Integration in Schools and Communities, 3*(1), Article 4.

Wilson, Alia. (2013, April 2). Sunnyvale: Ellis Elementary students put science in action. *MercuryNews.com*. Accessed at: http://www.mercurynews.com/sunnyvale/ci_23161358/sunnyvale-ellis-elementary-students-put-science-action

Chapter 11
Elementary School Engineering for Fictional Clients in Children's Literature

Elissa Milto, Kristen Wendell, Jessica Watkins, David Hammer, Kathleen Spencer, Merredith Portsmore, and Chris Rogers

To help orient readers for the remainder of the chapter, we open with a brief overview of one group's work on an engineering problem they framed for characters in a book they were reading for class. After this, we step back and share our perspectives on engineering and literacy in the elementary grades. We then provide evidence of the beginnings of students' practices, in particular with respect to (1) framing problems (2) planning, and (3) testing and realizing ideas. Finally, we offer suggestions for teachers interested in engaging their students in similar activities.

A Brief Example: Escape from Ember

Caroline and Samantha, fifth graders, have just read *The City of Ember* by Jeanne DuPrau. The book tells the story of Ember, an underground city where the only light and power come from a dying generator, and the only supplies come from an increasingly bare warehouse. Two of the book's characters find a way out, up through a deep chasm, but they face the problem: How can the inhabitants of Ember get to the top?

Caroline and Samantha are engrossed in the task of designing something to help, something they will prototype in the classroom.

Their first idea is a hot-air balloon with a basket, but they decide that although it would function in the real world and in the book, they could not try it the classroom because it requires an open flame. Their second idea involves a vertical gear and chain system with the bottom gear in Ember and the top gear at the surface. Like a

---

E. Milto (✉) • K. Wendell • J. Watkins • D. Hammer
K. Spencer • M. Portsmore • C. Rogers
Tufts University Center for Engineering Education and Outreach, Medford, MA, USA
e-mail: elissa.milto@gmail.com

ski lift, seats are attached to the chain to carry people up and then return to pick up more passengers. Caroline and Samantha abandon this idea because they feel they do not have the proper materials nor enough time (Fig. 11.1).

In the end, the girls choose a pulley system: The people from Ember would sit in a basket and use a pulley system to raise themselves. This, they feel, would work in all three settings (book, world and classroom).

As they complete their project, we ask the girls what has surprised them during the project.

> Caroline: And, what also surprised me is how we changed our ideas. Like we had two completely different ideas, and at first we kind of fought over it. I was like, 'No, I don't understand that,' then we're just kind of like decided.
> Samantha: I didn't get like the whole pulley- the, her idea, and she didn't think that mine would be a good idea, and then we thought about both of them realized that neither of them were [*sic*] a good idea, and just kind of just put this together...

In this example, we see the girls engaging in brainstorming, gathering information, considering client need, and making measured decisions based on a variety of

Fig. 11.1 Drawing of Caroline and Samantha's second idea for a solution

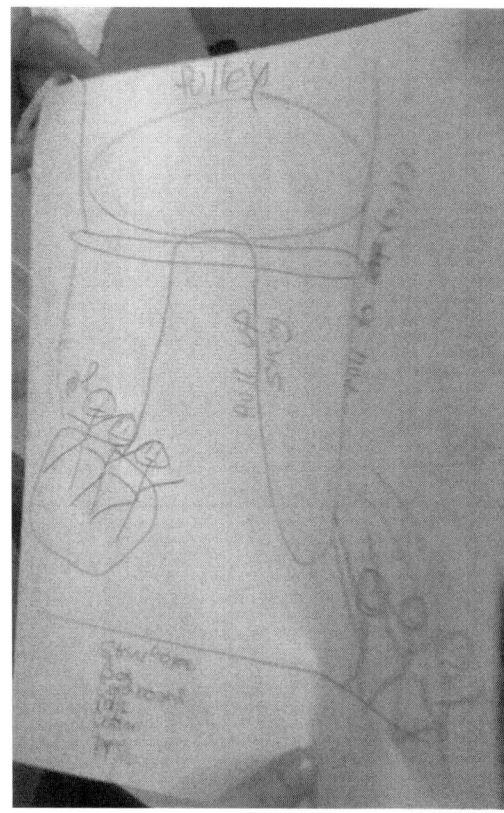

imposed and emergent constraints. They used *The City of Ember* text as the foundation for their design choices, gaining from it information about client need and design constraints. They considered solutions from a variety of perspectives and chose the one that made the most sense given the constraints of materials, time, and classroom and book settings. We see them engage in practices which Crismond and Adams (2012) characterize as typical of an experienced or informed designer.

In this chapter, we present this approach to engineering education in elementary grades: students' engineering for characters in books they are reading for class. The stories offer rich contexts for their finding and framing problems for clients, planning for solutions, and then building and testing their ideas.

Project Overview

Integrating Engineering and Literacy (IEL) is a research project at the Tufts University Center for Engineering Education and Outreach (CEEO), funded by the National Science Foundation, grant # 1020243.

Like our other work at the CEEO, IEL is motivated by our assumption that kids can engineer when given the chance. We don't expect students to be expert right away; we expect, as we show in this chapter, that they have resources and interests to begin. We are interested to understand and promote the beginnings of engineering thinking in young children.

IEL is not a curriculum; it is an approach to drawing engineering problems from readings. Each teacher chooses the text that his or her students will be using for IEL tasks. The teacher acts as a facilitator, helping students discuss and realize their ideas, while also promoting productive aspects of students' design thinking. Our focus here is on grades three through five, but we believe this approach could be used in other grades as well.

Over the last three years we have worked with over five hundred students, taught by eighteen teachers in eight different schools, collecting data in video of students' work and photographs or copies of artifacts including sketches, writings, and prototypes. Books used have included fiction, non-fiction, and anthologies.

Structure of IEL Activities

The general structure is as follows. A unit begins with students reading a book, independently, in small groups, or as a read-aloud. After reading, or sometimes along the way, the teacher facilitates students as they discuss the problems they see characters encounter. The teacher usually records the students' ideas, on the board or chart paper. Discussions often move between identifying problems and brainstorming possible solutions that would be feasible in the book and in the classroom. Students then work in groups to pick a problem, and they start working on scoping it and considering possible solutions; often students do this for several problems as part of choosing. Typically students spend time discussing solutions as a group and then begin building, testing and iterating, always—we hope—using the book as a reference to justify their ideas. Sometimes the students do not build, but work on a conceptual design. The heart of the process is that students have the opportunity to develop and find evidence for their own ideas about how to solve a messy, complex problem.

As we describe below, we have found two particular instructional moves helpful. One is for the teacher to specify in advance, when appropriate, that the goal is to have a working prototype to demonstrate in the classroom. The other is to have a midway full-class share-out session. The groups take turns presenting where they are and getting feedback on their ideas. IEL activities typically end with final presentations and feedback or some type of consolidation activity.

Professional Development

There are two main components in our Professional Development (PD) for IEL teachers. One is engaging the teachers themselves in engineering, and for many in the project this is their first experience. The second is working with the teachers to think about their students' learning. In that respect we pay particular attention to helping teachers recognize and think about supporting the beginnings of engineering in students' engagement. To these ends, IEL PD workshops consist of a mix of activities.

Generally, elementary school teachers have more experience in literacy than in engineering design, so we spend more time on thinking about the nature of engineering practice. Teachers work on their own IEL design projects to get a better understanding of engineering and the interdisciplinary nature of IEL activities. Interestingly, we have found that teachers' confidence in their literacy knowledge and teaching helps support them in learning about engineering. This reflects an overlap, for teachers as for students between objectives of literacy and of engineering, in particular of understanding the story and getting to know the clients and their situation.

We devote a significant amount of time to watching and discussing video clips of students' work in IEL activities, with the PD objectives of cultivating a stance of close attention to students' reasoning as well as abilities to recognize productive beginnings. (For us, these activities are also important for our research on student learning, and in this the teachers are often important collaborators.) Often there is value to what students are doing in what at first glance seems to be just "fooling around." Teachers are also given the opportunity to plan for the IEL experiences in their classroom. Part of the planning involves anticipating student thinking in literacy and engineering, their questions, and possible projects.

In the next section, we provide an overview of our perspectives on student learning in literacy and engineering, building on research and policy documents about productive engagement in both domains. We then present more specific descriptions of the beginnings of engineering we have observed in students participating in IEL projects, which we collapsed into three categories representing key aspects of their emerging practices.

Integrating Two Domains: Engineering and Literacy

Much of the power of this approach is in an overlap between two areas of education objectives: Literacy and Engineering. In this section, we discuss what we see as evidence of children's literacy and of nascent engineering.

Literacy

We take it as evidence of students productively engaging with text when, as they read or write about a fictional novel or an account of a historical event, they

- try to take on the perspectives of characters,
- note relevant aspects of the physical setting,
- wrestle with unfamiliar concepts and vocabulary, and
- use information in the text and their own world knowledge to construct an interpretation of the text.

The widely adopted Common Core State Standards (National Governors' Association Center for Best Practices 2010) also emphasize that students in middle and later elementary school should learn to attend to key details in texts and use textual evidence as they make inferences, identify important themes, and support claims about the text in collaborative discussions or written arguments.

IEL projects support students' literacy learning by providing an engaging reason for students to attend to multiple aspects of the text. In order to design something that would address a particular problem in the book, students must identify and understand problems faced by the characters. They must also think about the problem in the context of the overall story or historical account. Is this problem important? Who is involved with the problem, when does it occur, and why does it matter? By their structure, IEL activities effectively prompt students to put themselves in a character's shoes, look to the book to understand characters' wants and needs, and make inferences about the potential impact of proposed solutions. Many students use the book on their own initiative, as both an inspiration for their design and a resource of ideas for how it should work.

For example, as Samantha and Caroline designed their system for helping characters escape from the underground city of Ember, they focused on a problem the protagonists faced at the end of the story. If the protagonists did not figure out how to get people out of the city, food supplies would run out, the power would fail, and the city would be plunged into permanent darkness. As noted above in Fig. 11.2, the

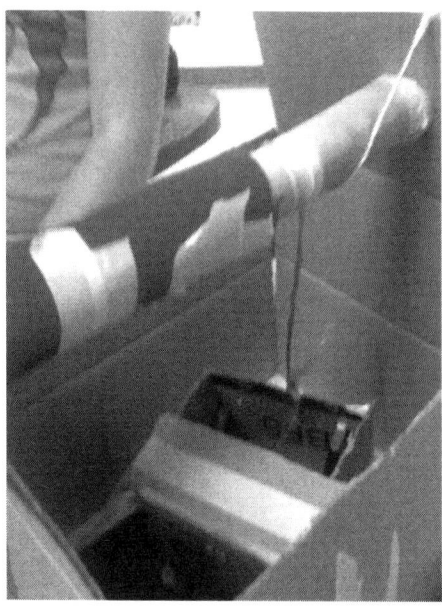

Fig. 11.2 Caroline and Samantha's final prototype: ember is symbolized with *black* felt at the *bottom* of the box. The *blue* felt symbolizes the light of the surface world

girls used felt on the bottom of the box to show the darkness of Ember. They represented the outside world by the blue color of the bar that scaled the chasm. These features of their design shows the girls' understanding of one of the main points of the book and their desire to communicate the difference of the two settings of the book. There is further evidence of their attention in how the girls discussed the characters' needs as they worked, using names and other specifics from the book to support their design choices.

We see this often. As students consider various problems and potential solutions, they discuss key events in the book, constraints of the setting, and character needs and personality traits. Students may take on the perspective of different characters, looking for ways to evaluate the relative importance of different problems and the feasibility of solutions. We encourage students to evaluate their own and others' designs using text-based criteria, helping them to think more deeply about the story and, at the same time, about the their clients and their situation. In short, students practice being good readers as they learn about the beginnings of good engineering.

This overlap in the intellectual work of engineering and literacy is evident in the alignment of IEL objectives with the Common Core State Standards' core ideas and literacy standards. Table 11.1 shows examples from CCSS for fourth grade.

Three Key Aspects of Engineering

For our purposes here, we take engineering to begin with the identification of problems for clients. Part of the challenge is "framing" or "scoping" problems, which generally involve vague and unstated goals and constraints (Jonassen et al. 2006),

Table 11.1 Integrated engineering and literacy reading and writing experiences and the related aspects of the Common Core State Standards

In an IEL classroom over the course of a school year, we hope to see all students:	Related CCSS core ideas and standards
Reading and Literature: Key Ideas and Details	
Students understand the text in order to define a problem and relevant constraints. Students use information that is stated explicitly and found through inference	CCSS.ELA-Literacy.RL.4.1 Refer to details and examples in a text when explaining what the text says explicitly and when drawing inferences from the text
Writing: Research to Build and Present Knowledge	
Students understand and draw on specific details from the text in order to justify problem constraints and design features	CCSS.ELA-Literacy.W.4.9 Draw evidence from literary or informational texts to support analysis, reflection, and research
Writing: Comprehension and Collaboration	
Students ask and respond to questions as part of design reviews. Questions and answers are drawn from an understanding of the text and an understanding of their own and other students' design choices	CCSS.ELA-Literacy.SL.4.1c Pose and respond to specific questions to clarify or follow up on information, and make comments that contribute to the discussion and link to the remarks of others
Writing: Presentation of Knowledge and Ideas	
Students address the text by designing solutions to characters' problems that are based on relevant information from the text. Students present their designs through presentations and other types of representations	CCSS.ELA-Literacy.SL.4.4 Report on a topic or text, tell a story, or recount an experience in an organized manner, using appropriate facts and relevant, descriptive details to support main ideas or themes; speak clearly at an understandable pace

in a way that both appreciates the clients' situations and supports ideas for solutions (Schön 1983; Dorst and Cross 2001). It then involves thinking of possible solutions, assessing their merits and shortcomings, and selecting ideas to pursue, which means building prototypes to test and refine.

We discuss these as "three key aspects" of engineering—framing problems, conceptual planning of solutions, and realizing and testing ideas. While we think of the process as beginning with identifying problems, we do not generally expect students to move through them in order. For example, progress in conceptualizing solutions can introduce new considerations for framing problems; results from pilot testing may provoke new work on conceptualizing solutions, and so on. That is, like models of the engineering design process (e.g. French 1998; Pahl and Beitz 1984), this is not a view of a linear sequence but rather of a set of kinds of activities.

Our motivation for this description is largely practical: the categories provide a serviceable framework for teachers (and us) to use in recognizing the beginnings of engineering with their students. It is not usually feasible for all students, within the typical time available for IEL projects, to experience engineering design all the way from scoping problems for characters to constructing functional prototypes. Describing three aspects gave us a serviceable way to structure our expectations for what we hoped to see at some point for all students, perhaps over several rounds of IEL projects, allowing that not all experiences will give evidence of all three.

Note that these aspects are not mutually exclusive. We may see evidence of more than one at a given time; they can overlap and support one another. Students and teachers are capable of engaging in all three aspects in one IEL project, but in our experience there is not generally time for that, and it is not necessary for students to engage in all three aspects in order for their work to be a productive educational experience. Further, we have found the more often students are given opportunities to act as engineers, the greater the number of engineering aspects their design processes include.

It is essential to be clear that our purpose with this framework is to guide our observations and assessments of students' work, not to define or constrain expectations for a given lesson. Precisely because we see productive engineering as iterating among different kinds of activities, we intend IEL lessons to afford students' shifting among them as makes sense. Thus the aspects we describe here may not correspond to expectations or plans for a lesson. Later in this chapter, we will discuss planning for IEL projects. Here we are focusing on evidence of nascent engineering in students' thinking and engagement.

Table 11.2 summarizes the three key aspects and shows their alignment with the Next Generation Science Standards (NGSS) disciplinary core ideas and performance expectations for engineering. In the following section, we describe and give examples of each of the Three Key IEL Aspects.

There are many aspects of the engineering process followed by professional engineers that we do not include in our analysis, for example life cycle analysis, a cradle-to-grave analysis, manufacturing and total quality management, and so on. We do not expect students to productize, estimate and adhere to budgets, deal with packaging and shipping aspects of the design, or many of the other aspects of product development. We are looking for the beginnings of engineering thinking.

Seeing the Engineering in Children's Thinking

In the U.S., there has been growing interest in engaging K-12 students in engineering in informal and formal educational settings. While engineering has been gaining visibility, and related activities and curriculum have been proliferating, there is limited research and understanding about how children engage in design. Some authors suggest that children, like beginning designers, have naïve versions of engineering practices—such as treating an engineering design problem as well-defined and proceeding to design a solution versus spending time problem scoping and delaying design decisions (Crismond and Adams 2012). There is also evidence from the study of children engaged in design that children may "skip" preliminary engineering design practices in favor of working with materials (e.g. Welch 1999; McCormick et al. 1994).

However, there is also evidence that even young children can engage in sophisticated engineering practices, such as planning or drawing (Portsmore 2008; Portsmore 2010). Evidence from children's work in IEL supports this perspective, that young children are capable of navigating through messy and ill-defined

11 Elementary School Engineering for Fictional Clients in Children's Literature

Table 11.2 Three key aspects of student work throughout integrated engineering and literacy experiences and the related aspects of the next generation science standards

In an IEL classroom over the course of a school year, we hope to see all students:		Related NGSS core ideas and standards
1. Frame problems	*Information gathering:* Using the literature as a resource, students pay attention to the client and to opportunities and limitations imposed by the classroom, teacher, or text *Multiplicity of problem elements:* Students consider multiple sets of constraints and criteria when setting up and solving a problem *Different perspectives:* Students consider different perspectives and contexts involved in the design, e.g., clients, users, manufacturers, salespeople, classmates, teachers	*NGSS-ETS1.A: Defining and delimiting an engineering problem* K-2-ETS1-1. Ask questions, make observations, and gather information about a situation people want to change to define a simple problem that can be solved through the development of a new or improved object or tool 3-5-ETS1-1. Define a simple design problem reflecting a need or a want that includes specified criteria for success and constraints on materials, time, or cost
2. Engage in conceptual planning	*Designing a solution:* Students articulate their own design ideas tailored for solving the problem at hand *Multiple solutions:* Students consider more than one way to solve the problem and use the fruits of brainstorming to adjust the problem scope *"Measured" decisions:* Students make thoughtful decisions about the solution based on the complex problem space; decisions are deliberate instead of random or ad hoc *Functionality:* Bringing math and science reasoning to bear on the problem, students think about how their design is really going to work *Ongoing planning:* Students engage in planning before and after building begins	*NGSS-ETS1.B: Developing possible solutions* K-2-ETS1-2. Develop a simple sketch, drawing, or physical model to illustrate how the shape of an object helps in function as needed to solve a given problem 3-5-ETS1-2. Generate and compare multiple possible solutions to a problem based on how well each is likely to meet the criteria and constraints of the problem
3. Realize and test their ideas	*Solution Validity:* Using physical tests and other types of evaluation, students obtain information about the validity of their solution *Idea refinement:* Students use these evaluations to refine their solution *Argumentation with evidence:* Students use evidence to argue for or against various ways to refine their solutions *Functionality:* Students continue to bring math and science reasoning to bear as they think about whether their design really works	*NGSS-ETS1.C: Optimizing the design solution* K-2-ETS-1-3. Analyze data from tests of two objects designed to solve the same problem to compare the strengths and weaknesses of how each performs 3-5-ETS-1-3. Plan and carry out fair tests in which variables are controlled and failure points are considered to identify aspects of a model or prototype that can be improved

problems, often in ways that are characteristic of more experienced or informed designers (Crismond and Adams 2012).

Some of the challenge, we suggest, is seeing nascent engineering in children's work. In the following sections, we show evidence of the beginnings of engineering, organizing our analysis by the three key aspects of (1) framing problems, (2) conceptual planning of solutions, and (3) realizing and testing ideas.

Framing Problems

When students frame engineering problems, they gather and use information to establish the scope of the problem they are solving, consider multiple sets of criteria and constraints, and take on different perspectives to determine what might count as a satisfactory solution.

Students gather information from the text, from the physical world, as well as from the teacher. This includes identifying the needs of the client and using these needs as problem constraints and criteria. For instance, in this chapter's opening vignette, the fifth-grade girls Samantha and Caroline took the needs of *The City of Ember* characters and the setting of the book into consideration as they planned their design.

They use this information to consider multiple sets of criteria and constraints. Samantha and Caroline thought about the situation within the book as well as within the classroom, thinking both about what the characters might do and what would be possible to prototype given the materials and time available.

Identifying criteria and constraints of a problem often involves thinking about the situation from multiple perspectives. Samantha and Caroline critiqued their ideas from the perspectives both of a classroom user and of a character in the book. They wanted to ensure that they and their classmates (and teachers) could physically manipulate what they built, but they also valued their design's fitting into the context of the book. They imagined the characters in *The City of Ember* using their design.

In these ways, we see evidence of the beginnings of engineering in elementary students. Here we present snippets of children's work from the project that illustrate problem framing in IEL classrooms.

Money Scooper

A fourth-grade class in a suburban school read *From the Mixed-Up Files of Mrs. Basil E. Frankweiler* by E. L. Konigsburg. In this book, two children, Jamie and Claudia, run away to the Metropolitan Museum of Art in New York City, where they live in hiding for several days. During whole-class discussions, the students and teacher listed problems encountered in the text, and then the students worked in groups on problems they chose from the list.

Sean and Zane chose the problem of helping the characters collect change from a fountain at the museum, which Jamie and Claudia did in the book when they needed money. The episode we consider here comes from a moment when Sean and Zane were starting to build one of their ideas for a solution, and a question about materials prompts them to return to framing the problem they are trying to solve.

In particular, the boys disagree about whether to include cotton balls in the money-collecting bag.

> Sean: We don't really need the cotton balls.
> Zane: Come on, let's keep that there, just to keep it quiet.
> Sean: No, we don't really need to keep it quiet. We don't need to keep the money quiet, though.
> Zane: Yeah, we do. Remember it was loud.
> Sean: That's probably someone else's (problem to work on).
> Zane: What?
> Sean: Keeping Jamie's money quiet, that's a whole 'nother thing [points to the paper across the room with the list of possible problems generated by the students]. That's a whole entire other problem.
> Zane: So? We can still do it. It'll be fine.

A few moments later, Sean takes on the perspective of the client as he describes how a difficult-to-use solution will not be helpful for the user. He says, "eventually when they're done taking the bath ..., then all that stuff is going to come with the money. And we're going to have to keep changing it over and over and over. That stuff isn't going to dry overnight."

Thus Sean and Zane were discussing the scope of their engineering problem: Is keeping quiet a relevant consideration of the problem they are solving? Attempting to narrow the scope of the problem motivated Sean to think more carefully about their proposed design and the cost to the clients—perspective taking—in ease of use. Zane was also taking the clients' perspective in suggesting that sound absorption is a valuable design feature. Both boys supported their positions based on the information from the text, as well as on their understandings of physical properties—that cotton balls would take time to dry, that they would dampen the sound of coins jingling.

The boys ended up resolving the disagreement by covering each of the cotton balls in duct tape, thinking the tape would keep the cotton balls dry while allowing them to dampen the sound of the coins.

Connecting to NGSS
The boys' awareness of relevant criteria and their work to scope the problem in this example illustrates NGSS 3–5 ETS1-1, defining a simple design problem reflecting a need or want that includes specified criteria for success and constraints on materials, time or cost.

Swan Swimmer

Another fourth-grade class, also suburban, read *The Trumpet of the Swan* by E. B. White. At the moment we examine, the students had generated a list of problems from the book, which was posted on chart paper. The teacher had asked the students to work in groups and think about several different problems and their potential solutions.

Alex and Jonathan have the idea to record their ideas about problems on Post-it notes, which they use to organize their discussion. They discuss both problems and potential solutions: helping the baby swans swim with a paddle boat, protecting the nest with a dome or security system, and providing water for the swans by filtering and suctioning the body of water.[1]

> Jonathan: [Looking at his Post-it notes] So I have an idea for um- … the raft idea.
> Alex: Let's- let's just narrow it- Let's do this first.
> Jonathan: I have a- I was thinking we could use a water bottle.
> Alex: Oh, and it would float around? Then how would they steer it?
> Jonathan: Ding ding ding! Paddle.
> Alex: But I don't know (unclear).
> Jonathan: Oh, true.
> Alex: But that's a good idea. Okay, let's try swimming first. Let's come- What do we have for swimming?
> Jonathan: We have the- this…
> Alex: Bike pedal, and my, thing.
> Alex: And… I have- like a wall around it maybe with some video cameras and stairs and a bear trap.
> Jonathan: Well, the thing is, we have to make these kinds of things.
> Alex: So protect nest, out of the question.
> Jonathan: I was thinking we could use a dome, like, out of, like, um- You know I don't know what to make it out of, but, a dome.
> Alex: Let's not- let's not do protect nest.
> Jonathan: Yeah okay, so that's out.
> Alex: Swimming.
> Jonathan: Um. Swimming, I- We already did swimming…. And I have this one that um, I was thinking they could find like a stick, and then they could find like, this, you know, like, how there are just cups floating around randomly, so I think they could just, like, use these. Use that and like…
> Alex: Oh.
> Jonathan: A lever. So a swan would push it.

---

[1] Ellipses (…) indicate that portions of the students' utterances have been omitted. In all cases, the meaning of the utterance is the same without or without the omitted text.

Alex: I came up with this. Like, we could maybe use a vacuum cleaner. It's like- so it filters the rocks on the bottom, and it gets water so the wa- the rocks won't go through. And just put water in a little bowl. Once you switch the lever, and this- You know the little toy cranes kids have?
Jonathan: The toy what?
Alex: Toy cranes that little kids have.
Jonathan: Yeah.
Alex: That aren't controlled by electricity. They pull the levers. One makes it raise; one makes it snap.

The boys' interaction shows evidence of their awareness of the need to define their problem. In lines 1 and 2, Jonathan seems to want to share details about his raft idea, but Alex recommends, "Let's just narrow it. Let's do this first." He appears to understand that limiting the scope of what they are working on is a task they need to accomplish before actually designing a device. Later, Jonathan joins the problem scoping effort by expressing concern about the idea of using video cameras, saying "we have to *make* these kind of things" (emphasis added). The boys decide that "protecting the nest" is too difficult a problem given the constraint of having to make something in the classroom.

Later they raise other constraints, including the materials available in the classroom and the lack of electricity in the world of the swans. These considerations affect their selection and scoping of the problem. The evidence shows their awareness that choosing and framing a problem involved negotiating several sets of constraints and criteria, including the available materials, the teacher's instructions, the context of the book, and the requirement for a functional prototype. We also like the snippet as an example of spontaneous, authentic literacy in the boys' idea to record ideas on Post-it notes.

Connecting to NGSS
Jonathan and Alex spend time defining and understanding the criteria and materials that help them define their problem. This is an example of NGSS 3–5 ETS1-1, defining a simple design problem reflecting a need or want that includes specified criteria for success and constraints on materials, time, or cost.

Conceptual Planning

The second key aspect of engineering we look for in students' work is conceptual planning of solutions. When students are engaging in conceptual planning, they are articulating ideas for possibilities, reflecting on their framing of the problem and drawing on their understanding of the world including of mathematics, science, and technology.

Peach Lifter

This snippet is from a fourth-grade class in an urban school. The class read *James and the Giant Peach* by Roald Dahl in small reading groups, the students writing problems they found on their own sheets of paper. Matt and Charles decide to work on the problem of the giant peach floating in the ocean. It and the characters were vulnerable to attack by sharks. Charles has the idea of devising a crane to lift the peach out of the water.

> Charles: How about- M, how about we create this type of crane thing to pick it up.
> Matt: I honestly wanna do that rope thing where they got to, like, wrap it around, and then they pick up- pick it up with umm, with like a stem kind of thing. And you can, like- it's a lever and it picks it up.
> Charles: Where are we going to get the lever?
> Matt: The lever- that can be the lever [sketches a straight line on the paper where they had been recording possible problems to solve], and then that can be the rope stuff.
> Charles: How about like we put something heavier than the thing [pointing to the sketch]. How about we-
> Matt: This is like- and then the- and that will be like the stem, and it'll be like a box thing, and this will be connected through [continues sketching his design as he talks]. So when-
> Charles: (overlap) How about, like, we tie something to something, and then put something heavier on it, so it- so that will go down, and the other, the peach, would go up.
> Matt: So, like, umm-
> Charles: So like a type of seesaw!
> Matt: Well, this [points to sketch] is my idea, and so you're thinking of this idea [begins to sketch a second design]. Like umm, platform thing, and … and then that can be longer, whatever. And so, you want the peach like that, and, like a net thing. And then you want like an enormous rock thing right here to like lift that.
> Charles: No, actually I want like this- I want like this. Can I-? [Takes the pencil and begins a third design sketch.] I want it like this, so watch. I want, a string tying to the peach, and then something heavier than the peach, lifting up.

The boys' conceptual planning continued into the next day, as they worked to better understand each other's ideas and their functionality, at one point grabbing a water bottle, pencil, and eraser for an impromptu test of a lever system (Fig. 11.3).

In these ways, their work is rich in evidence of this aspect of engineering, as well as of authentic literacy, in the their conversation, collaborative drawing, and manipulation of tangible materials to develop their ideas for lever system designs. We see them thinking of multiple possibilities (a crane and a "rope thing"—we suspect Matt was thinking of a winch), reasoning about mechanisms (e.g., the need to "put something heavier" than the peach if they hope to use a simple see saw lever), and conducting a simple experiment. There is also evidence of their working to under-

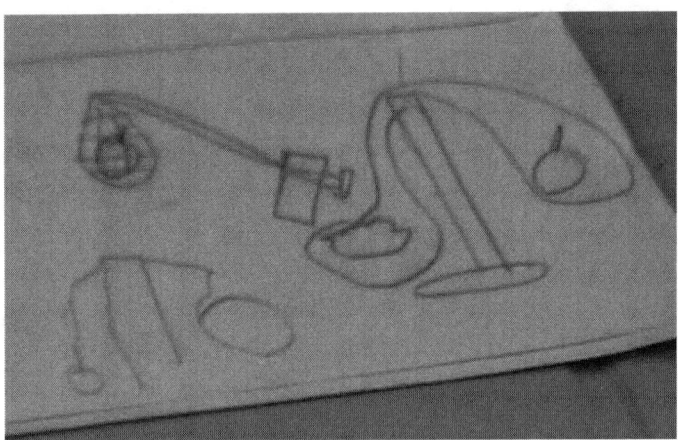

Fig. 11.3 Three sketches of peach lifter designs—on the *right* is the first sketch created by Matt to show his idea. At the *top left* is Matt's second sketch to show his understanding of Charles's idea. At the *bottom left* is Charles's sketch

stand each others' ideas (e.g., Matt in line 10, saying "this is my idea, and so you're thinking of this idea" as he sketches).

Connecting to NGSS
By comparing different solutions and their level of functionality in the world and the story, the group in this example demonstrated NGSS 2-5ETS1-2, generate and compare multiple solutions to a problem based on how well each is likely to meet the criteria and constraints of the problem.

Heat Sandwich

This example is from a fifth-grade classroom in a rural school district. The teacher chose an excerpt from *The Swiss Family Robinson* by Johann David Wyss, in particular because it describes the weather on the island and the challenges the family encountered. All students worked on the same problem, to design a shelter for the family that would keep them cool.

Jenn and Susan are working on their idea, which in this episode they are explaining to a member of our research team. The episode shows how they are drawing on their ideas about the physical world to inform their design. This exchange begins with Jenn explaining that felt (textile material) reflects heat.

> Jenn: So felt […] reflects back up any heat. Felt also reflects back cold as well. And I remember that cold air is less thick than heat, warm air.
> Researcher: Cold air is less thick than warm air?

Jenn: So we have a little chimney for the cold air to come in, but it'll be clogged up halfway with this stuff that's spread out so that the more cold air would be coming in than the warm air would.

Researcher: This is a really cool idea and I want to make sure I understand it…

Jenn: And heat ri-

Susan: And that way, and also it'll go out-

Jenn: And heat rises, so-

Researcher: Heat rises, so that's why you have the chimney on top? So heat will go out this way and cold air will come in that way.

Jenn: Cause heat air, I mean, well, cold goes down and heat goes up.

Researcher: Cold goes down, heat goes up. OK, so this is a way to have a, like, little path for them to go [hand gestures up and down]. And so, why do you have the cotton stuff on top of here?

Jenn: That's for insulation, just to make sure the air doesn't get through. I know, like, cause that's what you'd have to have, if this was made out of sticks like it's representing.

Researcher: This is sticks, not-

Jenn: This is supposed to be, like, tons of sticks and stuff, so.

Researcher: OK. Um, and you were saying something about- where's the felt going to go? It's on the bottom?

Jenn: Felt, there's some-

Susan: There's some on the bottom. We might do it on the side too, so it's hard harder for the heat (to get in).

Jenn: We also put some felt on our door so that the heat wouldn't get through the cracks.

Researcher: So you're trying to- what are you trying to do? Are you trying to keep it warm or cold?

Jenn: We're trying to keep it cold. We don't want the cold air since it's less dense to get out the cracks. And we want the heat air to go up there.

Researcher: You said that cold air is less dense, so-

Jenn: So it can fit through smaller things.

It may be difficult to recognize the nascent practices of engineering here, as Jenn and Susan draw on their understandings of heat and their design on three ideas:

"Heat" rises and cold air goes down.
 Hot air is "thicker" than cold air ("thicker" meaning something like viscous).
 "Heat" can be reflected.

Of course, there are difficulties with these ideas as science, but we are struck by the girls' direct use of them to inform their design of the structure. Thus they placed felt on the bottom of the shelter so that any heat from the ground will be reflected back to the ground rather than enter the shelter, and since they believe cold air is less "thick," they put cotton into the chimney in an effort to allow the cold air to come in, but to filter the "thicker" hot air from entering the structure. They also included insulation to keep the inside temperature constant and to block other air to enter the structure.

Without this conversation—were we only looking at the structure—we might have thought that they chose materials haphazardly and arranged them non-sensibly. With this discussion, we can see that they have put a great deal of thought into their design and used their understanding of the world to guide them. Correctness aside, Susan and Jenn are using their current best science ideas to plan carefully for a design solution that they think will function well. (Later, their measurement of the temperature gave them confirming evidence: They found it went down slightly in their structure!)

Connecting to NGSS
The girls in this example are using their understanding of the world to inform their design and to predict how it will function. Their activity connects to the Science and Engineering Practices promoted as part of NGSS.

Realizing and Testing Ideas

Finally, we show two snippets as examples of students' realizing and testing their design ideas. These come from our observations both of the students at work on their projects and of their presentations to their classmates.

Before we proceed to the examples, we pause with two caveats. First, not every IEL experience needs to include all three aspects. In particular, although every student in an IEL classroom should experience building a physical artifact at some point in the course of a school year, there is often not time for students to do justice to framing the problem, for example, and reach the stage of a working prototype. Indeed, much of the particular value of IEL is in the richness of the depiction children's literature provides of fictional clients, a context for cultivating empathy in design (e.g. Kouprie and Visser 2009).

Second, as we said above, we do not expect framing problems, conceptual planning, and realizing and testing ideas generally to occur in simple sequence. In practice, there is extensive overlap and iteration among these three aspects of engineering.

We turn now to the remaining two snippets.

Museum Backpack

This is from a fourth-grade suburban class that read *From the Mixed-Up Files of Mrs. Basil E. Frankweiler.* Ella and Laura constructed a prototype of their solution to the problem for Jamie and Claudia, the characters in the story who are hiding in the museum, of carrying twenty-five dollars in change they brought from home. Framing this problem, Ella and Laura discussed how Jamie and Claudia needed to be *inconspicuous*, a theme in the book that the class discussed (and a vocabulary word they were learning). This led them to design a backpack that would keep the

Fig. 11.4 Functionality of museum backpack being tested during a final presentation

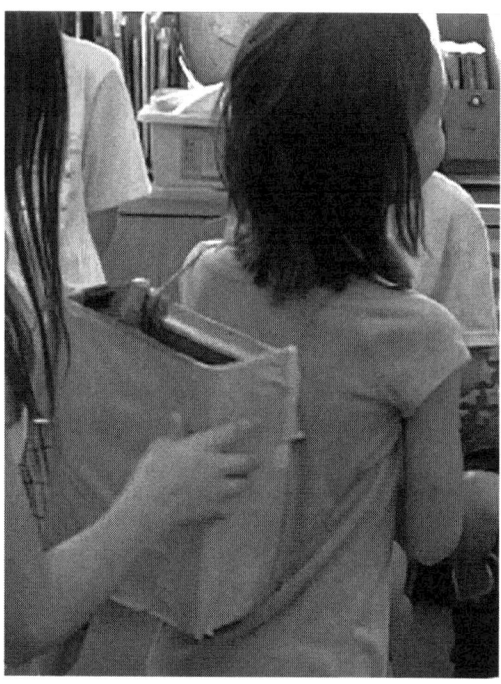

coins hidden and quiet. They have built and tested a prototype, showing in Fig. 11.4, which they are demonstrating to their classmates.

They have just shown how the backpack has space for clothes or food, with the false bottom hiding the space for coins, with insulation to dampen the noise.

>Kate: Won't that still make noise, the change?
>Ella: There's padding.
>Eric: Would people be able to see from the top?
>Laura: No, because there's going to be a cover here [gestures to the top of the backpack].
>Carola: I have a question. Why did you cover it with fabric?
>Ella: Oh, because if it was just cardboard, people would notice that cardboard was on their back and they would be suspicious.
>Teacher: What's the word?
>Ella: In-, con-, ceivable. Inconspicuous.
>Teacher: Good, so they want to be inconspicuous.
>Gabbie: Who's going to wear that?
>Ella: Both of them. Well, we didn't want to make it a purse because Jamie wouldn't-
>Laura: If Jamie was going to hold it that would be a little weird, so- [smiles]
>Ella: So girls could have worn this too.

The girls' work on their backpack, their peers' questions, and the girls' responses show the beginnings of engineering in several ways. We see Ella and Laura's care and thoughtfulness in choosing the components of the backpack, including selection of materials (cardboard for structure, padding for sound insulation, fabric for inconspicuousness). Their classmates ask pertinent questions, and the girls are able to respond, explaining features they have included such as a concealing lid, to address the concern Eric raised that people would be able to look down into the backpack. When Gabbie asks which of the two characters would wear the backpack, Ella and Laura explain that they chose a backpack rather than a purse, so that it would work for either character.

Connecting to NGSS

In Ella and Laura's presentation, we are able to see that they have considered multiple criteria in the development of their backpack and have made design choices based on the criteria that they have identified. Their activity shows they can meet NGSS 3–5 ETS1-1 performance expectation of students being able to define a simple design problem reflecting a need or want that includes specified criteria for success and constraints on materials, time, or cost.

Additionally, the girls are able to exhibit that they have tested their backpack prior to the presentation and have used identified criteria and constraints to inform their design. Their testing is in line with NGSS 3-5-ETS1-3 which sets the expectation that students plan and carry out fair tests in which variables are controlled and failure points are considered to identify aspects of a model or prototype that can be improved.

Water Filter

Our final example comes from a rural class of third graders who have read *If You Lived in Colonial Times* by Ann McGovern as part of their social studies curriculum. Adam, Owen, and Samuel are working together to construct a filter for the dirty water available to New England colonists. They have constructed a device with a paper towel tube lined in aluminum foil and filled with cotton balls. This tube empties into a basin made out of a box; the device is shown in Fig. 11.5.

Testing their prototype, they find water flows the way they planned, but there is leakage. Their first reaction is to cover everything with tin foil, but they find that does not solve the problem.

In the following excerpt the students are talking about their revised construction, having found there are still leaks, despite the additional foil.

> Owen: So maybe we should start with the new base, maybe in here?
> Adam: Well then we'll have, well, Owen, there's one problem with that. Then we'll have to take all the tin foil off, the pipes off...
> Owen: And we're not going to do too much.
> Samuel: I don't think that's enough.
> Adam: I think we should just stay with this.
> Samuel: Cause that's not really enough.

Fig. 11.5 Water filter

Owen: Well, so first of all we should see what we need to fix, right?
Adam: No, but that's for next time we make a pump, not, Colin, we're not making another one.
Owen: Guys, engineers don't always stay with what they made originally.
Adam: Well, yeah, but Ms. Mellin said we can't throw away our thing.
Owen: No, we're not going to throw it away, we can make a new one.
Samuel: Let's put the cotton balls on the top here and then it's just-
Owen: (overlap) OK, we'll just put all layers in it, 'kay? But we're laying more around it.
Teacher: What's the purpose of putting more layers on it?
Owen: Well, we're trying to keep it from leaking,
Adam: Owen, remember, from my personal experience of the past few hours, tin foil does not work as lining.
Owen: Tin foil does work. A bit.

The boys go through several cycles of testing and revision; here they discuss whether to change strategies entirely—whether the evidence from their tests rules out foil as a material to use as a lining. They are also weighing the advantages of starting over versus changing what they already have, given their results with this prototype, the constraints of time in the classroom, and the teacher's requests.

Eventually they solve some of the filter's structural issues to the satisfaction of some of the team members, and they move on to test the validity of its purification system, the cotton balls. They explore coffee filters and cloth as alternative particle strainers. But as they continue to test, they continue to be presented with information about the structural weakness of the filter tube. There is also evidence here of an essential feature of engineering, that tests often fail, and much of what happens is learning from failures and persisting with new ideas.

Connecting to NGSS
This group conducts many tests to assess the functionality of their water filter. The results of each test influences the changes they make to their design. This is an example of NGSS 3-5-ETS1-3, NGSS 3-5-ETS1-3, plan and carry out fair tests in which variables are controlled and failure points are considered to identify aspects of a model or prototype that can be improved.

Three Key Aspects Overview

We have chosen to present a series of quick snippets, rather than one or two extended accounts, in order to illustrate the range in evidence from students' work of the beginnings of engineering we have seen during IEL projects. We sorted them into three aspects, but as we noted, there is often evidence of more than one aspect of engineering. For example, Ella and Laura, the students who built the backpack, thought of new aspects of the problem framing and design concept as they built their prototype, such as realizing they had to cover the cardboard in fabric. Sean and Zane, the students who built the money scooper, were defining the problem and the needs of their clients as they were realizing and testing their scoop.

In this, we hope at once to have provided evidence of children's beginning abilities in engineering as well as to have illustrated the rich variety in how those abilities appear. Much depends on educators recognizing productive beginnings in children's thinking. Often, discussions about engineering in early grades build from views of children as lacking in abilities that instruction must provide. Part of the reason for this, we suggest, is that children's abilities can be difficult to recognize. Nascent engineering is not always obvious, such as for Jenn and Susan, who designed their shelter based on their rather different understandings of the properties of warm and cold air.

Another part of the reason for deficit views of children is in the structure of lesson implementation focused on achieving specific objectives more than on discovering the emergence of productive thinking. The complexity of students' engagement makes it difficult to control what and how they will think, which puts a greater priority on teachers' attention, interpretation and responsiveness to students' work as it unfolds. Some instruction will not tap into students' productive resources for engineering, because the students experience what they are doing as following directions, rather than as considering and designing to address the needs of their clients.

For these reasons, in IEL we emphasize students' intellectual agency within their activities, and we emphasize teachers' attention, interpretation, and responsiveness.

Considerations for Teachers

We have treated IEL as an approach, rather than a set curriculum, because we expect—and have seen— that there is great variety in how it can become part of a class's activities. For one, there is no specific library of IEL books; teachers can implement IEL with books that are already part of their curriculum. Nor are there fixed time requirements; a class might productively spend a half hour thinking about engineering problems in a text, or the class might spend many days developing, piloting, and revising ideas.

We have found that giving students the freedom to define their own problems and constraints allows for rich, productive environment that we believe is mutually beneficial to engineering and literacy. Teachers should work to create an environment where students feel comfortable "failing" in some ideas, learning from those failures, and trying again. In many respects, these values are at odds with the traditional values and structures of lessons in schools, in which failure is a terrible thing and students work toward specific, predetermined objectives.

Above we summarized how values within IEL align with standards, in particular in NGSS (Table 11.2). We have seen that in supportive classrooms where teachers make space for, recognize, and cultivate productive beginnings of students' thinking, students will naturally address standards of learning within IEL tasks. Conversely, we have seen constraining students too closely, such as with explicit, prescriptive links to particular standards, is counter-productive. "Good" students work to be obedient rather than agentive.

For example, we've found that when a formal discussion or lecture about the engineering design process takes place, the students tend to focus more on figuring out if they are doing the "right" step or if they are implementing the process correctly, rather than focusing on the design problem itself. Further, by having a diversity of problems and solutions, peers provide the questions and argumentation that helps the inventors think more critically about their project and actively promotes finding evidence to support the invention.

In what follows, we discuss considerations for teachers in implementing IEL. To ground this discussion, we focus on a fourth-grade class that read *The Trumpet of the Swan*, from which we drew the Swan Swimmer snippet above. Along the way, we will also comment on related moves other teachers have made in other classrooms. Ms. Jackson is the classroom teacher with six years of classroom teaching experience and no formal engineering training. The experience discussed below comes from her third year of doing IEL with her students.

Choosing a Book

Teachers in IEL classrooms have used a variety of books of differing genres and lengths. Ms. Jackson chose *The Trumpet of the Swan* because she knew the characters were accessible to the students, and she was able to anticipate several problems that they would find interesting to solve and for which they would be able to engineer functional solutions.

In general, we have seen that books most conducive to engineering are those that most engender students' empathy for the characters. Typically these are more realistic books about children their age. However, books with animal characters and historical or futuristic human characters can also support good engineering. With some books, such as *The Trumpet of the Swan*, we have found teachers need to clarify assumptions about the role and abilities of the characters, comparing and contrasting them to what the student engineers in the classroom are able to do.

One key decision to make, either group-by-group or for the class as a whole, is whether the students are engineering a device *for* the characters, or whether they are engineering a device that the characters themselves could make. If the latter, then students need to pay attention to which materials are available in the book, and what capabilities the characters have for manipulating materials. For example, if engineering an alarm that Louis, the main swan in *The Trumpet of the Swan*, could make himself, students need to know whether they can assume Louis can use his beak as a tool! When using magical or fantastical books, it is easy for students, and adults, to begin thinking of fantastical solutions since the context of the book is magical or fantastical. We have found that when using fantastical books, it makes a significant difference to the activity when teachers ask that solutions actually function in the classroom.

In some classes, students did not read full books but only excerpts, such as excerpts presented in anthologies, such as from *The Swiss Family Robinson* above. A challenge to these, as well as to some nonfiction, is that they do not provide as thorough a depiction of the characters and setting. Teachers may want to supplement the reading with discussion that could fill out the setting more thoroughly, about the clients and their needs.

Reading and Discussing

As Ms. Jackson and her students read *The Trumpet of the Swan*, they followed their usual format for book discussions. They stopped after chapters, augmenting their usual discussions by considering problems the characters were having. Ms. Jackson kept a list of the problems the students generated on a large sheet of paper.

Another teacher encouraged the students to list problems the characters encountered in each chapter as they read in small reading groups. They referred to their lists when they met with the whole class and discussed problems and how it would affect the characters and plot of the book.

Not only can book discussions foster a greater understating of the text and help students frame problems, but they can also help the students understand what is being asked of them in the engineering activities. Since the openness that accompanies IEL projects is often a new experience for students, supporting them as they figure out what they should be doing will help them understand more quickly, such as by giving them time to play around with brainstorming.

Reading and discussing the book most naturally generates problem framing, of the aspects we discussed, but it can also lead to conceptual planning. As the students are listing problems, some teachers make space for them to start talking about possible solutions. This can provide an opportunity to consider possible criteria and values for solutions, which can be especially valuable for fantasy books. Early discussions in classes reading *James and the Giant Peach* by Dahl and *Tuck Everlasting* by Natalie Babbitt included fantastical solutions such as shrink rays and time machines. This provided the teachers opportunities to discuss what they would do going forward. For example, one teacher listed ideas on the board in two columns: "can be built in this classroom" and "cannot be built in the classroom because it needs more technology." This afforded recognizing students' thinking while at the same time raised awareness of the practical constraints of the classroom, supplied as it was with only non-magical materials.

Eliciting Student Planning

As we described above, Ms. Jackson initiated planning by having students write ideas on Post-it notes. Each group of students picked several problems and brainstormed possible solutions. Some teachers have had students pick one problem and brainstorm multiple solutions. Prompts either on the board or on a sheet of paper can be helpful in outlining the process: What problem will you solve? How will you solve it? What materials will you need?

In the classroom reading *If You Lived in Colonial Times*, the teacher told the students that they needed to come up with a certain number of ideas for their solution. The students did so, but came up with some outrageous ideas because they had already devised some feasible ideas and needed to finish the task of listing solutions; this might be an example of how students can experience a task as inauthentic. (Of course, sometimes playing with outlandish ideas can inspire new insights and ideas.)

Giving students feedback on their plan before they become too attached to it is important. One way to do this is for each group of students to share their plan with the whole class or in smaller groups as a design review. For planning related to the story *America's Champion Swimmer: Gertrude Ederle*, the teacher had her students devise individual plans first and then had them bring those to the group.

Students often take charge of these discussions and hold each other accountable to the constraints of the classroom and the story by questioning the presenters about the feasibility of the designs and usability by the characters. Helping students think about the information and constraints presented in the book during feedback can help students flesh out their ideas for engineering solutions and keep them thinking about the text. These discussions can also be another opportunity for discussing the nature of the task.

Students may engage in planning throughout the entire design process. As they find elements of a design that do not work or change materials, they may need to go back to planning. Another aspect of planning is that planning looks different for each student and at different stages during their design process. Planning may include talking, drawing, or manipulating materials to convey ideas.

Choosing Materials

Material choice has a great impact on how students plan and the types of designs they create. It is not only important to think about the types of materials that will be available to the students, but also who decides what the materials are and when the students will know what material will be accessible to them. If identifying materials is part of the planning process, then teachers need to (1) build in lag time to be able to obtain materials (2) decide whether it is the teacher's or student's responsibility to bring in materials, and (3) decide what range of materials and tools they will make available (e.g., if students ask for wood and a saw, will that be possible?).

In most IEL classrooms, materials used for building consist of an array of inexpensive and recyclable materials such as paper egg cartons, towel rolls, tape, rubber bands, Popsicle sticks, and straws. Some teachers choose to include commercially available materials such as pre-made pulleys, Makedo kits, or Tinkertoys. While found materials are inexpensive, abundant, and encourage creativity, students may spend additional time constructing something that is readily available in a pre-fabricated form or more quickly constructed with a more costly alternative. For example, if students spend time building a pulley, rather than using a pre-constructed pulley, this will take time away from other aspects of their design. Teachers will need to consider where they want the students' efforts to be focused.

Sometimes the materials lead the students' design rather than it growing from the story and classroom constraints. In some classrooms it has been necessary to "outlaw" certain materials. In one classroom working from *The Mouse and the Motorcycle* by Beverly Cleary, one group of students incorporated a balloon into their design and suddenly other groups decided that they needed balloons too, although it made no sense for their designs. In a fourth-grade group reading *Esperanza Rising* by Pam Muñoz Ryan, students found a material that was highly attractive to them and added it to their design, resulting in a hot tub to be placed on the roof of a family of migrant workers' house!

Building

In Ms. Jackson's class, once she felt that an individual group had a coherent plan that they could explain to her, she allowed them to gather their materials from a table on the side of the room and begin building. They spent an hour each day over several afternoons building, testing, and iterating on their ideas—sometimes getting more or different materials based on input from their testing.

Each participating teacher has structured building time differently to work with their students' experience levels and their classroom schedule. In one fifth grade classroom, students had numerous design experiences and were able to plan and finish their prototypes in three hours. Less experienced students will need additional time to plan and to get feedback on their plans before moving forward with building.

Evaluation and Testing

When students were testing their designs for *The Trumpet of the Swan*, each group devised a different test or tests to address the function(s) of their prototype. They tested as they built, at crucial points in their design, rather than at the end after they completed their prototype. This allowed them to revise and redesign based on the feedback they got from the tests. Ms. Jackson walked around the classroom asking her students, "How are you going to prove to me that it works?" Sometimes the students would have an immediate answer and other times it would prompt a conversation about the functionality of the design.

In another classroom, students were working on similar projects to each other so the teacher told them that they would each need to use the same test (using a small electronic bug which jiggled about randomly) to simulate a dog in order to test the stability of a dog pen. This unfortunately resulted in some inauthentic testing that gave little genuine feedback: the bug was not to scale with the pens, and for at least some students represented a significant departure from the situation of their clients. In another classroom reading *The Miraculous Journey of Edward Tulane* by Kate DiCamillo, the class used a tub of water to symbolize the ocean during testing. The students developed devices to be used in the ocean, but, as some students complained, the walls of the tub would not exist in the ocean, and some designs made use of the walls for leverage. Students ended up being confused about which constraints (the walls of the tub or the openness of the ocean) they should consider as they moved forward with redesigns.

At this point, we would like to note that failure is a frequent and necessary element of design projects. Teachers should take care to foster an environment where students are comfortable with failure and learn to see that failure gives feedback that can be used to improve their designs.

We've seen a link between the language teachers use when talking about tests and the way students approach testing. Ms. Jackson was very clear when describing the task and when talking to the students, saying that what they built needed to work. She placed focus on a functional prototype, asking students how they would know it was working rather than asking students to explain specific tests before they began building. We've seen instances when students are tasked with explaining how they will test their prototype and devise tests to fulfill the requirement rather than to inform their design. This is especially detrimental when students are asked to define their test during the planning phase when they don't have a full understanding of how their prototype will work.

Using a magical book does not necessarily lend itself to magical prototypes if classroom expectations are clearly outlined to students. The boys that built the Peach Lifter were working within the magical *James and the Giant Peach*, but conducted several tests that gave them feedback regarding the functionality of their prototype. It is important that students have the chance to iterate on their design based on testing feedback.

Mid-design share-outs offer another type of evaluation of their design. The advantage of sharing at the mid-way point, rather than during a final presentation, is that students are able to incorporate the suggestions and ponder the comments of the other students. Including multiple design reviews is a common practice with professional engineers and we've seen that students reap the same benefits with their inclusion, as do professional engineers. In order to counteract the need for students to feel as if they must be able to answer every questions, it is helpful for students to be told that it's okay to answer "I don't know" when asked a question about how their prototype will work. If students feel comfortable not appearing that they have an answer to every question, a more fruitful discussion can occur.

Sharing/Presenting/Reflecting

At the completion of *The Trumpet of the Swan* task, Ms. Jackson's students presented their final prototypes to the other students. Each group of students shared their prototype and answered questions from the rest of the students. Presenting at the end of building is an exciting way for students to proudly share their design. Students are able to reflect on their process and see how other students have approached the problems in the book. A disadvantage of final presentations is that students are not able to change their designs based on feedback from other students, hence the need for periodic design reviews during the process.

Many teachers in their second and third years of IEL have chosen that their students do not participate in final presentations and have opted for mid-design share-outs. The advantage of the mid-design share out is that that students are able to alter their designs based on feedback from peers. This more closely mirrors the process in which experienced engineers participate.

Teachers have chosen alternative ways for the students to reflect and present in place of a final presentation. Students reading *Tuck Everlasting* used an iPad app to document their design processes and presented these to the class. Other teachers have incorporated a stronger writing component and asked students to write a final chapter describing what could have happened had their design existed in the book or to make a comic strip. Written products, movies, posters, or other compositions based on their projects can help students reflect on their engineering practice and support their literacy learning. Student engagement during writing projects often reflects their excitement for building and offers an extension to express thoughts about their projects.

Closing Remarks

>Alice: This is hard.
>Jen: This is *really* hard. I love it though!

While we have mostly focused on students' disciplinary thinking and engagement in engineering and literacy, what perhaps stands out most in IEL classrooms is how invested the students are in their projects. IEL projects can be challenging, but we have seen the students remain engaged and persist, even in the face of setbacks or failed tests. We've seen students stay in during recess or lunch to work on their projects. Some students continue to work on their designs at home—even after the final presentation. Parents have often mentioned IEL projects as the focus of their own conversations with their children. One student even used a computer program to model his design and talk about what features he might add beyond the feasibility of the prototype he developed in class. These examples highlight how IEL projects offer students an opportunity to not just learn about the steps of a design process, but to be excited about engineering as their own pursuit to develop new ideas and realize them.

As policy recommendations and standards are beginning to include more engineering as a part of K-12 education, IEL offers a way to leverage existing classroom literacy activities for authentic engagement in engineering. The interdisciplinary nature of IEL offers students a multiplicity of problems and solutions to solve, while deepening their engagement with books and other texts. By focusing on the productive resources students display for engaging in both domains, we have shown how students can enter into engineering practices within the rich context of literature.

Our approach to IEL has been to place student thinking in design as the centerpiece for both research and teaching. Therefore, we do not provide a prescriptive curriculum, but a more responsive approach to examining and anticipating student thinking and engagement. Our goal is to help teachers give students opportunities to explore, create, fail, iterate, and evaluate. While choices when setting-up IEL tasks are important, teachers should take care to allow students the freedom to build on their own ideas as they move through their engineering experience. By listening to and watching students, teachers will be able to find and promote moments of beginning engineering in their students. With each design experience, students will build on their emerging engineering and be able to incorporate more aspects of design thinking into their work.

In the coming year, we plan to work with students in grades one through six to find out what IEL looks like in grades other than three through five. We will look more closely at phenomena such as mid-design share-outs, the inclusion of LEGO bricks, and student planning. Additionally, based on a previous pilot study, we will continue to work with students that have reading and writing difficulties in an effort to better understand how we can better support this population, supporting their reading and writing, as well as their engineering efforts. Since responses to IEL by students, teachers, and administrators have been so supportive, we will work to widen our dissemination efforts, promoting IEL as an approach rather than a set curriculum. We would like to include a gallery that documents student ideas and work as related to specific texts.

References

Crismond, D. P., & Adams, R. S. (2012). A scholarship of integration: The matrix of informed design. *Journal of Engineering Education, 101*(4), 738–797.
Dorst, K., & Cross, N. (2001). Creativity in the design process: Co-evolution of problem–solution. *Design Studies, 22*(5), 425–437.
French, M. J. (1998). *Conceptual design for engineers*. London: Springer.
Jonassen, D. H., Strobel, J., & Lee, C. B. (2006). Everyday problem solving in engineering: Lessons for engineering educators. *Journal of Engineering Education, 95*(2), 139–151.
Kouprie, M., & Visser, F. S. (2009). A framework for empathy in design: Stepping into and out of the user's life. *Journal of Engineering Design, 20*(5), 437–448.
McCormick, R., Murphy, P., & Hennessy, S. (1994). Problem-solving processes in technology education: A pilot study. *International Journal of Technology and Design Education, 4*, 5–34.
National Governors Association Center for Best Practices & Council of Chief State School Officers. (2010). *Common core state standards for english language arts and literacy in history/social studies, science, and technical subjects*. Washington, DC: National Governors Association Center for Best Practices & Council of Chief State School Officers.
Pahl, G., & Beitz, W. (1984). *Engineering design*. New York: Springer.
Portsmore, M. (2008). *Exploring first grade students' planning in an engineering design problem and its relationships to artifact construction and success: A pilot study*. Unpublished qualifying paper, Tufts University, Medford.
Portsmore, M. (2010). *Exploring how experience with planning impacts first grade students' planning and solutions to engineering design problems*. Unpublished doctoral dissertation, Tufts University, Medford.
Schön, D. A. (1983). *The reflective practitioner: How professionals think in action*. New York: Basic Books.
Welch, M. (1999). Analyzing the tacit strategies of novice designers. *Research in Science & Technology Education, 17*(1), 19–34.

Chapter 12
Teaching Engineering Design in Elementary Science Methods Classes

Christine D. Tippett

> *Fusing engineering education with other subjects, such as mathematics and science, is an essential first step in promoting preservice teachers' potential to implement engineering education.*
>
> (Hudson et al. 2009, p. 165)

Introduction

One of the challenges currently facing science teacher educators in the United States and Canada (where I am located) is how best to prepare preservice teachers for the demands of a science curriculum that includes engineering design. Integrating engineering design into science education is not a new idea. *Science for All Americans* (American Association for the Advancement of Science 1989) included a chapter on the designed world that addresses agriculture, materials and manufacturing, energy sources and use, communication, information processing, and health technology. However, the publication of the *Next Generation Science Standards* (NGSS, Achieve, Inc. 2013a, b) places a greater emphasis on engineering design, and with the adoption of the *NGSS*, science teachers at all levels will be encouraged to include aspects of engineering in their classrooms.

Teachers of every subject at every grade level are expected to follow the curriculum mandated by their state department of education (in the United States) or their provincial ministry of education (in Canada), and recent science standards and science curriculum documents in both countries emphasize the inclusion of engineering design. In the United States, the *Framework for K–12 Science Education* (2012) and the *NGSS* (Achieve, Inc. 2013a, b) emphasize the relationships between scientific and engineering practices. Canadian science curriculum documents vary by province but are based on a document produced by the Council of Ministers of

---

C.D. Tippett (✉)
University of Ottawa, Ottawa, Ontario, Canada
e-mail: ctippett@uottawa.ca

Education Canada called *The Common Framework of Science Learning Outcomes K to 12* (CMEC 1997), which includes technology as one of four foundational pillars.

In the near future, many science education programs in the United States are likely to be based upon the *NGSS* (Achieve, Inc. 2013a, b), which delineates crosscutting concepts, disciplinary core ideas, and scientific and engineering practices. The emphasis the *NGSS* places on engineering is leading to a renewed awareness of engineering design in the Canadian education system. In-service and preservice teachers will need professional development that parallels the approach in the *NGSS* – engineering design as distinct from science inquiry yet closely connected. Where and how will teachers acquire the education and experience necessary to teach engineering design? There are three main possibilities: in-service professional development, school programs for K-12 students, or preservice education. Professional development for in-service teachers does help address the needs of experienced classroom teachers for whom the addition or inclusion of technology and engineering requires rethinking their current pedagogical practices, and there is a small but growing number of in-service professional development opportunities that highlight engineering design at the K-12 level. School programs aimed at K-12 students may offer the teachers of those students some opportunity to gain experience teaching engineering design. However, preservice teachers are expected to teach engineering design, too, during practicum placements and once they are certified and as a result, teacher education programs must provide opportunities for preservice teachers to experience engineering activities and learn how to teach engineering design.

In this chapter, I make a case for including engineering in elementary science methods courses, describe my initial attempts to incorporate engineering design, and outline a research program that will be based on the results of those attempts.

Why Include Engineering in Elementary Science Methods Courses?

The rationale for including engineering in elementary science methods courses has three underlying assumptions: (1) engineering should be taught in elementary school; (2) the subject of science is an appropriate place in which to situate engineering education; and (3) preservice teachers need to learn how to integrate engineering design in their science teaching. I address each of these assumptions in turn as I make the claim that science methods courses are an appropriate context for preservice teachers to learn how to integrate engineering and science.

Assumption 1 *Engineering should be taught in elementary school*

Many countries around the world, including Canada and the United States, have predicted a shortage of engineers and other STEM professionals (Charette 2013). Although there is some debate about whether this shortage actually exists

(Charette 2013; Engineers Canada, 2015; Sargent 2013), most people would likely agree that STEM literacy, or the lack of it, is a pressing issue in education. If today's K-12 students do not acquire an adequate foundation in science, math, and engineering, there will most definitely be a shortage of STEM professionals, and engineers and scientists are essential in developing and maintaining technological leadership and innovation, in manufacturing and other service areas, playing a vital role in "economic strength, national defense, and other societal needs" (Sargent 2013, p. 27). In addition, Charette (2013) notes that "improving everyone's STEM skills would clearly be good for the workforce and for people's employment prospects, for public policy debates, and for everyday tasks like balancing checkbooks and calculating risks" (p. 59).

It is hard to argue against teaching engineering, but when should engineering introduced? The answer appears to be as early as possible, with effective programs and approaches for students as young as pre-Kindergarten being described in the literature. "Learning engineering requires identifying opportunities to conceive of something new, comprehending how something works, and researching and applying knowledge to construct something novel and appropriate for others. Young children can engage in these activities and appear to be quite motivated and adept at doing so" (Brophy et al. 2008, p. 384). Children are naturally curious about how things work and they engage in informal engineering activities all the time (Museum of Science, Boston 2013). Petroski (2003) called children "born engineers" (p. 206), but noted that while children "experience the essence of engineering in their earliest activities" (p. 206) they rarely recognize that what they are doing is related to engineering. Sullivan (2006) pointed out that learning about engineering in the elementary grades can provide students with a more realistic picture of what engineering is and what engineers do, leading to more students considering engineering as a career and "helping them to recognize the complexities of contemporary issues, engage intelligently in the discourse of our times, and make informed choices that take future generations into consideration" (p. 6).

Assumption 2 *Science is an appropriate context in which to situate engineering education*

If we are to teach engineering in elementary schools, where should it be situated in an already crowded curriculum? Many researchers have suggested that engineering activities can act as a gateway for learning science, providing opportunities for the application of science concepts in ways that students find relevant and meaningful. As pointed out earlier, the idea of integrating engineering (or technological design) and science is not a new approach, but engineering is still frequently overlooked (or underemphasized) in STEM education, particularly at the elementary level (Brophy et al. 2008; ITEA 2009). However, teachers may already be using engineering design activities in their science classrooms, but often without identifying those activities explicitly, or perhaps even being aware of the distinction themselves (Brophy et al. 2008; Bybee 1998). Identifying activities appropriately, pointing out when students are engaged in engineering, and using "a common language across grade levels for both scientific and engineering practices and crosscutting concepts" (NRC 2012, p. 259) would highlight the engineering that is already being taught without taking time away from science instruction. Engineering design activities can help students to understand science concepts, especially if those concepts are explicitly pointed out at appropriate points during the design process (Brophy et al. 2008). Engineering activities can help to make science concepts relevant, and increased relevance enhances learning (Holbrook and Rannikmae 2009). For example, Redmond et al. (2011) found that using engineering activities that emphasized real-world applications of science and mathematics concepts had a significant positive impact on Grade 6 and 7 students' confidence in science and mathematics as well as their awareness of, and interest in, engineering as a career.

Assumption 3 *Preservice teachers need to learn how to integrate engineering design*

It seems that combining teaching about engineering with science education is an effective way for teachers to make the most of their instructional time. However, Brophy et al. (2008) conclude a review of promising integrated engineering instructional models for students in pre-kindergarten through Grade 12 by pointing out that many teachers lack the knowledge and experience to comfortably implement these models. They note that:

> Teachers are typically uncomfortable teaching content they do not understand well and thus they will often shy away from such content for fear of being unable to answer students' questions. This may be a particularly significant problem for K-8 teachers who are attempting to deal with engineering content and the processes of design and inquiry accompanying the learning of such content. (Brophy et al. 2008, p. 381)

Indeed, Hudson et al. (2009) suggest that the attributes of self-efficacy, confidence, and enthusiasm are foundational for teaching engineering and that those attributes can be affected by preservice experiences. Preservice teachers need opportunities to develop their understanding of the crosscutting concepts, disciplinary core ideas, and scientific and engineering practices, as well as experiences that will "help them understand how students think, what they are capable of doing, and what they might reasonably be expected to do under supportive instructional conditions" (NRC 2012, p. 257).

Assumption 4 *Engineering design should be integrated in elementary science methods courses*

Teaching engineering design to elementary students has the potential to make science concepts more relevant and preservice teachers need experiences that will build their confidence in and enthusiasm for teaching engineering. It is reasonable to assume that if integrating engineering with science is beneficial for elementary students that integrating engineering and science in elementary science methods courses will also be effective. Preservice teachers could participate in the kinds of activities that they might use with their own students and see firsthand how they might use an engineering design context to make science concepts relevant. Engaging in engineering activities during science methods courses would be likely to help preservice teachers to develop both the science and the engineering practices outlined by the NRC (2010). A combination of engineering and science activities could create opportunities for students in elementary science methods classes to discover that technology is more than computers and cell phones, to learn about natural phenomena, and to apply problem solving skills in contexts that mirror how technologists, engineers, and scientists solve problems, think critically, construct explanations, communicate information, and engage in reasoned argument in the real world (DiBiase 2001).

Incorporating engineering activities in elementary science methods courses could provide opportunities to explicitly consider the differences between engineering design and science inquiry, while similarities can also be examined (Bybee 1998; Capobianco 2012). The engineering design process could be contrasted against systematic scientific inquiry that seeks information about a natural phenomenon. Additionally, many engineering design activities lead naturally to science inquiries as questions arise during the problem solving process (Brophy et al. 2008). Incorporating engineering activities in science methods courses would allow teacher educators to emphasize engineering, technology, and science relationships and applications with preservice teachers (NGSS, Achieve, Inc. 2013a, b).

As promising as an integrated approach to engineering and science methods appears, to-date little research has examined such an approach for preservice elementary teachers. Berlin and White (2012), in their review of research on preservice teacher education programs that emphasized integration of science, mathematics, and technology, focused almost exclusively on math and science within STEM while neglecting engineering entirely, which suggests that few preservice teacher programs have attempted to integrate engineering with science methods courses and that even fewer of those attempts have been examined formally. Notable exceptions include studies by Capobianco (2012), Culver (2012), and High and Dockers (2007) who have worked with elementary preservice teachers in the United States, and by Hudson et al. (2009), who have worked with preservice middle school teachers in Australia.

Capobianco (2012) developed a science methods course, *Learning to teach science through design in the elementary school*, that would use five different engineering activities as the foundation for teaching and learning science. It was intended that preservice teachers would learn how to teach science inquiry and engineering design while considering the similarities and differences of the two processes. At the end of the one semester course, participants were better able to follow the engineering design process and to use key skills such as teamwork and communication. Participants reported an increased interest in design activities and an increased enthusiasm for including engineering design in their own classrooms.

High and Dockers (2007) examined the effects of additional instruction in engineering in an elementary science methods course. They found that the 23 preservice teachers in the innovative course had increased confidence in teaching engineering concepts compared to the 25 preservice teachers who took a regular science methods course, with no loss of confidence in teaching science concepts.

Hudson et al. (2009) investigated 17 preservice teachers' confidence in, and enthusiasm for, teaching engineering at the middle school level. Preservice teachers participated in two engineering activities during their science methods course and then taught similar activities to Grade 7 students. Results of a pre-post Likert scale assessment suggested that preservice teachers' confidence in teaching engineering increased significantly, and that they were more motivated to teach an integration of science and engineering.

Culver (2012) examined the views of 44 preservice elementary teachers enrolled in a more traditional science methods course during which they completed one engineering design task. After engaging in the task, participants felt that integrating engineering and science would be a realistic approach to take in elementary classrooms. These preservice teachers also felt that a methods course that emphasized the integration of engineering and science would be more beneficial than a separate engineering methods course.

Although the body of research is limited to-date, indications from these studies are that incorporating engineering activities in science methods courses can lead to increased preservice teacher confidence in teaching both science and engineering. There is an increased push for incorporating engineering design in science methods that arises from the publication of the *Framework* (NRC 2012), which points out that "science teacher preparation must develop teachers' focus on, and deepen their understanding of the crosscutting concepts, disciplinary core ideas, and scientific and engineering practices so as to better engage their students in these dimensions" (p. 257).

My Work

With the renewed emphasis on engineering design that has arisen with the publication of the *Framework* (NRC 2012) and the *NGSS* (Achieve, Inc. 2013a, b), I am interested in exploring approaches to more effectively integrate engineering in my own science methods courses. I work with preservice teachers who are enrolled in a 1 year post degree education program. The program includes one required science methods course that consists of 36 h of instruction occurring in weekly blocks of 3–3 1/2 h, depending on the number of classes scheduled during a particular term. I want to examine preservice teachers' beliefs about science, technology, and engineering while identifying activities and approaches that increase their confidence in teaching both science and engineering.

I am based in Canada, where education is a provincial responsibility and the national foundational document for science curriculum is *The Common Framework of Science Learning Outcomes K to 12* (CMEC 1997). Because I teach in Ontario, the provincially mandated curriculum document for elementary science that I use in my courses is *The Ontario Curriculum, Grades 1–8: Science and Technology* (OME 2007), which delineates three goals:

1. to relate science and technology to society and the environment

2. to develop the skills, strategies, and habits of mind required for scientific inquiry and technological problem solving
3. to understand the basic concepts of science and technology (OME 2007, p. 3)

Since Ontario teachers, like teachers across North America and Europe, are expected to follow locally mandated curriculum, they are expected to address technology and technological problem solving in their science instruction. While the answer to *Where and how do teachers acquire the education and experience necessary to teach engineering design?* might consist of three possibilities: in-service professional development, school programs for K-12 students, or preservice education, in Ontario, science methods courses are the obvious place in which to situate preservice teachers' engineering and design education because of the interrelated nature of science and engineering as conceptualized in foundational and curriculum documents.

Development of a Research Program

To-date, little has been written about engineering design in preservice classrooms, and even less has been written about what approaches might most effectively introduce or build on the differences in and relationships between science, technology, and engineering. Therefore, a thorough exploration of the multiple factors that may influence preservice teachers' ideas about, and confidence in, teaching science and engineering is warranted. Kelley (2012) notes "If researchers of STEM fail to thoroughly investigate the complexities surrounding teacher practices that integrate STEM, the recent efforts to infuse STEM education into the classroom will be void" (p. 35). Although Kelley is writing about the relationship between technology and engineering, his warning can be applied to the examination of any integrated aspects of STEM, or STSE (Science, Technology, Society, and the Environment).

In the remainder of this chapter, I focus on the development of a program of research, currently in preliminary stages, that will examine the consequences of embedding engineering design in elementary science methods courses. The question guiding my planning is *What is entailed in a systematic exploration of the impact of specific activities upon preservice elementary science teachers' conceptions of engineering?* The answer to this question will help shape the specific approaches to be taken during the next phase of research, in which guiding questions will likely include:

- What contextual challenges or constraints might be involved in teaching engineering design in science methods courses?
- What activities can be used to differentiate engineering design from science inquiry in a way that is meaningful and memorable for preservice teachers?
- What changes occur in preservice teachers' conceptions of engineering design when those activities are implemented?

For example, early indications are that the contextual challenges and constraints of teaching engineering design include preservice teachers' uncertainty about what engineering actually is. Based on these preliminary findings, I plan to gather more detailed information about preservice teachers' views of science, technology and engineering, using measures such as the Views on Science-Technology-Society (VOSTS, Aikenhead and Ryan 1992) and the Draw-A-Scientist Test (DAST, Chambers 1983), as well as by modifying the DAST and asking preservice teachers to draw an engineer.

The Trouble with Terminology: Is It Technology or Engineering?

Based on what I've observed and overheard in my science methods classroom over the past 5 years, both in British Columbia and Ontario, it appears that many of the preservice teachers I have worked have some uncertainty about what science, technology, and engineering are and how they might be related. That uncertainty exists, in part, because of varied and sometimes imprecise use of terminology, a problem pointed out in the *NGSS* where these key terms are carefully defined to align with current usage (Achieve, Inc. 2013a, b). In Canada, the foundational science curriculum document contains the following definition: "Technology, like science, is a creative human activity with a long history in all cultures of the world. Technology is concerned mainly with proposing solutions to problems arising from human adaptation to the environment" (CMEC 1997, p. 9). This definition is clear and concise, but according to current usage it is actually a definition of engineering design rather than technology (Achieve, Inc. 2013b, p. 103).

Another example of the trouble with terminology: I currently teach in Ontario, Canada where the mandated curriculum document for elementary science is *Science and Technology* (OME 2007). In this document, 'technology' is used to indicate the design process as well as the tools created through that process, although it most frequently refers to the process. At the secondary level, however, the mandated curriculum document is *Science* (OME 2008a, b), while the *Technology* program (OME 2009a, b) deals with communications technology, computer technology, construction technology, green industries, hairstyling and aesthetics, health care, hospitality and tourism, manufacturing technology, technological design, transportation technology – topics typically considered industrial or practical arts. This inconsistent use of the term technology in provincial curriculum documents has certainly lead to some confusion among the preservice teachers I work with in Ontario, with a fair number of them thinking that a methods course in Science and Technology will emphasize learning about educational technology (e.g., interactive white boards) or the use of ICT (e.g., tablets) in the context of science.

This confusion or uncertainty about technology and engineering has been reported by many researchers, including Constantinou et al. (2010) who examined the views of 183 elementary students, 132 middle school students, and 78 preservice elementary teachers. They found that regardless of age group, students had similar difficulties differentiating between science and engineering, with many participants overstating the role of experiments in science, confusing the use of engineering with manufacturing, lacking awareness of the importance of creativity, and conflating the use of technology with engineering design. Culver (2012) found that preservice teachers equated engineering with construction and engineering design with a process of trial and error. Hsu et al. (2011) found that teachers identified engineering with constructing and building, instead of a more comprehensive process that includes identifying a problem, planning solutions, constructing a model, analyzing, and repeating as necessary. Rose et al. (2004), found that adults have a narrow view of technology, perceiving technology to be ICT. Similarly, Cunningham et al. (2005) explored children's conceptions of engineering and technology. They found that students tended to link engineering with construction and technology with items that used electricity. Fewer than one third of students considered everyday objects such as bicycles and cups to be examples of technology.

To ensure a consist use of the terms science, technology, and engineering in this chapter, I compiled definitions from four influential North American standards and curriculum documents, as shown in Table 12.1. I compared these definitions, identified commonalities, and developed the following operational definitions:

Science is not just a body of knowledge that reflects current understanding of the natural world; it is also a set of practices used to establish, extend, and refine that knowledge. It is a human and social activity that is based on curiosity, creativity, imagination, intuition, exploration, observation, replication of experiments, interpretation of evidence, and debate over the evidence and its interpretations.

Technology results when engineers apply their understanding of the natural world and of human behavior to design ways to satisfy human needs and wants. Technology includes all types of human-made systems and processes – not just computers or electronic devices, which is a limited view often used in schools.

Engineering involves both knowledge and a set of practices. The major goal of engineering is to solve problems that arise from a specific human need or desire. To do this, engineers rely on their knowledge of science and mathematics as well as their understanding of the engineering design process. The term "engineering design" has replaced the older term "technological design," consistent with the definition of engineering as a systematic practice for solving problems, and technology as the result of that practice.

Observations About Teaching Engineering Design

In this section, I describe how I have approached the topic of engineering design in recent elementary science methods courses. Although the teaching sequence shown in Fig. 12.1 is the sequence I followed the last time I taught, the anecdotal

Table 12.1 Definitions of science, technology, and engineering from four foundational North American science education documents

Source	The common framework of science learning outcomes K-12 (Council of Ministers of Education Canada 1997)	The Ontario curriculum grades 1–8: science and technology (Ontario Ministry of Education 2007)	A framework for K-12 science education: practices, crosscutting concepts, and core ideas (National Research Council 2012)	Next generation science standards: Vol. 2. Appendixes (Achieve, Inc. 2013b)
Science	Science is a human and social activity with unique characteristics and a long history that has involved many men and women from many societies. Science is also a way of learning about the universe based on curiosity, creativity, imagination, intuition, exploration, observation, replication of experiments, interpretation of evidence, and debate over the evidence and its interpretations (p. 9)	Science is a way of knowing that seeks to describe and explain the natural and physical world. Occasionally, theories and concepts undergo change but, for the most part, the basic ideas of science – ideas such as the cellular basis of life, the laws of energy, and the particle theory of matter – have proven to be stable (p. 4)	In the K-12 context, science is generally taken to mean the traditional natural sciences: physics, chemistry, biology, and (more recently) earth, space, and environmental sciences (p. 11). Science is not just a body of knowledge that reflects current understanding of the world; it is also a set of practices used to establish, extend, and refine that knowledge. Both elements – knowledge and practice – are essential (p. 26)	Science is a way of explaining the natural world. [It] is both a set of practices and the historical accumulation of knowledge. An essential part of science education is learning science and engineering practices and developing knowledge of the concepts that are foundational to science disciplines (p. 96)

(continued)

Table 12.1 (continued)

Source	The common framework of science learning outcomes K-12 (Council of Ministers of Education Canada 1997)	The Ontario curriculum grades 1–8: science and technology (Ontario Ministry of Education 2007)	A framework for K-12 science education: practices, crosscutting concepts, and core ideas (National Research Council 2012)	Next generation science standards: Vol. 2. Appendixes (Achieve, Inc. 2013b)
Technology	Technology, like science, is a creative human activity with a long history in all cultures of the world. Technology is concerned mainly with proposing solutions to problems arising from human adaptation to the environment. Since there are many possible solutions, there are inevitably many requirements, objectives, and constraints. Hence, the chief concern of technologists is to develop optimal solutions that represent a balance of costs and benefits to society, the economy, and the environment (p. 9)	Technology is … a way of knowing, and is also a process of exploration and experimentation. Technology is both a form of knowledge that uses concepts and skills from other disciplines (including science) and the application of this knowledge to meet an identified need or to solve a specific problem using materials, energy, and tools (including computers). Technological methods consist of inventing or modifying devices, structures, systems, and/or processes (p. 4)	Technology … include[s] all types of human-made systems and processes – not in the limited sense often used in schools that equates technology with modern computational and communications devices. Technologies result when engineers apply their understanding of the natural world and of human behavior to design ways to satisfy human needs and wants (pp. 11–12)	Technology describes all the ways that people have modified the natural world to meet their needs and wants. Technology does not just refer to computers or electronic devices (p. 103)

(continued)

Table 12.1 (continued)

Source	*The common framework of science learning outcomes K-12* (Council of Ministers of Education Canada 1997)	*The Ontario curriculum grades 1–8: science and technology* (Ontario Ministry of Education 2007)	*A framework for K-12 science education: practices, crosscutting concepts, and core ideas* (National Research Council 2012)	*Next generation science standards: Vol. 2. Appendixes* (Achieve, Inc. 2013b)
Engineering	Not given	Not given	Engineering involves both knowledge and a set of practices. The major goal of engineering is to solve problems that arise from a specific human need or desire. To do this, engineers rely on their knowledge of science and mathematics as well as their understanding of the engineering design process (p. 27)	The term "engineering design" has replaced the older term "technological design," consistent with the definition of engineering as a systematic practice for solving problems, and technology as the result of that practice (p. 103)

observations that I share are an amalgamation from a number of previous courses; in other words, the responses I relate here may not be associated with this particular teaching sequence. I've chosen to present preservice teacher reactions this way because I am reflecting informally on what I have seen across cases rather than reporting the results of a formal research program. My accumulated observations are, in fact, the platform upon which I am designing an upcoming project.

I have used each of the activities presented here more than once although each time I teach about engineering, I adapt the activities and the sequence to reflect my current understanding of terminology and effective activities, as well as to accommodate the needs of my students. For example, the last time I taught, I was especially consciously of distinguishing between technology and engineering explicitly and frequently.

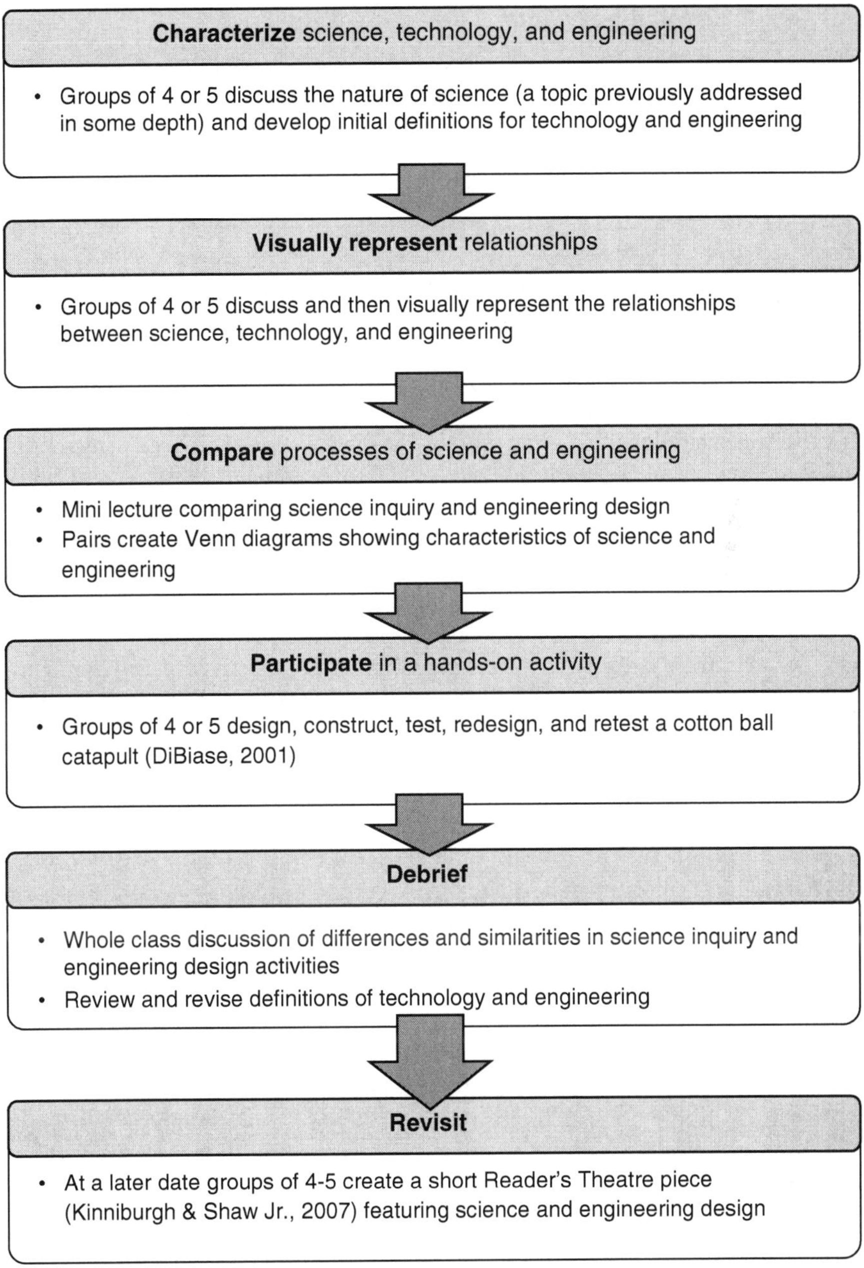

Characterize science, technology, and engineering

- Groups of 4 or 5 discuss the nature of science (a topic previously addressed in some depth) and develop initial definitions for technology and engineering

Visually represent relationships

- Groups of 4 or 5 discuss and then visually represent the relationships between science, technology, and engineering

Compare processes of science and engineering

- Mini lecture comparing science inquiry and engineering design
- Pairs create Venn diagrams showing characteristics of science and engineering

Participate in a hands-on activity

- Groups of 4 or 5 design, construct, test, redesign, and retest a cotton ball catapult (DiBiase, 2001)

Debrief

- Whole class discussion of differences and similarities in science inquiry and engineering design activities
- Review and revise definitions of technology and engineering

Revisit

- At a later date groups of 4-5 create a short Reader's Theatre piece (Kinniburgh & Shaw Jr., 2007) featuring science and engineering design

Fig. 12.1 Focusing on engineering design: a typical teaching sequence

A Typical Teaching Sequence

After several classes that focus on science with an emphasis on inquiry, investigation, and answering questions, I introduce the topics of technology and engineering design. Preservice teachers participate in a hands-on engineering design activity that involves constructing a cotton ball catapult (Dibiase 2001), I supplement this hands-on experience with additional activities that are intended to highlight the differences and similarities of science inquiry and engineering design, such as a mini lecture on the nature of science inquiry (answering questions) and engineering design (solving problems) and small group creation of Reader's Theatre (Kinniburgh and Shaw 2007) pieces featuring science and engineering design.

I typically begin the class with an explicit unpacking of terminology. First, I ask students to discuss **science** – we have spent several weeks positioning science as a creative endeavor of inquiry as we seek answers to questions about our natural environment, and small group discussions usually yield similar definitions. Then, I introduce **engineering**, which usually ends up defined as applied science – a definition that may be contested by some readers, but it serves our needs at this point – which is the process of designing solutions to problems. That leaves **technology** as … It is often at this point that things get tricky! Preservice teachers' suggestions normally include information communication technology (ICT), instructional technology like whiteboards and assistive software, occupational skills and the industrial arts, or problem solving and design – which presents some difficulty, given how we have just defined engineering.

I ask preservice teachers to continue their small group discussions and come up with a definition of technology that accommodates our definitions of science and engineering. My observations suggest that during this small group activity, preservice teachers begin to appreciate the difference between using technology the noun, which is its current usage, and technology the process, which is an outdated usage that has been replaced by engineering or engineering design (Achieve, Inc. 2013b).

In the next activity, small groups create a visual representation of the relationships between science, technology, and engineering, based on the definitions that they have just constructed. Results here are varied, with representations including webs, diagrams, flow charts, and occasionally a pictorial metaphor such as a faucet with running water. Regardless of form, for the most part these representations do show the kind of relationships that can be seen in Fig. 12.2. However, when I do show Fig. 12.2, there is always someone who points out that the arrows really should go both ways, suggesting a high level of understanding of the ways in which science, engineering, and technology influence, and are influenced by, each other.

The next step is to review the phases of scientific inquiry and contrast those phases with the engineering design process. In a previous class focusing on the Nature of Science, we will have discussed the lack of a single scientific method. The phases in science inquiry can be conceptualized, as shown in Fig. 12.3, as a set of common actions or activities (Aiken 1991). There is no single linear progression, but during a typical science inquiry, each of these phases is likely to be undertaken

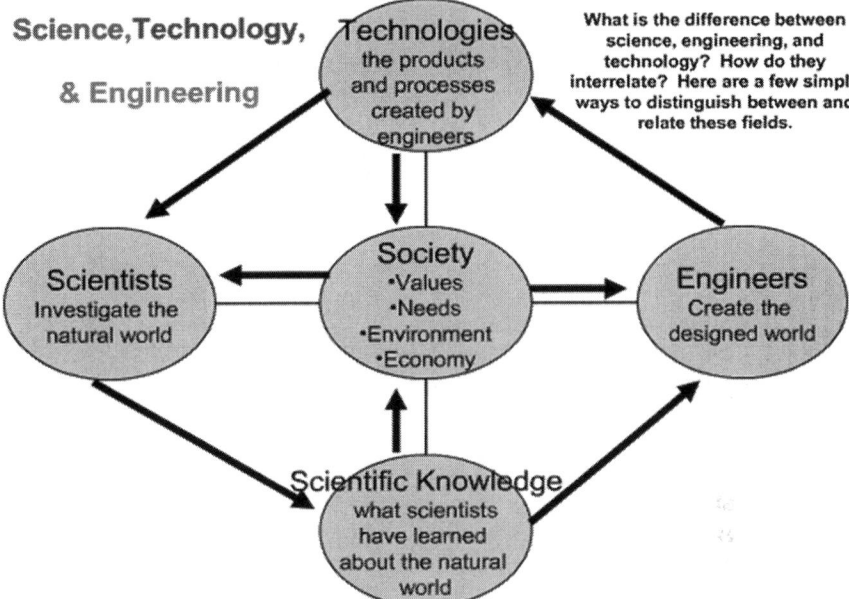

Fig. 12.2 The relationship between science, technology, engineering, and society (Retrieved from http://www2.chicousd.org/dna/libraries/For_Staff.html)

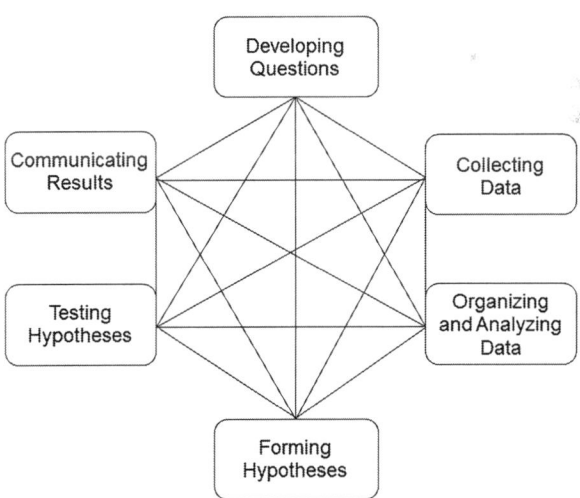

Fig. 12.3 Phases in science inquiry. *Lines* represent two-way paths between phases

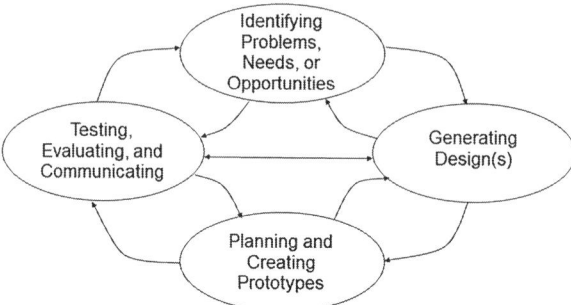

Fig. 12.4 Phases in the engineering design process. The testing and planning phases are typically iterative

at some point. Similarly, the engineering design process consists of a set of common actions that may occur recursively throughout the process, as shown in Fig. 12.4.

In the last activity before they undertake a hand-on engineering design task, I ask preservice teachers to work in pairs to complete a Venn diagram comparing science and engineering.

The activity that serves as the hands-on focal point of my instruction around the concept of engineering design with preservice elementary teachers is the Cotton Ball Catapult (DiBiase 2001). I've used the activity a number of times because it requires easily found materials, has a clear problem solving emphasis (as opposed to an investigative aspect), and follows a recursive sequence. The Cotton Ball Catapult activity was developed to be representative of an interdisciplinary STS activity, with technology defined as involving problems of adapting to our environment (DiBiase 2001) which, using the terminology established earlier in this chapter, is equivalent to engineering design. In the activity, students are presented with a problem along with materials that can be used when creating a solution to the problem and constraints to be considered during the problem solving process. The problem is to build a catapult that will fling a cotton ball the farthest, with the additional constraints including using all of the provided materials, and only those materials, while observing safety considerations (e.g., no thumb tack in the cotton ball). The steps in the activity, as I present it, are: given the problem, design and build a prototype catapult; test the prototype against other groups' creations; critique the prototype, based on test results; redesign, rebuild, and test a new and improved model.

During the Cotton Ball Catapult activity, preservice teachers typically appear to be highly motivated and engaged. Indicators include the participation of most (usually all) group members during the planning and creating stages, and particularly high participation during the rebuild stage. Groups are focused on the task at hand, with most discussion centering on the problem. The energy level during the trials is also high as preservice teachers compete to see whose catapult is best. The iterative nature of the task, requiring rebuilding and retesting, is often identified by preservice teachers as key in maintaining or heightening engagement. In the second trial of catapult designs, most designs are more successful than the initial prototype, even if success is still limited compared to the 'best' catapult.

Following the retest of the catapults, I debrief the preservice teachers, asking them to define terms, compare the phases involved in science inquiry and engineering design, and to think about the various phases as they were portrayed in the catapult design process as compared to how previous inquiry based activities that we have done. When preservice teachers are getting ready to leave at the end of class, I normally hear comments like *That was a fun class*, and *Boy, time went quickly*.

A week or two later I implement the final activity intended to emphasize engineering design and contrast it with science inquiry. I introduce the idea of Reader's Theatre (Kinniburgh and Shaw 2007) and after reading through a couple of examples as a whole class, I ask preservice teachers to work in small groups to create their own short Reader's Theatre piece about science and engineering. This activity is typically undertaken with a great deal of enthusiasm, although I can't be sure whether the enthusiasm is entirely due to the Reader's Theatre approach, or if some of the enthusiasm is carried over from the Cotton Ball Catapult activity.

At the end of a recent term I did a content analysis of preservice teachers' science notebooks, looking for entries that addressed science and engineering. Figures 12.5 and 12.6 are examples of science notebook entries, used here with permission. During this particular term, each of 37 preservice teachers made 5 notebook entries, for a total of 185 entries. When I read through the notebooks, I found 8 entries that were about the Cotton Ball Catapult activity specifically or engineering design in general (22 % of students, 4 % of entries). This small number wasn't too worrisome – there were numerous activities and experiences that the preservice teachers could have chosen to reflect upon, and 4 % of the overall entries meant that 22 % of the preservice teachers had selected engineering design. Unfortunately, a closer inspection of those entries revealed some misconceptions: the concepts of science inquiry and engineering design were explained correctly but then applied incorrectly when recounting the Cotton Ball Catapult activity, the Cotton Ball was described step-by-step but with no links to engineering design or science inquiry, or the Cotton Ball Catapult activity was presented as an example of the process of scientific inquiry. Only two entries accurately captured the nature of engineering design as problem solving contrasted with science as inquiry.

Implications for Teaching Science Methods Courses

Science notebook entries, though analyzed informally, suggest that preservice teachers' notions of science inquiry and engineering design are difficult to refine. "Without sufficient opportunities to 'Do S&T', students … may be unlikely to develop realistic conceptions of 'Characteristics of S&T'" (Bencze 2010, p. 46). Therefore a single engineering activity in a science methods course is unlikely to be sufficient to address any misconceptions or to help preservice teachers construct accurate understandings of the engineering design process. Science methods courses

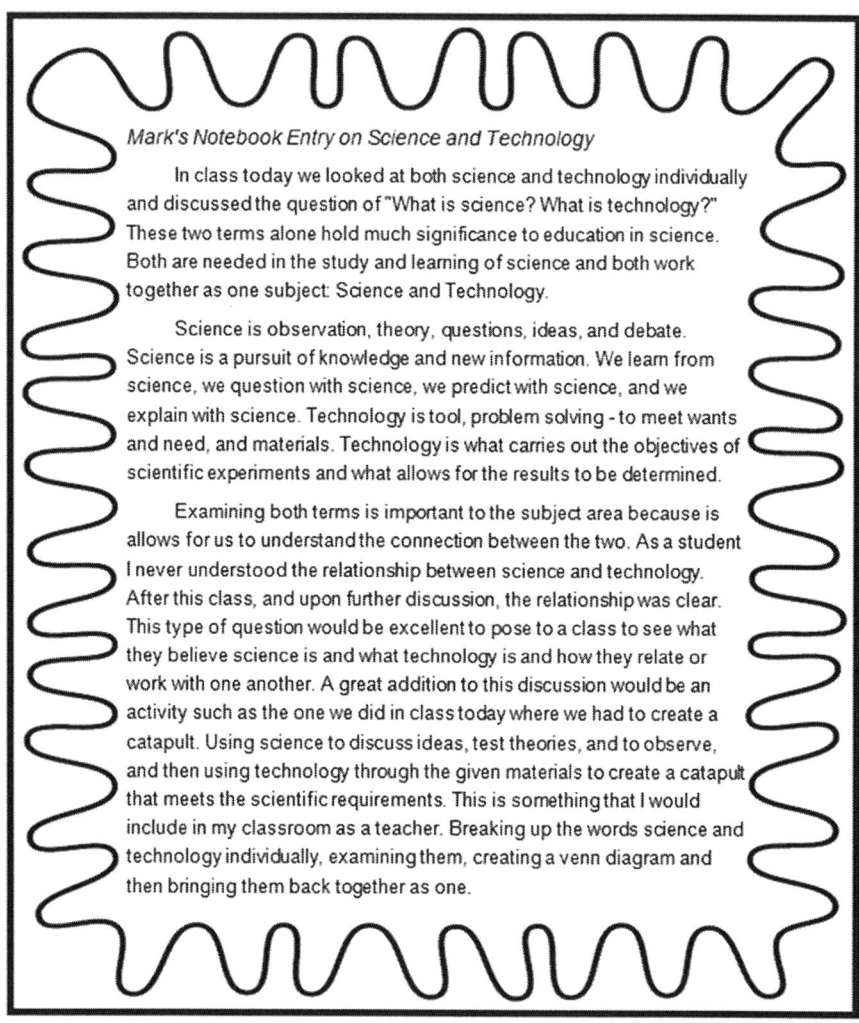

Fig. 12.5 Mark's science notebook entry about science and technology

likely need to include multiple hands-on activities that emphasize engineering design.

Although it is clear that preservice teachers have some specific preconceptions (and misconceptions!) about engineering and technology, it is not as clear exactly what those preconceptions are. Assessing preservice teachers' conceptions of science, technology, and engineering and their ideas about the relationships between those conceptions at the beginning of term would allow science teacher educators to take a constructivist approach to teaching engineering design. Assessment could be done more formally with a pencil and paper measure, or more informally as a discussion.

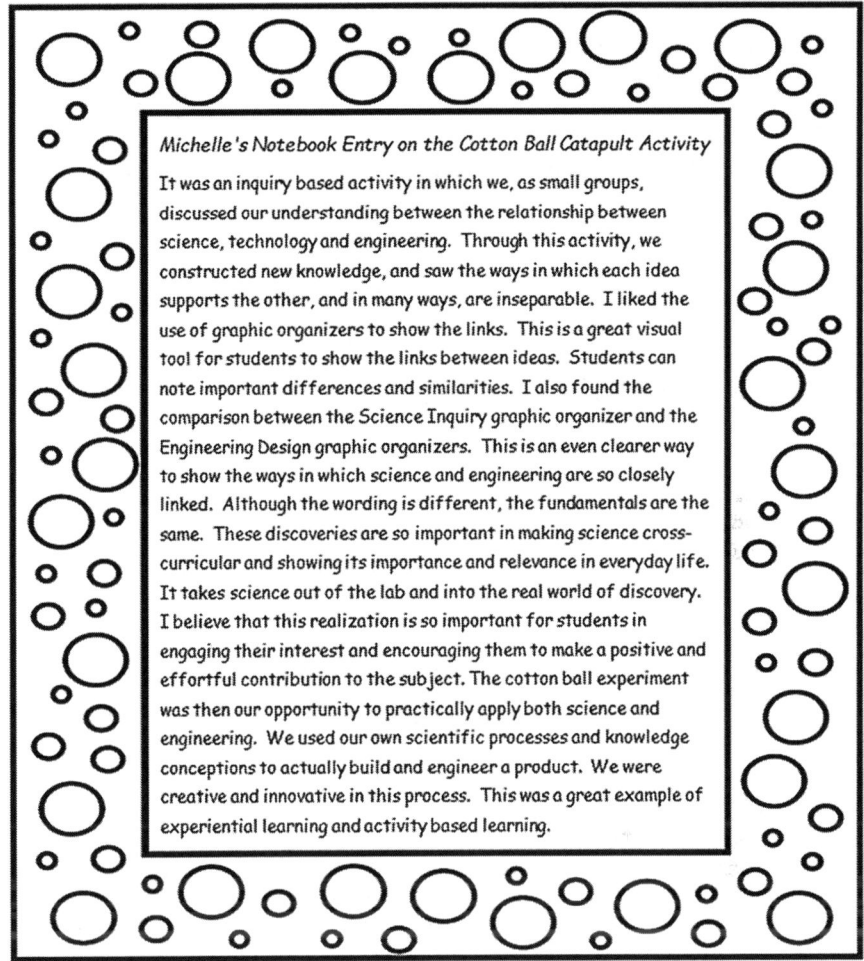

Fig. 12.6 Michelle's science notebook entry on the cotton ball catapult activity

Terminology is a definite area of concern, and imprecise or inconsistent use can contribute to misconceptions. To maintain consistency with current usage, ***technology*** should always be used as a noun. In addition, preservice teachers' notions of technology the noun tend to be limited to ICT or devices that use electricity. Explicitly positioning ICT as one small aspect of technology might help address these limited notions.

Although the following list of needs was developed with elementary and high school science teachers in mind the list is equally appropriate for science teacher educators:

- understand the scientific ideas and practices they are expected to teach
- have an appreciation of how scientists collaborate to develop new theories, models, and explanations of natural phenomena

- understand what initial ideas students bring to school
- know how to best develop an understanding of scientific and engineering practices, crosscutting concepts, and disciplinary core ideas across multiple grades
- understand how to move students along a developmental progression of practices, crosscutting concepts, and core ideas,
- possess science-specific pedagogical content knowledge, such as the ability to recognize common prescientific notions that underlie a student's questions or models
- be able to choose the pedagogical approaches that can build on those notions while moving students toward greater scientific understanding
- understand the scientific and engineering practices, crosscutting concepts, and disciplinary core ideas; how students learn them; and the range of instructional strategies that can support their learning

know how to use student-developed models, classroom discourse, and other formative assessment approaches to gauge student thinking and design further instruction (National Research Council 2012, p. 256)

Further Research

While engineering and technology are unlikely to be given equal status with science in the science classroom, there is space available for incorporating engineering into science instruction. In many cases, science teachers and science teacher educators are already including aspects of engineering and technology in their science classes but without acknowledging or perhaps even being aware of the distinctions between science and engineering. While the *NGSS* (Achieve, Inc. 2013a, b) and the *Framework* (NRC 2012) both newly emphasize engineering design, other curriculum documents, such as Canada's *Common Framework of Science Learning Outcomes K to 12* (1997) and Ontario's *Science and Technology K-8* (2007) already address the relationships between science, technology, and engineering.

There is a real need for evidence to support anecdotally promising approaches to teaching engineering design in science methods courses. Currently, there are only a handful of research projects examining in-service professional development that highlights engineering design or documenting engineering design experiences for elementary and high school students. Even fewer projects are focused on preservice teachers, and with the complexities involved in investigating the incorporation of engineering design in science methods courses, there is great potential for future research. Questions might include:

- How have the *Framework* (NRC 2012) and the *NGSS* (Achieve, Inc. 2013a, b) helped science teacher educators to improve elementary methods courses?
- How can elementary science methods courses most effectively incorporate engineering design for elementary preservice teachers?

- What experiences lead to increased preservice teacher confidence in teaching engineering in elementary science classrooms?
- How might teacher educators foster preservice teachers' ability to design and implement engineering activities in the science classroom?
- What level of teacher preparation is necessary for teachers to adopt and adapt existing engineering curriculum materials?

My Own Next Steps

Based on the early evidence reported above I am planning a formal research project examining preservice teachers' conceptions of engineering and how those conceptions might be affected by activities and experiences in a science methods course. This project will involve most, if not all, of the following components:

- Pre and post assessments of preservice teachers' conceptions of science, technology, and engineering, using measures such as the Views on Science-Technology-Society (VOSTS, Aikenhead and Ryan 1992), the Draw-A-Scientist Test (DAST, Chambers 1983), and/or the Draw-an-Engineer Test (DAET, Diefes-Dux and Capobianco 2011)
- Formal observations of preservice teachers' participation in multiple engineering design activities such as the Cotton Ball Catapult as well as some of the activities described by Capobianco (2012)
- Content analysis of end-of-course reflections or notebook entries
- Analysis of a questionnaire on the engineering design activities, with the goal of highlighting the activities that preservice teachers find most memorable and meaningful as well as identifying the activities that lead to more accurate views of science, technology, and engineering
- Semistructured interviews with a small group of preservice teachers, with the goal of further exploring activities that effectively differentiate between science inquiry and engineering design
- The research project will address the second question above, effective incorporation of activities, and may give some insights into the third and fourth questions, confidence in teaching and ability to incorporate engineering design.

Concluding Remarks

Preservice teachers need multiple, on-going opportunities to learn about engineering design and how to incorporate it in their classroom. The best place to situate engineering activities is in established science methods courses so that preservice teachers have opportunities to develop the practices outlined in recent standards, such as the *The Common Framework of Science Learning Outcomes K to 12* (CMEC 1997), *The Ontario Curriculum, Grades 1–8 Science and technology* (OME 2007), *A Framework for K-12 Science Education: Practices, Crosscutting Concepts, and Core Ideas* (NRC 2012), and the *Next Generation Science Standards* (Achieve, Inc. 2013a, b).

When teaching engineering design in science methods courses, teacher educators must be aware of local contextual constraints such as curriculum documents' use of technology as a noun and a verb. Differences between engineering design and science inquiry should be explicitly considered by preservice teachers, and those differences are best examined in a science methods course.

Many engineering activities can lead naturally to learning science concepts as questions arise during the problem solving process. Providing a balance of engineering and science activities in elementary science methods courses could create opportunities for preservice teachers to discover that technology is more than ICT, and to learn science concepts as they engage in the engineering design process to solve problems, think critically, and construct and communicate explanations. Incorporating carefully planned engineering activities in elementary science methods course could allow preservice teachers to develop the science and engineering practices outlined by the NRC (2012) and offer space for science teacher educators to emphasize engineering, technology, and science applications (NGSS, Achieve, Inc. 2013a, b). Of course, in order to determine what 'carefully planned' might entail, we need a great deal of additional information in order to understand our preservice teachers' current understandings of and beliefs about science, technology, and engineering; to identify engineering activities that will facilitate growth in understanding of science and engineering knowledge and skills; and to provide opportunities for preservice teachers to apply their knowledge while practicing skills.

Preservice teachers need opportunities to develop their understanding of engineering design and to investigate how science and engineering are related. They need hands–on engineering design experiences in order to be adequately prepared to integrate engineering design into their science classes. An appropriate place for these opportunities and experiences is the science methods course, but more research is needed so that we can identify best practices for elementary science teacher educators.

References

Achieve, Inc. (2013a). *Next generation science standards: Vol. 1. The standards*. Washington, DC: The National Academies Press.
Achieve, Inc. (2013b). *Next generation science standards: Vol. 2. Appendixes*. Washington, DC: The National Academies Press.
Aiken, F. (1991). *The nature of science*. Portsmouth: Heinemann.
Aikenhead, G. S., & Ryan, A. G. (1992). The development of a new instrument: "Views on science-technology-society" (VOSTS). *Science Education, 76*(5), 477–491.

American Association for the Advancement of Science. (1989). *Science for all Americans*. New York: Oxford University Press. Available at http://www.project2061.org/publications/sfaa/online/sfaatoc.htm

Bencze, J. L. (2010). Promoting student-led science and technology projects in elementary teacher education: Entry into core pedagogical practices through technological design. *International Journal of Technological Design Education, 20*, 43–62. doi:10.1007/s10798-008-9063-7.

Berlin, D. F., & White, A. L. (2012). A longitudinal look at attitudes and perceptions related to the integration of mathematics, science, and technology education. *School Science and Mathematics, 112*(1), 20–30. doi:10.1111/j.1949-8594.2011.00111.x.

Brophy, S., Klein, S., Portsmore, M., & Rogers, C. (2008). Advancing engineering education in P-12 classrooms. *Journal of Engineering Education, 97*(3), 369–387. doi:10.1002/j.2168-9830.2008.tb00985.x.

Bybee, R. W. (1998). Bridging science and technology. *The Science Teacher, 65*(6), 38–42.

Capobianco, B. (2012, January). *Investigating essential features for successful implementation of an elementary science methods course on engineering*. Paper presented at the annual meeting for the Association of Science Teacher Education, Clearwater. Retrieved from https://stemedhub.org/resources/680

Chambers, D. W. (1983). Stereotypic image of the scientist: The draw-a-scientist test. *Science Education, 67*(2), 255–265.

Charette, R. N. (2013). The STEM crisis is a myth. *IEEE Spectrum, 50*(9), 44–59. doi:10.1109/MSPEC.2013.6587189.

Constantinou, C., Hadjilouca, R., & Papadouris, N. (2010). Students' epistemological awareness concerning the distinction between science and technology. *International Journal of Science Education, 32*(2), 143–172. doi:10.1080/09500690903229296.

Council of Ministers of Education Canada. (1997). *The common framework of science learning outcomes K to 12*. Retrieved from http://publications.cmec.ca/science/framework/

Culver, D. E. (2012). *A qualitative assessment of preservice elementary teachers' formative perceptions regarding engineering and K-12 engineering education*. Master's thesis. Retrieved from http://lib.dr.iastate.edu/etd (Paper 12888).

Cunningham, C. M., Lachapelle, C. P., & Lindgren-Streicher, A. (2005, June). *Assessing elementary school students' conceptions of engineering and technology*. Paper presented at the American Society for Engineering Education Annual Conference, Portland.

DiBiase, W. J. (2001). Constructing a cotton-ball catapult: An interdisciplinary STS learning experience. *Science Activities, 38*(1), 11–16.

Diefes-Dux, H. A., & Capobianco, B. M. (2011). *Work in progress: Interpreting elementary students advanced conceptions of engineering from the Draw-an-Engineer Test*. Frontiers in Education Conference, F3J-1-F3J-2.

Engineers Canada. (2015). Engineering labour market in Canada: Projections to 2025. Retrieved from http://www.engineerscanada.ca/sites/default/files/Labour-Market-2015-e.pdf

Holbrook, J., & Rannikmae, M. (2009). The meaning of scientific literacy. *International Journal of Environmental and Science Education, 4*, 275–288.

Hsu, M.-C., Purzer, S., & Cardella, M. E. (2011). Elementary teachers' views about teaching design, engineering, and technology. *Journal of Pre-College Engineering Education Research, 1*(2), 31–39. doi:10.5703/1288284314639.

Hudson, P., English, L. D., & Dawes, L. (2009). Analysing preservice teachers' potential for implementing engineering education in the middle school. *Australasian Journal of Engineering Education, 15*(3), 165–174.

International Technology Education Association (ITEA). (2009). *The overlooked STEM imperatives: Technology and engineering*. Reston: Author.

Kelley, T. R. (2012). Voices from the past: Messages for a STEM future. *Journal of Technology Studies, 38*(1), 34–42.

Kinniburgh, L., & Shaw, E. (2007). Building reading fluency in elementary science through readers' theatre. *Science Activities, 44*, 16–22.

Museum of Science, Boston. (2013). *Why teach engineering to children?* Retrieved from http://www.eie.org/sites/default/files/engineering_for_children.pdf

National Research Council. (2010). *Preparing teachers: Building evidence for sound policy*. Committee on the Study of Teacher Preparation Programs in the United States, Center for Education. Division of Behavioral and Social Sciences and Education. Washington, DC: The National Academies Press.

National Research Council. (2012). *A framework for K-12 science education: Practices, crosscutting concepts, and core ideas*. Washington, DC: The National Academies Press. Retrieved from http://www.nap.edu/catalog.php?record_id=13165

Ontario Ministry of Education. (2007). *The Ontario curriculum, grades 1–8: Science and technology*. Toronto: Queen's Printer for Ontario.

Ontario Ministry of Education. (2008a). *The Ontario curriculum, grades 9 and 10: Science*. Toronto: Queen's Printer for Ontario.

Ontario Ministry of Education. (2008b). *The Ontario curriculum, grades 11 and 12: Science*. Toronto: Queen's Printer for Ontario.

Ontario Ministry of Education. (2009a). *The Ontario curriculum, grades 9 and 10: Technological education*. Toronto: Queen's Printer for Ontario.

Ontario Ministry of Education. (2009b). *The Ontario curriculum, grades 11 and 12: Technological education*. Toronto: Queen's Printer for Ontario.

Petroski, H. (2003). Engineering: Early education. *American Scientist, 91*, 206–209.

Redmond, A., Thomas, J., High, K., Scott, M., Jordan, P., & Dockers, J. (2011). Enriching science and math through engineering. *School Science and Mathematics, 111*(8), 399–408.

Rose, L. C., Gallup, A. M., Dugger, W. E., & Starkweather, K. N. (2004). The second installment of the ITEA/Gallup poll and what it reveals as to how Americans think about technology. *Technology Teacher, 64*(1), S1–S12.

Sargent, J. F., Jr. (2013). *The U.S. science and engineering workforce: Recent, current, and projected employment, wages, and unemployment*. Congressional Research Service report no. R43061. Retrieved from https://www.fas.org/sgp/crs/misc/R43061.pdf

Sullivan, J. F. (2006). Broadening engineering's participation – A call for K-16 engineering education. *The Bridge, 36*(2), 17–24.

Chapter 13
Infusing Engineering Concepts into High School Science: Opportunities and Challenges

Rodney Custer, Arthur Eisenkraft, Kristen Wendell, Jenny Daugherty, and Julie Ross

K-12 schools across the United States find increasing encouragement to teach engineering as an important component of STEM (science, technology, engineering, and mathematics) education. Engineering (a) provides authentic educational problem solving contexts for mathematics and science; (b) may increase the number of students interested in STEM areas, particularly from underrepresented populations (Brophy et al. 2008); and (c) might facilitate the technological literacy of all students (Erekson and Custer 2008). Given the increased emphasis on engineering at the K-12 level, the National Academy of Engineering (NAE) convened a Committee on K-12 Engineering Education resulting in a report that stressed the contribution of engineering to the development of an effective and interconnected STEM education system (Katehi et al. 2009). Several engineering-oriented programs from curriculum development to teacher professional development have emerged.

Many of the K-12 engineering education efforts have pursued an integrative approach whereby engineering is infused into the existing curriculum, whether it is within science, technology, mathematics or other courses. At the same time, new national assessments and educational standards are including engineering strands,

---

R. Custer
Black Hills State University, Spearfish, SD, USA

A. Eisenkraft
University of Massachusetts-Boston, Boston, MA, USA

K. Wendell
Tufts University Center for Engineering Education and Outreach, Medford, MA, USA

J. Daugherty (✉)
Purdue University, West Lafayette, IN, USA
e-mail: jldaughe@purdue.edu

J. Ross
University of Maryland-Baltimore County, Baltimore, MD, USA

requiring new curriculum as well as effective teacher preparation to deliver such curriculum. For example, the National Research Council 2012 report, *A Framework for K-12 Science Standards*, included engineering as one of four strands and identified cross-cutting concepts in engineering, as well as science. This led to the release of the *Next Generation Science Standards* (NGSS), which includes engineering in each of its three dimensions –Science and Engineering Practices, Disciplinary Core Ideas, and Crosscutting Concepts. Recognizing the importance of engineering concepts and noting that the high school curriculum is constrained in the amount of time that can be specifically assigned to engineering, we have been investigating how to infuse engineering concepts into science through the efforts of the National Science Foundation funded Discovery Research in K-12 project; Project Infuse.

Project Infuse was funded to research teacher learning through an innovative approach to professional development that is engineering concept-driven. Project Infuse teachers and researchers have been engaged in the development and refinement of an engineering concept base, the development of an assessment instrument to measure learning gains of the concepts, and approaches to infusing engineering into instruction. This chapter describes the approach taken by Project Infuse in these endeavors and highlights the key issues involved with infusing engineering concepts into physics, which include what engineering concepts are appropriate for high school physics and how these concepts can be infused into instruction. Since the goal of infusing engineering concepts is to facilitate both the learning of science content and engineering, one of the findings from the project has been the importance of embedding engineering concepts into science-based scenarios and content. This is opposed to simply "doing" engineering-types of activities without a significant understanding of what engineering is and of engineering practices and core concepts.

The focus on engineering concepts provides a touchstone for the teachers to understand engineering and entry points for its inclusion into their teaching. The rationale for a concept-driven approach to professional development is grounded in cognitive science, and teacher professional development research. Cognitive science research indicates that conceptual understanding is necessary for situating information, content and ideas into a particular context; in this case engineering into physics. Conceptual understandings allows learners to apply what they have learned to new situations and learn related information, and provides for the creation of a connected web of knowledge (Bransford et al. 2000). Concepts also organize knowledge into meaningful instruction (Donovan and Bransford 2005). Teacher professional development research indicate that professional development should take into account teachers' conceptions of teaching and of the learning process and allow for active learning and reflective participation (Burbank and Kauchak 2003; Clarke and Hollingsworth 2002; Loucks-Horsley et al. 2003). Understanding engineering concepts and then reflecting on students' learning of these concepts is the underpinning element of Project Infuse's teacher professional development approach.

Specifically, the project is engaged in several research and education activities that include: (a) refining the conceptual base of engineering for secondary level learning, (b) preparing a set of professional development activities that will develop teachers' understanding of engineering concepts and engage the teachers in a process of curriculum concept infusion, and (c) developing the instrumentation to collect data on the teachers' understandings of the engineering concepts and how this impacts their teaching. It is important to note that Project Infuse is fundamentally a research study, which is designed to better understand the effectiveness of an engineering concept-based approach to professional development. This chapter describes this approach in further detail and explains the different paths to engineering infusion explored by the teachers. Some preliminary observations are shared from the professional development experiences with a cohort of physics high school teachers.

An Engineering Concept-Based Approach

In order to identify the engineering concepts that could serve as the basis for infusing engineering into science, an extensive process was undertaken. Studies have been conducted to identify key concepts and the National Research Council (2012) report, "A Framework for K-12 Science Education" identified cross-cutting engineering concepts important to science. Custer et al. (2010) study included focus groups with engineering educators and engineers and an in-depth analysis of a broad range of engineering-related literature to identify core engineering concepts. This process resulted in 13 concepts (analysis, constraints, design, efficiency, experimentation, functionality, innovation, modeling, optimization, prototyping, systems, trade-offs, and visualization). Rossouw et al. (2010) conducted a Delphi study and a panel meeting to identify engineering concepts and contexts that can be used for developing curricula. Three rounds were conducted resulting in five key concepts (design [as a verb], systems, modeling, resources, and values) and 16 sub-concepts (optimising, trade-offs, specifications, invention, product lifecycle, artefacts ['design as a noun'], structure, function, materials, energy, information, sustainability, innovation, risk/failure, social interaction, and technology assessment).

In addition to identifying key engineering concepts for K-12 education, the project team developed a systematic process for defining the core concepts. A definition of concepts is essential in order to provide a clear and precise foundation on which to develop curriculum and assessment, as well as professional development. A variety of texts were reviewed to identify definitions for the concepts, including introduction to engineering textbooks (used primarily with freshmen engineering students), standards documents, and philosophy of engineering literature (e.g., Bucciarelli 2003; Mitcham 1999; Koen 2003; Florman 1996). Definitions were documented if they were specific to the engineering domain but in a broad conceptual way (not to a specific engineering discipline). The definitions were recorded verbatim, as well as any supporting text that further elaborated the concept. This information was presented to the project leadership team, who decided to focus on a smaller set of primary concepts that are central to engineering, important at the secondary level, and can provide strong links to science education. Four primary concepts emerged and sub-concepts were identified under these concepts serving to highlight key components. The concepts and sub-concepts are:

- Design (constraints, trade-offs, optimization, prototyping)
- Analysis (life-cycle, cost-benefit, risk)
- Systems (structure, functions, interrelationships)
- Modeling (visualization, prototyping, mathematical models)

This work coincided with the publication of *A Framework for K-12 Science Education: Practices, Crosscutting Concepts, and Core Ideas* and the Next Generation Science Standards, which includes engineering standards. The four engineering concepts identified for the project are evident in the framework and standards. Design is a central concept of the engineering component. The committee of the framework indicated that they were "convinced that engagement in the practices of engineering design is as much a part of learning science as engagement in the practices of science" (pp. 1–5). Two of the eight practices that are considered to be essential elements of science and engineering are: (a) developing and using models (modeling) and (b) analyzing and interpreting data (i.e., analysis). The committee also identified seven crosscutting scientific and engineering concepts, one of which focused on systems: "defining the system under study – specifying its boundaries and making explicit a model of that system – provides tools for understanding and testing ideas that are applicable throughout science and engineering" (pp. 4–1, 2).

These four engineering concepts are also crucial in physics and other sciences. However, the scientific use of these terms is decidedly different from the engineering uses. In science, design speaks to experimental design; analysis is done without regard to the utility of the findings (e.g. theory building as crucial); systems refer to arbitrary, theoretical ways in which we focus our analysis; and models speak to abstract representations such as the model of the atom. Given that these four terms are used in both engineering and science, we realized that distinctions should be made apparent to students to help them better understand both the terms and the context in which they are applied. To do this, we asked participating teachers to use Venn diagram representation to show their understandings of how each term is used in engineering, in science, and where there is overlap. The Venn diagrams we show below were developed with the teachers and should be considered preliminary in nature. All of the terms chosen by the teachers can be interpreted in multiple ways and require elucidation. For example, when teachers talked about analytical criteria as being impartial in science, they were expressing that although data can be interpreted in different ways by different scientists, the collective scientific community strives for impartiality in analysis. In future publications from our project, all of these shorthand phrases will be elaborated to provided clearer meanings. The teachers' Venn diagrams are presented here in a linear format for ease of reading.

Design

1. **SCIENCE**
    (a) Design an approach to answer questions about the natural world.
    (b) Knowledge for knowledge sake (not product driven).

(c) Generalization.
(d) Design experiment → Predict results.

2. **ENGINEERING**

   (a) End result is a product, process, or system.
   (b) Within defined criteria.
   (c) Optimization.

3. **SCIENCE AND ENGINEERING**

   (a) Iterative cycles.
   (b) Build models.
   (c) Data driven.
   (d) Collaborative.

Analysis

1. **SCIENCE**

   (a) Staring with the simplest case and trying to generalize.
   (b) Laws and theories first and last.
   (c) Impartial and objective criteria (Footnote: The teachers chose the term "impartial" to contrast scientific analysis with engineering analysis, but we recognize the potential for bias in scientists' analytical work).
   (d) Answering a question just to know the answer.

2. **ENGINEERING**

   (a) Clear purpose → specific case or product.
   (b) (Material, measurements, constraints)

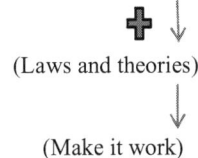

   (Laws and theories)
   ↓
   (Make it work)

   (c) Subjective and objective criteria.
   (d) Answering a question to make the product work.

3. **SCIENCE AND ENGINEERING**

   (a) Iterative to define.
   (b) Data driven.
   (c) How and why things happen.

Models

1. **SCIENCE**

    (a) Scientists create unified universal models.
    (b) Strive for simple models.

2. **ENGINEERING**

    (a) Engineers create specific models for each design.
    (b) Models are used for optimization.

3. **SCIENCE AND ENGINEERING**

    (a) Models change the way we think.
    (b) Used for communication and explanatory understanding.
    (c) Models have limitations.

Systems

1. **SCIENCE**

    (a) Scientists choose system boundaries.
    (b) Idealization.
    (c) Descriptive.

2. **ENGINEERING**

    (a) Engineering constraints dictate system boundaries.
    (b) Always practical [solution-serving].
    (c) Modularity.

3. **SCIENCE AND ENGINEERING**

    (a) Account for interrelationships among components.
    (b) Systems work together.

The Teacher Professional Development

Teachers participated in a series of summer institutes and school year activities for 2 years. During the summer institutes, teachers are engaged in three major types of activities, including: (a) conceptual development, (b) curriculum infusion, and (c) classroom preparation. The teachers engaged in activities designed to build their conceptual understanding of engineering, including case study analysis, historical reviews, concept mapping, and engagement with engineering design analysis activities. These experiences were explicitly designed to actively engage teachers in constructing their knowledge of engineering concepts and provide them with an active context within which to identify and reflect on engineering concepts (e.g., optimization, design, constraints, trade-offs, and efficiency). Curriculum infusion activities included revising existing curriculum modules and developing engineering infused lessons. The teachers were engaged with the outcomes of the concept infused lessons and participated in a process of critical reflection and analysis of the materials. Finally, classroom preparation activities focused on applying what has been learned in the summer institutes to their classrooms. The primary activity that occurred for the teachers during the school year between the two summer institutes was the delivery of the concept-infused lessons to students. In preparation for this process, the summer institute included reflection on the pedagogical knowledge, skills, techniques, and challenges associated with developing and implementing engineering infused lessons.

Why Engineering Infusion?

Infusing engineering into the existing curriculum is a significant issue faced by high school science teachers. Many approaches have been proposed to introduce high school students to engineering concepts and practices. While some approaches treat engineering as a separate stand-alone course or sequence of courses, this project focuses on infusion of engineering concepts into science classrooms; weaving the learning of engineering content throughout the fabric of a science class. This approach has a number of benefits. First and foremost, it can balance the importance of teaching engineering content with the reality that the high school curriculum is constrained in the amount of time that can be spent on engineering concepts. However, it can also allow students to develop a more sophisticated understanding of how science and engineering are related to one another and are practiced in the real world. Infusion can increase student interest and engagement in science by providing real-world applications for science learning. If implemented well, it facilitates student learning of science because students are required to analyze and synthesize scientific concepts in order to apply them to an engineering design challenge (Fortus et al. 2004; Kolodner et al. 2003). Rather than just "doing" engineering activities, infusion encourages the grounding of engineering design decisions in a conceptual understanding of science.

While there are benefits to the infusion approach, there are also significant challenges that teachers face as they incorporate engineering concepts into standards-based curricula and instructional activities. Currently, science teachers demonstrate a very broad range of exposure to engineering concepts; from none to sophisticated understanding. For high quality infusion to occur, a teacher must first learn the engineering content at a sufficient depth to be prepared to make good decisions in the classroom. For example, teachers need to be able to determine which engineering concepts are good matches for infusion in a specific lesson or unit. If the concepts are not well matched, the engineering content will be superfluous or worse, may distract from science learning. Determining how much infusion to include for a given science class is also a challenge. If too much focus is on engineering concepts, students may lose sight of the science content resulting in decreased science learning. Conversely, if engineering concepts are tangential to science learning or at insufficient depth, students may not show learning gains in engineering. Striking an appropriate balance takes experience and sound judgment by teachers.

Teaching open-ended engineering design challenges also requires a significant shift in pedagogical approach and style for many science teachers. Many teachers have never participated in formal training in how to work with student teams in a project-based collaborative learning environment and are therefore uncomfortable doing so. This environment also requires a different approach to assessment of student learning, further taxing teachers' comfort levels.

Finally, many teachers find it challenging to structure an engineering activity to fit within allotted time constraints. Very often, the time required for project completion ends up requiring more than has been allotted and the activity is ended prior to completion or causes other lessons/units to be compressed to compensate. Neither approach is optimal. Because engineering design challenges are open-ended by nature, it takes practice and experience to learn how to constrain an activity in a way that maximizes learning without taking undo time.

Approaches to Engineering Infusion

The following section provides guidance on how to manage these challenges and provides specific examples of how engineering infusion can be done in the classroom. Through the work of Project Infuse, the project team has observed four distinct approaches to the infusion of engineering concepts into a high school physics course. The first includes engineering as a part of a lesson and the second as framing of an entire unit. The third involves exploration of engineering case studies. The fourth is the occasional day or two devoted to a design challenge.

Engineering as a Part of a Lesson

Any physics lesson can be enhanced by using an engineering application as a means to engage students at the introduction of the lesson, a way to extend the lesson through the application of the physics concepts, or a combination of both. As an example, consider a physics lesson about shadows from *Active Physics* (Eisenkraft 2010), which utilizes the 7E instructional model (Eisenkraft 2003). In the 7E model, the lesson begins with something to "engage" the students followed by an opportunity for the teacher to "elicit" the students' prior understanding. Next, the students perform an investigation in which they can "explore" the science concepts. The investigation provides data that the students can use to "explain" their observations with teacher assistance. The teacher can then "elaborate" these conclusions by incorporating them into a larger context including additional, related phenomena and theories. Finally, the students can "extend" or transfer their learning to a new domain and apply their new knowledge. Throughout the lesson, the teacher is able to "evaluate" the students – during the engage, elicit, explore, explain, elaborate and extend, as well as with a summative evaluation.

As an example of the 7E instructional model in a single class where engineering can be infused, we can dissect a lesson on shadow formation (Eisenkraft 2010). In the shadows lesson, students learn that light travels in straight lines and observe how shadows are produced. They develop a scientific model of how shadows are formed. This lesson, without engineering infusion, begins by the teacher noting that when she was outside on her way to school, she observed her shadow. Here in the classroom, it appears that the shadow has disappeared. She asks the class, "Where did my shadow go?" "What do I need for a shadow?" The purpose of relating this observation is to "engage" the students and begin to "elicit" their prior understandings. The students are then asked if a mouse's shadow can be as big as an elephant's shadow. The teacher listens to the students' responses without correcting the students, nor trying to get the "correct" answer. The teacher is asking these questions to find out about students' prior understandings. The students then "explore" shadows. They are given a candle, a large post-it and a small post-it and are asked to investigate if they can make the shadows of the two post-its identical in size (like the shadows of a mouse and elephant.) Student groups are usually successful within 10 min. The teacher then asks students to "explain" how they were able to change the size of the shadows. They complete a claim-evidence statement, "To make the shadow larger, I can _____." To further "explain" why this occurs, students need to use the fact that light travels in straight lines. What is their evidence for this? Some will mention the rays of light emerging from clouds while others may mention the light beams at a laser light show. A ray diagram model can be created to explain the shadow region based on the fact that light travels in straight lines (Fig. 13.1).

Students then draw two ray diagrams. They can vary either the size of the post-it, the distance from the light source or the distance from the screen. They are then asked if this ray model of light is consistent with the results of their investigation. The ray diagram model can also be used to determine the size of the shadow using the mathematics of similar triangles or ratios. Upon closer examination of the shadow region, students will observe that the shadow does not have a sharp line between light and dark but has a thin, gray region. The "elaborate" portion of this lesson involves students trying to extend the ray model to account for this gray region. They can do this by recognizing that the candle is actually an extended light source. All of the light does not come from a single point as in the original diagram but some light comes from the top of the candle flame and some comes from the bottom of the candle flame. Finally, the shadow lesson offers students the opportunity to "extend" the physics to explain the shadow of the Earth on the Moon during a lunar eclipse or the shadow of the Moon on the Earth during a solar eclipse.

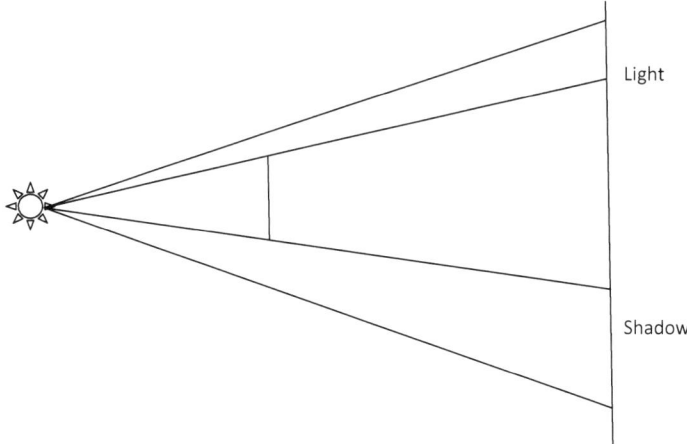

Fig. 13.1 A ray diagram model

In an engineering-infused version of this physics lesson, the "extend" might look a bit different. In one option, the teacher can pose an engineering design challenge related to shadows during theater productions. Considering a theater lighting designer as a client, can the students design a system that would enable the 'shadow monster' in a play to be a certain size, even when the height of the actor is not known until the day of the show? Specifically, if the monster must be 8 ft high, but we don't know if the actor is 5 ft or 6 ft, what do we do? The teacher might scaffold students' work on this design challenge by having students return to the lab to investigate what shadows look like when three different colored spotlights illuminate an actor.

Another option for infusing engineering via the "extend" of the physics lesson concerns wind generators. A wind generator can produce a shadow. As the blade moves, the shadow moves. This repetitive light and dark across a house can be bothersome. If the strobe effect of light and dark is severe enough and just the right frequency, it can possibly cause epileptic attacks. The engineer would have to maximize electricity generation while keeping the rotation of the blades above or below a certain value.

A final engineering-infused "extend" would challenge the students to design and build a sundial. The task of sundial design and construction uses the physics of shadows from the lesson but can also be structured so that students work with the engineering concepts of design, systems, analysis and models. To develop further their understanding of engineering design, students can be asked to consider constraints (size, cost, protection from elements), which leads to choices of materials, as well as considerations of aesthetics and structure (why are sundials usually not a straight stick, but are often a triangle with a curve?) (Fig. 13.2 by liz west, Sundial, 2007).

To foster students' thinking about the concept of engineering systems, the teacher might require one person in each design team to build the base of the sundial,

Fig. 13.2 A sundial

another person to build the gnomon (shadow stick) and another person to determine the best location and orientation. Students would conduct engineering analysis as they carried out the math and physics required to determine the correct placement of hours on the base, the path of the sun at different latitudes, the path of the sun at different times of year and whether these require changes in the structure of the sundial. Finally, students would consider engineering models both in terms of their geometrical model of the sundial as well as the physical prototype they build to check to see if it works as intended.

Another approach to infusing engineering at the level of the stand-alone physics lesson is to "engage" students with engineering applications at the beginning of the lesson. The teacher could have shared any of the engineering cases described above – shadows of wind generators, sundials, or theater design – when the lesson first commenced. Students would then have a rationale for learning about shadows. This approach gives teachers a way to enhance a stand-alone physics lesson with engineering infusion without the greater time commitment of a design-and-build challenge. Of course, the same engineering application could be studied during the "engage" of a physics lesson *and* followed with an actual design-and-build challenge during the "extend." But this is not necessary for the physics lesson to be enhanced – at least to some extent – by engineering infusion. Although an engineering-infused physics *course* should provide students with at least one extended opportunity to design, build, and improve solutions to an engineering problem during the academic term, individual physics *lessons* within the course can meet some objectives for engineering infusion without requiring students to participate in physical prototyping.

Engineering as Framing of an Entire Unit

The shadow lesson is actually one section of a longer instructional unit set in the real-world context of entertainment productions. As in most project based learning (PBL) students are introduced to a real-world challenge before beginning the unit's lessons, and they complete this challenge as their final learning and assessment experience of the unit. The real-world challenge in "Let Us Entertain You" is to design a sound and light show to entertain students your age. The sound must come from musical instruments, human voices, or sound makers built by the students, and the light must come from a laser or convention lamps. Students must also provide an explanation of the physics principles involved in the show. This assignment is described as an engineering design challenge. In this project based learning approach, all the sections of the unit (including shadows) are exposures to physics principles that can be used in this design challenge. Moreover, students are introduced to five phases of engineering design work in the introduction to the unit. In the Let Us Entertain You unit, these phases are called Goal, Inputs, Process, Outputs, and Feedback. These labels correspond to the problem scoping, research, design, testing, and optimization practices of engineering design found in many other models of engineering design processes (e.g., Atman et al. 2007).

- Goal: define the problem; identify available resources; draft possible solutions; list constraints to possible actions.
- Inputs: complete the investigations in each section; learn new physics concepts and vocabulary
- Process: evaluate work to date; compare and contrast methods and ideas; examine possible trade-offs to help reach goals and maximize efforts; create a model from your information; design experiments to test ideas and the suitability of the model.
- Outputs: present mini-challenge and intermediary steps or products; present main challenge based on feedback to mini-challenge
- Feedback: obtain response from target audience leading to modification of the goal; identify additional constraints, require restarting the input and process stages.

The goal is to design and build the light and sound show (students engage in *problem scoping* as they make sense of this goal). The sections where students learn the physics content are considered inputs (or engineering design *research*). The students then work on their light and sound show as the process phase (also considered the *conceptual design* and *detailed design* phases). The show is presented to the class as output (an experience that could also be called *testing* or *analysis*). The other student teams and teacher provide feedback to the group presenting (an opportunity for *optimization*).

The students are reminded of the sound and light show challenge at the completion of each section. After they learn about the effect of the length and tension on strings and the pitch of the sound, they are asked how this knowledge can be used in their light and sound show. Similarly, they are asked to apply their physics knowledge when they learn about sounds from tubes. Once the sections on creating sounds are completed, the students return to the engineering design cycle when they are asked, as a mini-challenge, to create the sound portion of their show. The mini-challenge provides the students with the opportunity to attempt the challenge, to gauge its difficulty, and to get feedback from students. The students then return to the sections and learn about reflection and refraction of light and applications of these concepts in mirrors and lenses. They return to the engineering design cycle at the completion of the 10 sections in the unit and design and build their show.

Note that each of the sections can have engineering infused in its 2–3 days of lessons as was shown with the shadow section. It is easy to imagine engineering infusion in a lesson about stringed instruments or funhouse mirrors. The engineering infusion of the unit as a whole is different. In project based learning, the engineering design challenge creates the structure and raison d'être for all of the sections. It also provides motivation and an opportunity to apply all the physics concepts (e.g., reflection and refraction of light, lenses, mirrors, effect of string length and tension on pitch, standing waves in strings). In addition, the elements of design, systems, analysis and models also exist. The design is now the design of the entire light and sound show. The systems include the sound system and the light system and their relative components. The analysis is inherent in the size of the effects, and the coordination of the components. The show is, in fact, a model for a larger show that would be used in an auditorium. Scaling this classroom model to a large audience is not a trivial task.

Engineering as Exploration of Case Studies

Another opportunity exists for infusion of engineering concepts in a high school physics curriculum occurs with the use of case studies. A case study can be a small reading or a brief video about an engineering design challenge that can introduce students to certain aspects of engineering. In the case study, the students are reading (or viewing) and interpreting rather than being involved in a hands-on aspect of engineering. Of course, some teachers may be able to use the case study as a jumping off place to involve students in a hands-on activity.

During the Project Infuse professional development workshops, we used a series of written case studies created by Cory Culbertson, a high school engineering teacher, specifically for our project. One of these case studies is included here to provide a fuller understanding of this curriculum addition (Fig. 13.3).

Through the case study, students are drawn into a new appreciation of architecture and the role of engineering and aesthetics. The discussion questions help them see the importance of design and analysis. The reading and discussion questions could be expanded to also include descriptions of the role of systems and models. This was not necessary since systems and models were the focus of other case studies that were presented. In this example, one can see how learning about engineering

Keeping Fallingwater from Falling

Frank Lloyd Wright is America's best-known architect, and perhaps no other building better represents Wright's iconic style than Fallingwater in Mill Run, Pennsylvania. Built in 1936, Fallingwater was commissioned by the Kaufmann family to be a vacation house with a view of the beautiful waterfall on their property. Instead of building the house *next* to the waterfall, Wright boldly decided to place the house *over* the waterfall. Its broad flat lines echo the stone ledges around the waterfall. Fallingwater is considered an architectural masterpiece for the way it not only complements the natural surroundings but even enhances them.

Frank Lloyd Wright's Fallingwater

However, Wright's design stretched the limits of structural engineering in his day. The home's wide terraces stretched over the falls, supported only at one end, in an arrangement known as a cantilever. While graceful, the cantilevered terraces were under enormous stress at their support points. Wright chose to use reinforced concrete as his building material. Reinforced concrete is a durable and versatile material that allowed him to create the forms he envisioned. However, concrete does best when stressed in compression, not tension. The cantilevered terraces had areas of high tension that posed a real challenge for Wright's engineers. Their design called for inverted T-shaped beams to support the terraces, with a small number of steel bars in the top of the beams to resist tension forces. Careful analysis of the building's loads allowed them to create a design in which the tension loads were within the safe limits of the concrete and steel. However, Wright felt that their analysis was too conservative. He also sharply disagreed with an outside engineering firm that recommended a large increase in the amount of reinforcing steel in the terraces.

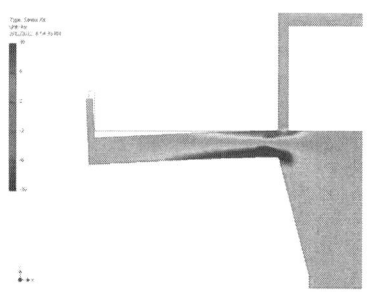

Computerized stress analysis of a cantilevered terrace similar to that at Fallingwater. Red shows areas of high tension. Blue shows areas of high compression. (C.Culbertson)

Soon after construction, the terraces began to sag noticeably. Later, cracks began to appear at the stress points. It appeared that the concrete would not be able to resist the tension forces indefinitely. Stopgap repairs were made for decades, but it was clear that the problem was in the original design. In 1995, Robert Silman Associates was hired to investigate. By using new computer modeling techniques and radar images of the interior of the concrete, RSA discovered that the terraces were dangerously close to failing. It was time to finally correct the problem. The challenge was not only to fix the damaged structure but to find a way to reduce the stresses in Wright's original design without altering its historic appearance.

RSA's engineering **team came up with a solution**

Fig. 13.3 Case study developed by Cory Culbertson

that placed a hidden support structure inside Wright's famous terraces. The key to the new design was the use of the same reinforcing material that Wright had originally undervalued - steel. Unlike concrete, steel is able to resist high tension forces. In fact, suspension bridges make dramatic use of steel tension cables to support their high-flying structures.

Unlike a suspension bridge, though, Fallingwater could not have its new steel cables strung up for the world to see. They had to be hidden inside the existing terrace floors and walls. This presented the engineers with a new problem: as the angle of the support cables became closer to horizontal, the amount of vertical load that they could support decreased significantly. At such a shallow angle, the steel cables would be stressed almost to their limits. Once again, Fallingwater engineers found themselves doing careful load analysis to keep their structure within safe limits of material stress.

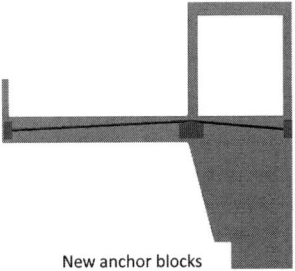

New anchor blocks

Placement of one of the steel cables within the terrace. As the angled cables are tensioned, the reaction forces pull the outer edge of the terrace inward and upward. (C. Culbertson)

The final design took advantage of the best properties of steel and concrete - steel to resist tension forces, and concrete to resist compression forces. The steel cables were carefully anchored into blocks buried inside the terrace floors. Then, they were slowly pulled into tension. The engineering team watched carefully as the lip of the main terrace rose by almost an inch. By careful placement of the new reinforcements, the engineering team had provided a much greater margin of safety and a longer life for the house.

Of course, visitors to Fallingwater will not see any of this. Instead, they will see Frank Lloyd Wright's house as it originally appeared when it was built. Only now, this architectural masterpiece will be able to majestically float over its waterfall for many more years to come.

Discussion Questions
1. Identify some points where engineering analysis played a critical role in the history of Fallingwater.
2. If engineering analysis is largely based on mathematics, how can there be disagreements over whether it is accurate or not?
3. Design is often described as an iterative process. However, few buildings are reworked until they are satisfactory, as Fallingwater was. Does the concept of iterative design apply to large structures and other one-time designs? If so, how?
4. Steel-reinforced concrete was a relatively new material in Frank Lloyd Wright's time, and it was not as well understood as it is now. Does Wright deserve blame for using a material that was still somewhat unproven? Describe more recent examples of situations in which somewhat unproven materials have been used.
5. Wright's vision for Fallingwater sometimes conflicted with the more conservative desires of both his clients and his structural engineers. Does the engineering design process have room for highly innovative, visionary people? What benefits and costs are associated with highly innovative design?

Fig. 13.3 (continued)

can be as engaging as being involved in engineering in the lessons on shadows or the design and creation of a light and sound show.

Design Challenge as Benchmark Experience

Many physics teachers introduce the stand-alone design challenge to their students as a means of acquainting them with engineering design. One such challenge is the bridge building activity where students must build a bridge from a limited set of materials and are judged by how much weight their bridge can hold. Another such challenge is the egg drop. An egg is enclosed in materials and dropped against a hard surface. Alternatively, an unprotected egg is dropped onto a cushioned surface of students' design. The egg that survives the highest fall is deemed the winner.

Design challenges of this sort are fun, exciting and require cooperation among team members. Sometimes these design challenges include engineering principles while other times they are merely exercises in trial-and-error. In the egg drop, we expect students to apply the principles of impulse and force to build the best cushion. The designs can be analyzed using the relevant equations and measurements. With the bridge building design challenge equations or measurements of stress and strain can be used to inform and analyze the students' designs. However, analysis with respect to equations or measurements is not often presented as an integral component of the challenges. Teachers can draw students' attention to design concepts and relevant science content knowledge with thoughtful questioning. The following list of questions can help deepen the potential for an egg drop challenge to be a benchmark learning experience to which teachers and students refer all year long.

- How does the cushioning process affect the (a) impulse, (b) force, and (c) time of the egg-ground collision?
- Predict the necessary changes required in the cushioning materials if the drop height were doubled.
- What is the effect of placing the egg in the cushioning material versus having the egg land on the cushioning material?
- How would the concept of iterative design help you succeed in the egg drop challenge?
- What analysis of forces could you do to predict where the egg is most likely to break?
- In what ways could your egg protector be considered an engineering system?

As a benchmark experience in our work with teachers, we have used the marshmallow design challenge where teams can use up to 20 pieces of spaghetti, 1 m of masking tape, and 1 m of string to build the tallest free standing structure that can hold a marshmallow at its top. The teams have 18 min to complete the challenge. As analyzed in a popular TED talk by Tom Wujec (http://www.ted.com/talks/tom_wujec_build_a_tower.html), most teams build their structure carefully and methodically. They take care not to break any spaghetti because they may need all 20 pieces.

Only as the clock signals that time is running out do they place their marshmallow at the top of the structure for the first time. To their chagrin, the tower collapses. In contrast, successful teams – including most of the kindergarten groups to whom Wujec has posed the challenge – fail fast and often by placing the marshmallow on top earlier and learning about the structural strength of the spaghetti. Wujec suggests that kindergarteners do so well on the marshmallow challenge because they understand this principle of failing fast and often. On the other hand, perhaps the kindergarteners do not recognize the fact that they are working under the constraint of limited spaghetti. They may be working under the assumption that if they break all of their spaghetti within the 18 min, they can simply ask for more.

To better understand the marshmallow challenge, the teachers in our summer workshop devised a preliminary research study where 100 middle school students were asked to complete the marshmallow challenge under slightly different conditions. One group completed the challenge as described. Another group was told that the most successful teams put the marshmallow on top early in the process. A third group was given 2 min to play with a few strands of spaghetti and marshmallow before the challenge began. The fourth team used 2 min to brainstorm ideas before beginning. We observed little difference in success in the marshmallow challenge due to these different interventions. This result suggests that no matter how the marshmallow challenge is set up, what may be most important for its potential as a benchmark experience is the reflection questions that a teacher poses afterwards. The questions below can deepen students' thinking about the marshmallow challenge and amplify its impact throughout the year.

- If, instead of spaghetti noodles, you were given a quantity of angel hair pasta or lasagna noodles equivalent in mass to 20 strands of spaghetti, how would your design change? What are the structural characteristics of the different pasta noodles, and how do these characteristics affect the marshmallow challenge?
- How would your design change if the mass of the marshmallow were doubled?
- Predict the changes to your design if a minimum height requirement were imposed.
- How could methods of iterative design help you create a successful tower for the marshmallow?
- What analysis of forces could you do to improve your chances of building the tallest successful tower?
- How could you construct models of your tower? How could modeling your tower improve your understanding of it and your chances for making it succeed?

Design challenges of this sort can be used in two distinct ways. They can be stand-alone exercises that can provide a glimpse of the engineering process for students. In contrast, they can be used as referents during the school year as different physics topics are studied. In this approach to infusion, a single engineering design problem – and the experience of solving it – becomes a "benchmark" engineering experience to which teachers and students return throughout a unit, semester, or year to illustrate particular science and engineering concepts. The initial solving of the design problem itself may take only one class period, but the conversations derived from the design problem may reinforce science learning for a much longer period of time – and this is where the engineering *infusion* really takes place, much more so than in the sometimes disconnected experience of simply building a physical device.

A design challenge that has this benchmark experience potential is one in which students are given a plastic cup, three felt tip pens, a small motor and masking tape. They have to create a "motorized artist" – something like a robot that can draw. This challenge can be given the first week of school. A day or a week after creating the artists, students might create an instruction manual detailing how to put it together, and in the process discuss the engineering concept of visual models. It can be referred to during the study of motion where students will now have to describe the movement of the pen. The students might return to the artwork produced by their devices and characterize the distance, total displacement, vector directions, and other factors of the pens' travels. As they do so, they might discuss the value of one style of artwork over another and discuss the engineering concepts of cost-benefit analysis and trade-offs. It can be referred to again during the study of forces as an example of the force needed to move the pen and the force of friction of the pens and the surface. They might examine another team's device and predict its motion before turning it on. After doing so, they might discuss the engineering concept of analysis. During the study of electricity, the power of the motor can then be analyzed. In this way, the motorized artist design challenge moves from trial-and-error to engineering as each element of physics can lead to a more informed design change.

Connections to the Next Generation Science Standards

The NGSS is structured so that each learning objective combines three dimensions of the Framework – crosscutting concepts, disciplinary core ideas and science and engineering practices. A distinction is made between science and engineering practices. In the *project based learning approach of* creating a light and sound show (Eisenkraft 2010), we provide opportunities for all of the science and engineering practices. You can see how these occur in sections (similar to the shadow section) and the unit as a whole with the examples below.

Engineering practice	Light and sound show
Asking questions and defining problems	
Unit level	Create a light and sound show to entertain your friends
Section level	How can we use diffraction to increase the volume of a wind instrument?
Developing and using models	
Unit level	Create a model for a light and sound show
Section level	Build a model of a stringed instrument

Engineering practice	Light and sound show
Planning and carrying out Investigations	
Unit level	Create a light and sound show
Section level	How can you form large images from lenses?
Analyzing and interpreting data	
Unit level	What is the maximum number of people that can enjoy your light and sound show?
Section level	Determine how the size of the shadow varies with the angle of the light source using the ray model of light
Using mathematics and computational thinking	
Unit level	What is the maximum volume of the instruments you built?
Section level	Compare the image size for different lenses
Constructing explanations and designing solutions	
Unit level	Insure that your light and sound show meets all the required criteria
Section level	How do you choose which wind instrument to build?
Engaging in argument from evidence	
Unit level	How can different musical instruments all produce the same note?
Unit level	Explain how you know that a longer tube will produce a lower pitch regardless of the material used
Obtaining, evaluating, and communicating information	
Unit level	Present your light and sound show and survey the class regarding its creativity, use of physics principles and entertainment value
Section level	Demonstrate three different pitches from your string instruments

Engineering as a Disciplinary Core Idea

The Framework also introduces engineering as a disciplinary core idea. The first core idea has to do with "Engineering Design." Engineering design can be introduced in each project based learning unit and throughout the year. It can refer to one type of engineering design cycle – goal, input, process, output, feedback, goal… – three times in each unit. Students work on their projects throughout the unit. They are practicing engineering rather than "trial-and-error" because each iteration of their design is a result of the application of new physics principles. They design under constraints and try to optimize their solutions.

The project based learning unit in creating a light and sound show certainly reflects aspects of this core idea and component ideas. For aspects of HS-ETS1-2, challenges such as creating a light and sound show requires students to design a solution by breaking it into smaller, manageable problems. Both of these and others also meet aspects of HS-ETS1-3 by prioritizing criteria and trade-offs. This project may fall short in responding to HS-ETS1-3 in that it does not treat major global challenges. Another unit challenge which teachers investigated during their summer workshops involved electrical use in developing countries. The science concepts were applied to the solution to the challenge. It is important to recognize that major global challenges often have aspects of science in their solution, but none of the solutions depend only on science. All major global challenges are dependent on politics, economics and history. Science plays a role but it is naïve to think that solutions are solely chosen on the basis of science and engineering principles. The project based learning unit in creating a light and sound show certainly reflects aspects of this core idea and component ideas (see HS framework components below).

The framework outlines specific aspects of engineering design.

HS-ETS1-1. Analyze a major global challenge to specify qualitative and quantitative criteria and constraints for solutions that account for societal needs and wants.

HS-ETS1-2. Design a solution to a complex real-world problem by breaking it down into smaller, more manageable problems that can be solved through engineering.

HS-ETS1-3. Evaluate a solution to a complex real-world problem based on prioritized criteria and trade-offs that account for a range of constraints, including cost, safety, reliability, and aesthetics, as well as possible social, cultural, and environmental impacts.

The challenges that we focused on in the summer workshops depended on the interdependence of science, engineering and technology. The major focus was infusion into a physics curriculum. It then becomes fairly obvious that the design features and use of engineering practices are always in the context of physics content and scientific practices. The Framework's specific aspects of engineering design demands that students see the influence of engineering, technology and science on society and the natural world. What we have tried to stress is that engineering's impact on society is not restricted to major global issues but appears in everyday solutions to everyday problems.

Conclusion

Infusing engineering into a physics curriculum is a necessary step if one assumes that the high school curriculum cannot accommodate a full engineering course. How to infuse engineering concepts into a physics course in a meaningful way requires careful planning. In our work, we have uncovered multiple strategies when combined can be used to infuse engineering into the physics classroom. The inclusion of engineering into the shadow lesson is one such strategy. On a larger scale, the adoption of project-based learning (e.g. the light and sound show) also allows for inclusion of engineering principles across multiple weeks of instruction. In this case, the engineering becomes a way to structure and provide engaging context for an entire unit of instruction.

A third engineering infusion strategy that can be used in combination with the others is the addition of case studies as an ancillary to the physics instruction. They can be inserted when related topics are taught in physics (e.g. the Frank Lloyd Wright building can be assigned during a unit of forces) to connect those concepts to engineering concepts. A fourth approach that can be used in combination with other strategies is the use of smaller scale design challenges. The values of these are enhanced if the students are able to use physics principles during their design. Without the opportunity to use physics, the design challenge may simply be trial-and-error and would not be considered engineering infusion.

References

Atman, C. J., Adams, R. S., Mosborg, S., Cardella, M. E., Turns, J., & Saleem, J. (2007). Engineering design processes: A comparison of students and expert practitioners. *Journal of Engineering Education, 96*(4), 359–379.

Bransford, J. D., Brown, A. L., & Cocking, R. R. (Eds.). (2000). *How people learn: Brain, mind, experience, and school*. Washington, DC: National Academy Press.

Brophy, S., Klein, S., Portsmore, M., & Rogers, C. (2008). Advancing engineering education in P-12 classrooms. *Journal of Engineering Education, 97*(3), 369–388.

Bucciarelli, L. L. (2003). *Engineering philosophy*. Delft: Delft University Press.

Burbank, M. D., & Kauchak, D. (2003). An alternative model for professional development: Investigations into effective collaboration. *Teaching and Teacher Education, 19*(5), 499–514.

By liz west (Sundial). (2007). CC-BY-2.0 (http://creativecommons.org/licenses/by/2.0), via Wikimedia Commons.

Clarke, D. J., & Hollingsworth, H. (2002). Elaborating a model of teacher professional growth. *Teaching and Teacher Education, 18*(8), 947–967.

Custer, R. L., Daugherty, J. L., & Meyer, J. P. (2010). Formulating a concept base for secondary level engineering: A review and synthesis. *Journal of Technology Education, 22*(1), 4–21.

Donovan, M. S., & Bransford, J. D. (2005). *How students learn: Science in the classroom. Committee on how people learn: A targeted report for teachers*. Washington, DC: National Research Council.

Eisenkraft, A. (2003). Expanding the 5E model. *The Science Teacher, 70*(6), 56–59.

Eisenkraft, A. (2010). *Active physics*. Armonk: It's About Time.

Erekson, T. L., & Custer, R. L. (2008). Conceptual foundations: Engineering and technology education. In R. L. Custer & T. L. Erekson (Eds.), *Engineering and technology education* (57th yearbook, Council on Technology Teacher Education, pp. 1–12). Woodland Hills: Glencoe.

Florman, S. (1996). *The introspective engineer*. New York: St Martin's Press.

Fortus, D., Dershimer, R. C., Krajcik, J. S., Marx, R. W., & Mamlok-Naaman, R. (2004). Design-based science and student learning. *Journal of Research in Science Teaching, 41*(10), 1081–1110.

Katehi, L., Pearson, G., Feder, M., & National Academy of Engineering and National Research Council. (2009). *Engineering in K-12 education: Understanding the status and improving the prospects*. Washington, DC: National Academies Press.

Koen, B. V. (2003). *Discussion of the method: Conducting the engineer's approach to problem solving*. New York: Oxford University Press.

Kolodner, J. L., Camp, P. J., Crismond, D., Fasse, B., Gray, J., Holbrook, J., et al. (2003). Problem-based learning meets case-based reasoning in the middle-school science classroom: Putting learning by design (TM) into practice. *The Journal of the Learning Sciences, 12*(4), 495–547.

Loucks-Horsley, S., Love, N., Stiles, K. E., Mundry, S., & Hewson, P. W. (2003). *Designing professional development for teachers of science and mathematics*. Thousand Oaks: Corwin.

Mitcham, C. (1999). *Thinking through technology: The path between engineering and philosophy*. Chicago: University of Chicago Press.

National Research Council (NRC). (2012). *A framework for K-12 science education: Practices, crosscutting concept, and core ideas*. Washington, DC: National Academies Press.

Rossouw, A., Hacker, M., & de Vries, M. J. (2010). Concepts and contexts in engineering and technology education: An international and interdisciplinary Delphi study. *International Journal of Technology and Design Education*. Retrieved from http://www.springerlink.com/content/4u32551r6h44kx42/fulltext.pdf

Chapter 14
How Do Secondary Level Biology Teachers Make Sense of Using Mathematics in Design-Based Lessons About a Biological Process?

Charlie Cox, Birdy Reynolds, Anita Schuchardt, and Christian Schunn

In the fall of 2011 five secondary level biology teachers in the northeast United States implemented an experimental instructional module that challenged their students with a design problem. This challenge required students to perform both mathematical analysis and the engineering application of biological concepts in order to reach a resolution. Specifically, given the parental genotypes of two gecko parents, students were tasked to: (a) mathematically represent the relative frequency of all possible offspring genotypes; and (b) design a systematic breeding program for the geckos that would consistently produce a rare and highly desired genotype as a result. Presented here is a study of how the participating teachers made sense of the mathematics and engineering design applied to the biological process of inheritance, and their reflections on their own implementations of the instructional module. Emergent themes dealt with the limitations of mathematics in teachers' own biology education, their lack of experience with either engineering or design, and their efforts to help students address similar circumstances.

The Organization of This Presentation

A presentation of this study needs some explanatory background in order to be understood by a wide range of readers, and that requires the introductory section to collect and sort a good deal of information from a variety of sources. The first part of this

---

C. Cox (✉)
Penn State University, University Park, PA, USA
e-mail: cxc655@psu.edu

B. Reynolds • A. Schuchardt • C. Schunn
University of Pittsburgh, Pittsburgh, PA, USA
e-mail: schunn@pitt.edu

section is a review of biology as it is being taught at the secondary level, comparing its characteristics to those of chemistry and physics. This is followed by a description of the policies that will soon profoundly affect science instruction at that level.

For practicing K-12 teachers who no doubt have already begun to contemplate how the latest policies will affect their pedagogy, much of the introductory section serves as an assurance of due diligence on the part of the researchers with regard to practitioner concerns. For others outside the profession, the researchers' intent is for them to consult the introductory material in order to bring themselves "up to speed" with those concerns.

After that, an example response to those concerns is detailed through the content of an experimental instructional module aimed at integrating mathematics and engineering practices with a typical secondary level biology topic. In this section the design challenge that forms the basis of the module is described. It is constructed to require students to draw on their mathematical resources and to make an engineering application of biological concepts in order to arrive at a resolution. Many approaches to a resolution are possible and either competitiveness or collaboration (at the level of individuals, teams, and the entire class) can be emphasized where deemed advantageous by the teacher implementing the module.

Finally, teachers who participated in the study reflect on and react to their implementations of the module in their classrooms and the professional development that informed those implementations. This section concludes with the insights that emerged from teachers' experiences with the module. It is likely that this section and the preceding one will be the ones of most interest and use to K-12 practitioners.

Current State of Secondary Level Mathematics and Biology vis-à-vis One Another

Many secondary school biology teachers are hesitant to put mathematics into service, either as a descriptive method or predictive tool, because topics in any of the sciences at that level are separate and distinct from those in the mathematics classroom down the hall. This is reflected in the lack of mathematics' incorporation in science textbooks (Cantrell & Robinson, 2002).

Furthermore, both mathematics and biology can be taught as a collection of abstractions, without application to observable processes. That is, secondary level biology students can be handed a sequence of well-defined concepts (e.g., DNA, genes, chromosomes) associated with well- defined relationships and processes (e.g., transcription, dominance, random assortment), but no student can actually see any of these without microscopes, so the concepts and processes remain abstract. Meanwhile, the same students encounter similarly well-defined abstractions in their mathematics courses, with no demonstration of these applications to events or objects in their day-to-day lives. The dissociation of mathematics from biology at the secondary level neither indicates what students will likely encounter if they choose to pursue biology as a major or possible career nor promotes how a student's interest in biology could lead to finding engineering or mathematics useful at all.

While not willfully ignored, opportunities for mathematics to be applied to a biology process can easily be neglected. At least in part this is because biology does not afford neatly describable and predictable demonstrations of foundational concepts the way physics and chemistry do. From one organism to the next, "wet" anatomy and physiology might not always appear or behave exactly the same way, and certainly do not perform processes consistently to the same extent that, say, precipitate formation does for chemistry.

This is due to biological processes' stochastic nature being much more evident in class demonstrations than it is for processes in physics and chemistry, and it is related to the amount of conditions that can be observed. If a biology lab could address thousands of parents and offspring, then it would be reasonable to expect students to discover recurrent ratios of genotypes in the offspring, because the large numbers would approximate predictable results. In comparison to chemistry, however, while not every single particle that could form a precipitate will do so, the enormous number of tiny particles that are typically present yield enough of the expected performances so as to render that outcome consistent, predictable, and verifiable from observation.

Likewise, biology labs tend to deal with much larger scales and much smaller samples of observations than do secondary level chemistry and physics. Consider one pair of parent organisms that can have only so many offspring in a semester. Because the parental alleles that are inherited as offspring alleles separate and combine randomly (and there can be tens of thousands of different genes for a species), students can go only so far with determining, recording, and comparing offspring genotypes in that semester. After all, it took Mendel several years and acres of plants before the data he collected yielded their information.

Thus, while useful probabilistic expressions might not spring to mind in chemistry and physics classes at the secondary level, they are entirely appropriate for dealing with the otherwise overwhelming enormity of data associated with combination and permutation in inheritance processes. Unfortunately, if students don't use probability in other science classes, such as physics (that can be linked easily with mathematics), and if they don't encounter probability, permutations, and combinations in a mathematics class, that means biology teachers have to introduce those concepts at the same time they're introducing the inheritance process so that students can get a grasp of the topic. And, if biology teachers do not typically bring mathematics into their classrooms (because mathematics and science are segregated), the extent of the students' grasp is severely curtailed.

Another aspect impeding application of mathematics is the rate at which biology advances can leave gaps between teachers' understanding and the current state of scientific thinking (Cakir & Crawford, 2001). As a consequence, teachers may be inclined to instruct students through memorization of simpler concepts than those being contested in the field. Kleickmann et al. (2013, p. 94) raise the point that, "… the available formal professional development programs tend to consist of short-term workshops that are often fragmented and noncumulative," (referring to a

German study, but generalizing to other countries and citing American studies). Not only that, but if biology teachers try to weave mathematics and engineering and design into their presentations, that effort entails all the additional content knowledge and pedagogical content knowledge that they themselves need to learn, implement, and maintain about those fields. So, if the effect of professional development is questionable within teachers' expected purviews, it seems unreasonable to expect much benefit when the subject matter is unfamiliar, as mathematics and engineering might be unfamiliar to biology teachers.

The current situation is that it is easy to find instructional implementations of engineering in secondary physics and chemistry (e.g., Robinson & Kenny, 2003), and it is easy to find other implementations that apparently do not distinguish one discipline in the sciences from another when introducing mathematics and engineering (Ralston, Hieb, & Rivoli, 2013). Yet secondary level programs focused specifically on biology continue to lack the resources to offer students a range of interesting real-world problems of the sort that actual biologists could address in their professional practice (e.g., modeling the logistics of preserving an endangered sub-species of tiger).

Look at the lessons to be learned from the revival of engineering design in higher education, resulting from years of studies conducted through grants from the National Science Foundation (NSF), with an intensive concentration in the 1980s. There is a reason why engineering professions want people with design skills at the entry level, including research and creative application of scientific principles, and it proved counter-productive for higher education curricula to downplay those skills in favor of other subject matter. If anything, it makes perfect sense to expose such skills to students at the secondary level wherever that can happen, but especially in science courses including biology, in order for them to make an informed choice about careers that might interest them and that they might wish to pursue.

In other words, not only do individual students benefit, but also so do the biology and engineering professions; in the case of the professions the advantage is an influx of people who want to practice in those fields because they are familiar with and perhaps even enjoy what those fields require. Furthermore, early exposure to engineering might lessen the strain on introductory levels of those programs at university, the current popular location for students to resolve whether they have made good career decisions or not (reducing the time and resources students spend as undergraduates when otherwise they would have to start over after concluding that engineering was not a good initial choice for a major).

What Secondary Level Instructional Interventions Will Need to Include

The integration of mathematics with science in P-12 education is currently accelerating toward a critical state of concern for teachers in all science disciplines. This comes as a result of the Next Generation Science Standards (2013, hereinafter

NGSS) resuming where the National Academy of Engineering and National Research Council's (2009) framework left off, that is, in the actual presentation of "crosscutting concepts" (interrelationships among science disciplines) and explicit connections to other subjects in the Common Core State Standards (2011). While updating existing curricula with which they are familiar will be of genuine concern to teachers across the country, there will be the additional complication of addressing engineering concepts and practices, as well (including engineering design), with which most science teachers will not be familiar. Therefore, identifying specific issues that these teachers might face, investigating strategies for mathematics and engineering integration in specific disciplines, and disseminating these strategies at scale and through the literature contributes to the shared effectiveness of all teachers' efforts in this undertaking. But how will this take place?

The Learning Research and Development Center (LRDC) at the University of Pittsburgh launched the Biology Levers Out Of Mathematics (BLOOM) study in order to design and develop instructional modules for both integrating mathematics with secondary level biology and exploiting opportunities for engineering design that had previously lain dormant in the biology classroom. The example of a BLOOM module to be detailed herein presents a design challenge to students and demonstrates where BLOOM can help secondary level biology teachers in using this design challenge in order to get a handle on working with NGSS (2013) performance expectations MS-LS3-2 (middle school) and HS-LS3-2 and HS-LS3-3 (high school):

- Develop and use a model to describe why... sexual reproduction results in offspring with genetic variation.... Emphasis is on using models such as Punnett squares, diagrams, and simulations to describe the cause and effect relationship of gene transmission from parent(s) to offspring and resulting genetic variation.
- Make and defend a claim based on evidence that inheritable genetic variations may result from: new genetic combinations through meiosis...
- Apply concepts of statistics and probability to explain the variation and distribution of expressed traits in a population.... Emphasis is on the use of mathematics to describe the probability of traits as it relates to genetic and environmental factors in the expression of traits.

It will be shown below that the BLOOM module implemented in this study not only addresses these expectations, but in the case of the Punnett square, it also encourages students to replace that cumbersome device with a more sophisticated and powerful mathematical expression, thus incorporating a *crosscutting concept* intended for HS-LS3-3 whereby "algebraic thinking is used to examine scientific data and predict the effect of a change in one variable on another," (NGSS, 2013). In addition, this module emphasizes the *science and engineering practices* for secondary level life sciences of "asking questions and defining problems" and "developing and using models" by requiring students to prepare a presentation about the path they took to their final results. They have to define the sequence of the path and defend it step by step, such that it can be replicated without ambiguity. That justification includes the use of the mathematical expression they develop, while working together as an entire classroom of participants, in order to supersede the Punnett square.

Module Content: Biology–Mathematics Connections Needed for the Design Challenge

Posit a biological process that can be represented by a mathematical expression and furthermore assume that this mathematical expression can be derived from analyzing previous results of the process. Then it is not difficult to manufacture an engineering problem based on manipulating the variables in the expression in order to determine the results, without having to enact the process in actuality.

The BLOOM module presented here addresses inheritance, a biological process that lends itself to mathematical representation through algebraic expression. Consider Mendel's Law of Segregation of Alleles in the case of some animal whose genes each have two alleles; for one of these genes each allele may be either type *A* or type *a*. For this gene alone, each parent could then be one of these genotypes: *AA*, *Aa*, or *aa*. In any of those three possible instances, every parental gamete will contribute one of those two alleles to an offspring. A parent that has *aa* genes will have an *a* in each of its gametes, another parent with an *AA* gene will have an *A* in its gametes, and yet another parent with *Aa* will have either an *A* or an *a* in its gametes.

When applying these principles as an engineer might, one can predict the range of possible outcomes for an offspring having any pair of those parents and determine which of those outcomes, if any, are more likely to occur than others. Furthermore, the prediction of likely proportions of permutations can be extended to multiple genes (the complete genome for any organism being far beyond the convenient range currently served by instruction about the Punnett square).

One trick in applying mathematics to biology is to establish and maintain sensible mapping of biological processes onto mathematical expressions and vice versa. For example, the ratios of expected genotype occurrence in offspring (1/4:2/4:1/4) from a mating of heterozygous parents (both have genotype *Aa*) have meaning for respective allele permutations of *AA*, *Aa*, and *aa* occurring in the offspring. Put another way, when both parents have the same *Aa* genotype, there are likely to be twice as many *Aa* genotypes in their offspring in proportion to either *AA* or *aa*. Or, to present it a third way, given a large enough sample of offspring from these *Aa* parents, one would expect 1/4 to be *AA*, another 1/4 to be *aa*, and the remainder to be *Aa*.

Why? Well, if one starts with the male parent (informally call him "dad" in order to make it easier to keep track) being *Aa* and contributing either one of those alleles to a gamete, with the same being true for the female parent (call her "mom"), that means the offspring genotypes from the combined parental gametes could be:

- *A* from mom and *A* from dad = *AA* for an offspring
- *a* from mom and *a* from dad = *aa* for an offspring
- *A* from mom and *a* from dad = *Aa* for an offspring
- *A* from mom and *A* from dad = *aA* for an offspring

Except, wait a minute: *Aa* and *aA* are the same. It doesn't matter with which parent the allele originated. So the convention is to label both of them *Aa*, and now it is apparent that there are likely to be twice as many of those as either of the other genotypes.

Detailing the Design Challenge in the Module

For the engineering problem, this BLOOM module presented students with a challenge to detail a breeding plan over several generations of mating for producing rare kinds of geckos, as requested by a fictional zoo (the client). Starting with a given amount of pretended funding, students could "buy" geckos with known genotypes and then breed them to get offspring (neither the parent geckos nor their offspring were real) for which the possible genotypes and the likely ratios of those particular genotypes out of any given set of offspring could be calculated.

Keeping in mind the intended alignment with the NGSS (2013) performance expectation to develop and use a model of gene transmission and variation, one follows directly to the derivation of a mathematical expression that represents the inheritance process and facilitates the sequencing of the breeding plan. But, in order to make those calculations, the students first had to generalize their mathematical expressions to take into account not only any number of genes but also parents with any permutation of alleles.

Ostensibly, this was necessary because the zoo clients had specific criteria for what they would accept, and the criteria involved analysis of parental genotypes and prediction of offspring genotypes for multiple genes simultaneously. In fact, this generalization is related to the NGSS (2013) performance expectation regarding statistics and probability as explanatory vehicles for the variation and distribution of expressed traits in a population.

Once the offspring from a mating were predicted, then students could "sell" them, and use the "profits" to "buy" more expensive geckos with correspondingly more exotic genotypes that could then themselves be bred, producing another round of offspring, and so on until the zoo's criteria for rare animals (expressing some permutation of recessive or incompletely dominant or co-dominant alleles) was met or exceeded.

Say that there are three traits under consideration: size, pigment, and pattern. If the zoo asks for two of those traits to be consistently expressed by recessive genes, then students need to buy whatever common geckos they can afford with their limited initial capital, perhaps setting some funds aside to purchase a particular breeder at greater expense because it is known to have one of the desirable recessive genes. Once they produced true breeders for that expression, that is, parents who could produce only offspring with similar genotypes, students could sell their excess stock (by then including some geckos of greater value than those originally purchased), and reinvest in another breeder known to produce a different recessive expression. This aspect of the challenge is congruent with the NGSS (2013) performance expectation to make an evidence based claim regarding inheritable genetic variations.

Take as a simple example the following: the *A* or *a* allele expresses skin pigment, and *B* or *b* expresses a skin pattern. If a student buys a common gecko male of genotype *AABB* and a female of the more expensive genotype *aaBB* then the student can expect *AaBB* offspring, most of which can be sold for further investment later. But once the student has a female and a male both known to be *AaBB* to breed, that means the offspring from mating them can be expected to be about 1/4 *aaBB* (for which the phenotype is a distinctive lack of skin pigment) which are the more expensive true breeders for the recessive *a* allele. One of them can be kept and the rest sold.

If the student keeps an *aaBB* male and purchases an equally expensive *AAbb* from the profits to date, then it is apparent that all the offspring will be *AaBb*. Mating two of those *AaBb* geckos is likely to produce quite a few common ones (*AABB*, *AaBB*, *AABb*, and *AaBb*, no one distinct phenotypically from any other), some true breeders for *a* and some true breeders for *b*, and sooner or later a true breeder for both *a* and *b* (expected phenotype ratios of 9:3:3:1). Over time, the student will create enough true breeders to satisfy the zoo's needs.

Note that there was additional complexity beyond that of manipulating genotypes in that some phenotypes are not associated with one genotype exclusively, and it was necessary for students to determine how to get true breeding genotypes that could produce only similar genotypes in their offspring, thus perpetuating the phenotype, as well. This required a biologically-based distinction among recessiveness and the various kinds of dominance (simple, incomplete, co-dominance) in the relationships of alleles available from each parent. When that was established, the range of expressions possible in the offspring could be calculated.

To students, the apparent intent of the challenge was for them to purposefully breed geckos with known genotypes (or acquire geckos with known genotypes) in order to arrive at a particular genotype acceptable to the zoo, according to a precisely determined plan that they derived themselves, and for which they would need mathematics to predict and keep account of each stage, turning a profit for their efforts. The actual educational intent was for those students to work out for themselves how the laws of combination and expression worked and could be represented mathematically and then manipulated, regardless of the organism involved.

Now, consider that in its appendix devoted to engineering design the NGSS (2013) directs secondary level teachers to provide students with opportunities for:

- Defining the constraints in the problems they face
- Developing multiple iterative solutions by first analyzing complex problems in search of simpler pieces that then can be resolved and synthesized as solutions to the larger challenge
- Establishing criteria for assessing and evaluating trade-offs in the resources they have at their disposal for dealing with their problems

Upon review it may be seen that these are exactly the components of the design challenge.

Some Logistical Aspects of the BLOOM Module: First Appearances to a Teacher

In general, the duration of a BLOOM module can vary between 2 and 4 weeks of daily 45-min classroom sessions. The instructional intent of implementing a BLOOM module is that students must generate and graph data or derive some algebraic expression that describes a biological process and gives them a way to predict outcomes of that process. The module used for this study addressed inheritance.

One of the initial guiding questions for this study was if and how deriving that representation would work with what participating teachers had previously done regarding inheritance, prior to BLOOM. After all, the BLOOM module breaks from tradition for inheritance content presentation in several significant ways:

- Meiosis is not the introductory topic. Instead the BLOOM module starts with fully formed male and female gametes.
- Genotype is treated with little mention of phenotype for three quarters of the module until the concept of phenotype is not only necessary to introduce, with respect to solving the design challenge, but also explicable at last from the genotypic information constructed as a foundation theretofore. This reduces extraneous cognitive load (Sweller, 2011) that would otherwise occur when simultaneously defining both genotype and phenotype while maintaining the distinction of one from the other.
- The Punnett square, a centerpiece of the usual instructional approach, is instead summarily dropped as an unwieldy prediction generating widget that rapidly loses biological meaning in exponential complexity. Instead, students are asked to derive compact and more powerful mathematical expressions.
- Teachers allow students broad leeway to approach a well-defined but ill-structured problem in gecko breeding, as engineers and biologists might encounter. Being ill-structured, the problem has the appearance of being wicked (Rittel & Webber, 1973), and so is a departure from typical problem solving for most students and teachers. Actually, the problem used in the BLOOM implementation is relatively well-defined in order to function less wickedly than what engineers potentially encounter, and instead acts more in the manner of a puzzle, for which there are several ways for the pieces to be assembled but only a finite range of so many pieces and their beginning and end states. Yet it is not a familiar textbook biology problem by any means.

Research Questions about Biology Teachers Using Mathematics and Engineering

This paper's focus is not primarily any of the module's instructional effects. Instead, this investigation concerns how a number of individual biology teachers made sense of the BLOOM module that was being iteratively developed through rapid prototyping and then implemented in their classrooms. Cox (2009) describes the path that novel subject matter takes in higher education, from initial agreement among faculty about definitions for the subject matter to final legitimization as explicitly advertised subject matter in a course catalog; there was a similar process at work with the BLOOM module. An initial agreement about what mathematics expression was appropriate for mapping inheritance had to be negotiated among the BLOOM developers and presented to the participating biology teachers during their professional development and subsequent classroom implementation. These teachers presented the interplay between mathematics and inheritance to their students, and both teacher and student reactions tempered what the BLOOM developers kept and modified in subsequent iterations of the module. In this way, it was discovered what applications of mathematics and what presentations of the mathematics proved robust enough to not only survive confrontation with teachers' and students' biology understandings, but also to augment those understandings.

Of additional importance, observation of teacher efforts was not limited to resolution of only the mathematic content knowledge required for the modules (representing a legitimization of mathematics' place in biology content). What this study also attended to were any shifts in teachers' pedagogical content knowledge (Ball, Thames, & Phelps, 2008; Davis & Krajcik, 2005; Shulman, 1986) as those occurred during the practice of teaching biology through mathematical applications.

Finally, as previously mentioned, the emphasis on mathematics was complexified with an introduction of engineering concepts and practices, and how those affected content knowledge and pedagogical content knowledge were observed, as well.

How did biology teachers describe what happened in their classrooms during their implementation of the BLOOM study instructional material? How did the focused use of mathematics affect the nature or extent of their individual pedagogical resources and their use of those resources for teaching biology?

Unit of Analysis and Anticipated Critical Dimensions of Phenomena

Although Elmore (1996, p. 16) was not involved with this study, his characterization of teachers who maintain "ambitious and challenging practice in classrooms" pertains nicely to our participants as teachers who are "motivated to question their practice on a fundamental level and look to outside models to improve teaching and learning." As the study's unit of analysis, there were five participating teachers, with each having one or two daily sections of a secondary level biology course (ranging from grades nine through twelve), and each section consisting of from 10 to 25 students, depending on absenteeism. Because the BLOOM module was being developed in a rapid prototyping manner, making it available every semester over the year to date, three of these teachers had also participated in previous implementation rounds. The other two were newly recruited in an effort to expand BLOOM implementations within the same geographical area (participants represented public school districts and parochial schools in the northeast United States). All participating teachers were female in this round.

During this study, participants met as a group only three times, for about 3 h each time, first at one professional development session before the implementation started and then another during the implementation, with a reflection session after the implementation had concluded. Contact time was thus a critical dimension that affected what participants could achieve as a group.

A further critical dimension of the implementation was that of the difference between intended and enacted duration of the module. While the BLOOM project team considered the module to require a 2–3 weeks implementation schedule, various factors dragged this out to from 4 to 5 weeks in the field. In addition to interruptions at each school from conflicting events that had been set months beforehand, including standardized testing, there were unpredictable amounts of time required for students to reach conclusions on their own as the module materials encouraged teachers to do.

Methodology

This study involves an empirical approach to gather phenomenological data, relying heavily on: observation of the participants encountering the BLOOM module in professional development sessions; observations of participants implementing the BLOOM module in their classrooms; and interviews with participants immediately after their class sessions, as well as two delayed interviews afterward. These last two interviews contributed the most data to this study.

Although Rossman and Rallis (2003, p. 98, citing Seidman, 1998) describe a phenomenological sequence of interview as having three components, it was prudent here to combine the first two into one longer interview with each participant, covering both the professional history and the implementation of the BLOOM module (see Appendix A). This corresponds to a naïve description as detailed in Moustakas (1994, pp. 13–15, citing Giorgi, 1979, 1985), an anecdote or narrative that a participant living the experience (i.e., teacher enacting an implementation, in this study) tells about the experience, a recounting of events without delving for explanation or justification.

The second interview (see Appendix B) was then devoted to a dialog between the BLOOM developer and each participant, regarding the participant's individual reflections and interpretation of the implementation. In order to facilitate the crucial act of triangulation known as member-checking (Lincoln & Guba, 1985), each participant was presented with the data analysis relative to her interviews and observations and asked to interrogate the researcher's interpretations, especially those which rang false or unconvincing. This is where the previously empirical orientation of data collection and analysis explicitly gives way to a heuristic manner, in what van Manen (1997, p. 99) characterizes as the hermeneutic conversation where the researcher and participant tackle the question, returning theme by theme to ask again and again, "Is this what the experience is really like?"

Data Analysis

The teachers participating in this implementation did not regularly convene as a group, having only three professional development sessions as described under the previous heading of *Unit of analysis and anticipated critical dimensions of phenomena*. On one hand, these sessions were purposefully structured improvements over the typical format as described by Kleickmann et al. (2013, p. 94, "short-term workshops that are often fragmented and noncumulative"), including a post-implementation meeting (p. 92, "Several studies suggest that teaching experience needs to be coupled with thoughtful reflection on instructional practice, with non-formal learning through interactions with colleagues, and with deliberate formal learning opportunities."). On the other, participants did not attempt to discuss ongoing implementations with one another outside of professional development sessions, engendering little in the way of community. As a consequence, each of the participating teachers will be discussed in turn as an individual. In order to maintain their confidentiality, each has been assigned a pseudonym: Alice, Betty, Carol, Dorothy, and Emma.

Alice Would Have Liked to See More Math Years Ago

Alice teaches in a parochial school that, while not inner city, occupies a neighborhood of older wood framed homes built cheek by jowl, dotted with factories and warehouses succumbing to dilapidation, and laced throughout with a maze of meandering streets. She has participated in previous implementations of the module, and is familiar with the changes that have accompanied the iterations. She also has the most experience in the classroom of all the participants, so her reaction to the module's increasing sophistication is of great interest, in that her naïve description of any implementation has likely given way to repeated reflection long before this study, and whatever sense she is going to make of it has already been accomplished.

It is possible that hers is the transition described by Drake and Sherin (2009) whereby only after repeated usage of materials, can teachers establish the level of trust they place in the designer's intent and the materials' utility, as opposed to the initial confrontation when affordances and constraints still need to be discovered. Indeed, Alice made a point of listing what particulars from the BLOOM module she intended to incorporate in her future presentation of inheritance, including leaving meiosis for the conclusion.

For her, mathematics is a medium necessary for analysis and presentation of data, and there clearly is not enough of it in general biology today. She is one of the participants who has consistently maintained Elmore's (1996, p. 16) "ambitious and challenging practice in classrooms" and motivation "to question their [own] practice on a fundamental level and look to outside models to improve teaching and learning." She said in this study's first interview that, "A teacher has to be open to seeing differently or kids won't look at [content] another way." Thus, when the BLOOM study first recruited her to work with a mathematically intensive module, she responded enthusiastically.

As with all of BLOOM's participants, her attitude is in direct contradiction to her own biology education in secondary school, where mathematics dared not speak its name. In the secondary level biology classes that she took as a student, genetics was ignored. However, she did not follow a direct path to becoming a biology teacher, in that she first chose a related field for her initial teaching practice and then returned to university some years later.

By then, biochemistry had been introduced into the curriculum. For her, the place of mathematics in biology was to be taken for granted from that time on, and she believes that more biochemistry and its accompanying mathematics is needed in the biology curriculum where she teaches. Likewise, any preparation for physiology studies must include mathematics because "everything for physiology has an equation."

On one hand, Alice's lack of exposure to mathematics at her own secondary level of biology parallels that of all our participants, as will be shown. On the other, her experience at university seems to differ significantly from that of the other participants, so the insight to be gained from her interview probably is not going to be entirely the same as for the other participants. This is evident in other aspects, as well; consider her answer to a question about textbooks, to the effect that the one she is using provides a graphing exercise for each of its numerous labs, something she has emphasized in other responses as being crucial for biology students to practice. No other participant gave more than a brief dismissal regarding the state of the textbook in use (note that the textbook publisher varies from school to school in this study). Was she actively looking for affordances that others had already quit trying to find?

Betty Will Not Give Up on Her Students' Exposure to Mathematics

Betty teaches at a public school that might not exactly be run down, but certainly has been used roughly for many years and shows its age. Student absenteeism is much worse there than at any of the other participants' schools, and this disrupts attempts at team-based projects such as those in the BLOOM module. Betty does her best to shift students from team to team in order to make progress every day, and has implemented several versions of the module previously, but still finds her students taking more weeks to get done than those of other participants.

It is likely her conscienteeism about reaching every student that slows her down. In response to the absenteeism, she is determined that, when a student actually does decide to show up he or she will be brought along to the level of those in attendance every day.

Although Betty is an experienced teacher here, she is still dealing with the cultural differences in this setting compared to the student/teacher model of relationship she grew up with in her native country. With regard to the apparent grudging respect she gets from the students, she feels that innovations such as the BLOOM module, that places the responsibility for research and discovery of knowledge needed to grasp the content squarely on the student, are paths worth exploring in order to engage her classes.

Unlike most of the other participants, her secondary level education explicitly addressed the mathematics with which biology teachers should be equipped. There was no hesitation on her part in dealing with that aspect of the module. She and Alice actually addressed the design challenge together during the professional development, and they seemed to follow the derivation of the mathematical expression with little instruction.

Among the participants it was also this pair who first attended to multiple genes in each parent as they set about sequencing the breeding for the design challenge. This is not to say that the BLOOM module's exclusive focus in its initial phases on genotype aligned with Betty's strategies of how genotype/phenotype interaction should be taught, but rather that she was willing to deal with the potential for cognitive discomfort on the part of her students in order to discover any possibly beneficial effects from the module's implementation.

She was keen to find any increase in evidence-based generalization and inductive reasoning among her students, especially involving the use of analogies in order to transfer inheritance concepts to something other than geckos. She was persistent in finding and making opportunities for students to phrase their biology questions as comparisons to topics they already knew, and this practice predates her work with BLOOM. For example, when interviewed for the first time, she had just that morning led her students through the similarities of compound interest (familiar to some students, and generally engaging due to its financial nature) and calculating population growth.

But in order to get to that stage, Betty sees at least two prominent obstacles: segregation of subjects; and level of expertise perceived necessary. In the former, students have been conditioned to expect rigid and impervious boundaries between subjects, such that the mention of mathematics in a biology class is an intrusive anomaly. In the latter, students have not had to formulate mathematical expressions in service of their own problems, and so expect that only a mathematician would have the expertise to do so. Simply because the BLOOM module attacks those misconceptions does not ensure that students will either embrace an integration of mathematical subject matter with that of biology or attempt what they had previously classified as exclusively expert behavior and beyond their abilities.

Carol Was Wary of the Mathematics at First

Carol's school is one of recent vintage, and situated on its own campus just outside a commercial strip of its suburban community. Easy going and affable, Carol also participated in the various versions of the BLOOM module, and developed a forthright attitude in dealing with the BLOOM researchers, which they encouraged. From time to time, she augmented the BLOOM materials with worksheets and information that she felt her students needed, but this decreased with each iteration, either due to her concerns being addressed from one version to the next or perhaps attributable to her increasing trust in the materials.

Carol does not have any issues with the mathematics (algebra and probability) itself, but was not always confident about the extent to which she resorted to it in the past, as when she asked in the first interview, "Is measurement math?" Likewise, she does not object to exploiting opportunities for mathematics in her teaching. But she is very careful to watch for students "getting lost in the math," because the integration of the two subjects is an uncommon occurrence for them to face, and she feels that not everyone can handle that. Of course, "nobody pushes cross curriculum" at her school at any level (individual teachers, departments, administration), and unless the state's impending biology standards do, it is unforeseeable that anyone will.

She speaks of mathematics as an "enhancement," perhaps for those "math-oriented" students who need a challenge beyond the day-to-day biology content, and often introduces her opinions about mathematics' place in biology with caveats. For example, in response to the National Research Council's (NRC, 2012, p. 64) statement about mathematics' dual communicative and structural functions, she begins, "*If the student is able to handle math* to make logical deductions, then it is a wonderful tool to explain biology." [emphasis added] When she does entertain the use of mathematics in an assessment item, it is only with her "advanced kids."

It is not surprising that Carol would adopt this prudent wariness. She is an experienced teacher and no doubt has seen a highly touted reform or two run its course and vanish. Nor does her own background as a student give her any compelling reason to throw in with the BLOOM module before it has proven itself to her satisfaction. She does not recall "math pushing me" or any intensive concentration on mathematics over the period from her secondary level biology courses through university and on into pre-service teaching. Furthermore, the textbooks she works with currently provide no such emphasis.

Prior to the BLOOM module, the Punnett square performed adequately as her touchstone for inheritance related mathematics. "We have this grid that can show you real easy what these combinations are." Oddly enough, her observation about deriving a general mathematical expression to replace the Punnett square was that the "denominator [of the expression] slowed you down when it wasn't 16," that is, when at least one parent was not heterozygous for two alleles. This raises the question of whether it really was easier for students to use the Punnett square when the denominator was 16 (a dihybrid cross) as opposed to the general expression.

This particular situation, often depicted as the cross of two parents $AaBb \times AaBb$, is a well known litmus test that separates mechanistic or intuitive approaches from precise calculational ones when generalizing from one gene to dealing with two or more genes (Moll & Allen, 1987; Tolman, 1982). Were the students who were slowed down by the general expression neglecting to attend to the biological process in order to focus on mathematics, or did getting the math to work bring biology any more into focus for them than plugging allele designations into the Punnett square?

In fact, Carol had already developed another approach to the Punnett square on her own, that enabled students to make the transition from one gene to two and even three or more. She first had them isolate the Punnett square for each individual gene, producing however many two by two squares as there were genes. Then, in each quadrant of the first gene's grid, a second gene's entire grid was inserted. The results in each subdivided unit of the first gene all have the same alleles from the first gene but vary by the alleles of the second gene (as shown in Fig. 14.1). So, from a pair of 2×2 grids a third grid emerges as 4×4, with 16 units total. If one then inserts another 2×2 grid, for gene Cc, say, into each unit of the 4×4 grid, the result is a subdivision of each of those 16 units into 4 new units, such that an 8×8 grid emerges, with 64 units total, all of which have two alleles from each of the three genes.

This approach had occurred to the BLOOM developers, as well. On one hand, the Punnett square is not robust enough to withstand accounting errors, and it might help students that this technique makes it difficult to fill in the grid incorrectly. On the other hand, the formation of parental gametes that would appear at the heads of rows and columns in the typical Punnett square is ignored, thereby deleting one of its actual redeeming features. Furthermore, the acreage required to accommodate generating the permutations of multiple genes burgeons just as rapidly in Carol's approach to Punnett squares as in any other, no matter how accurate one is about keeping track of them all. Upon reflection, Carol seemed satisfied that her implementation of the BLOOM module had in fact exceeded the limitations of the Punnett square as it is typically constructed.

Given all that, Carol was still only tentatively in favor of the BLOOM module's mathematical emphasis. While she felt that such aspects as calculating increasing dollar values for correspondingly rarer gecko offspring indicated an acquaintance with inheritance, she was not happy with the ambiguity of topics that eluded resolution, as in whether there were three or four different products of a monohybrid cross (genotypic AA, Aa, aa versus algebraic AA, Aa, aA, AA). In addition she would like to extend the use of the manipulables into a modeling of meiosis, rather than setting them aside at that crucial phase. She also discussed how earlier versions of the BLOOM module defined the target genotype and phenotype more explicitly and were better suited for classes with lower abilities. She intended to implement both earlier and later versions of the module in the future, with her honors classes getting the later version.

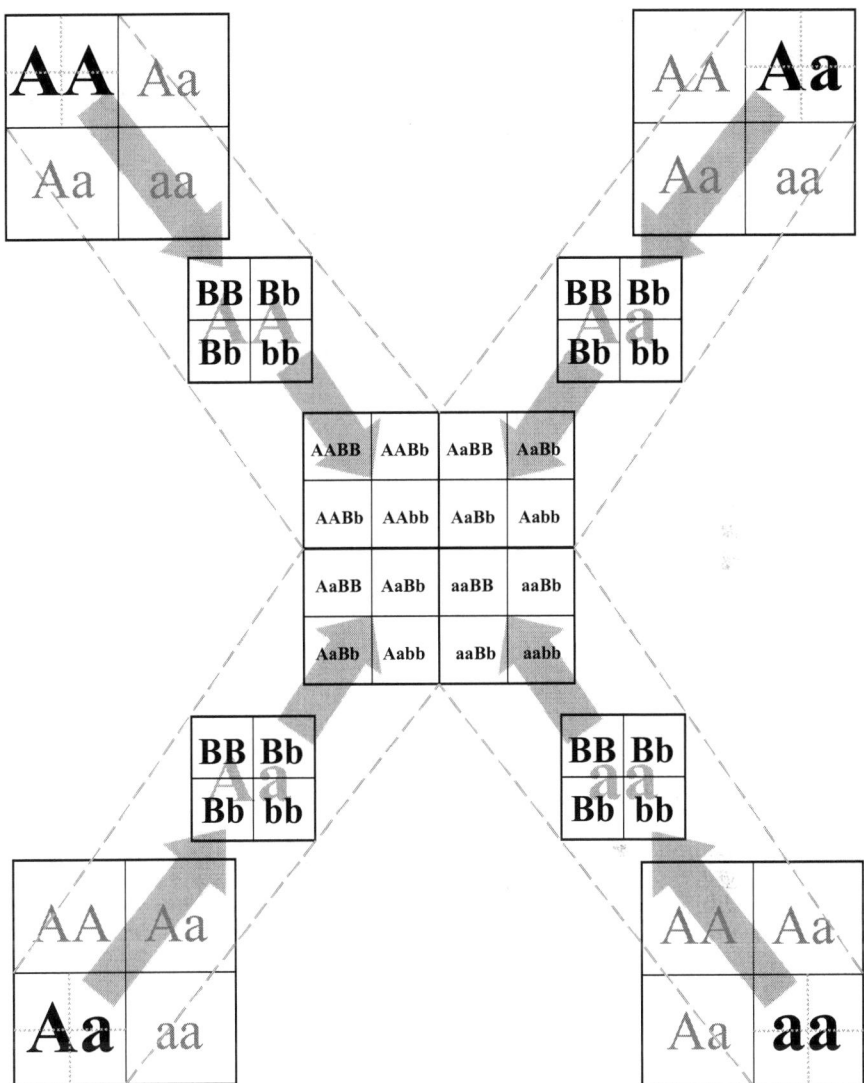

Fig. 14.1 Carol's Approach for permutations of more than one gene: Consider Punnett square for each gene by itself, insert the entire Punnett square for another gene into each unit of the first, subdividing it (here the entire Punnett square for Bb × Bb is inserted into each unit of the Punnett square for Aa × Aa)

Dorothy Emphasizes the Mathematically Rigorous Aspects of Statistics When She Can

Dorothy's suburban parochial school alone boasts a sealed concrete floor finish and the modern exposed roof deck and ductwork in lieu of a ceiling tile grid. Certainly, the equipment in her room has met the least suffering at student hands, and her current charges are not ones to leave a lab area in tatters by any means. While there is no doubt who is in charge of the classroom, the atmosphere is almost collegial with give-and-take as she engages her small groups of 15 or so students.

Her orientation to biology teaching is pronouncedly more quantitative than that of other participants, and it is clear that she is familiar with statistical methods of analysis and their terminology. Regarding the opportunity to show students what biologists actually do, she says, "Real researchers have tests," elaborating on this to implicate measurement, comparison, and statistical significance as components of answers to research questions in a biology course. "Every time a student does an experiment there is a statistical test," is how she describes the honors students' work. Although she is not yet this rigorous with her lower level students, she wants this to be more the case for all of her biology classes.

But Dorothy's own secondary level biology involved no mathematics. And it was not until she was working on a thesis at university that she needed statistics; it just was not necessary for weekly lab reports. Concluding that it requires her initiative to bring mathematics into the secondary biology classroom, she not only helps her regular students with statistics, but also participates in summer workshops that focus on that subject matter.

Responding to the enormous scope of student interests in research has helped her work past a common subterfuge that teachers adopt when confronted with a situation about which they are ill informed. "I'm not gonna lose control if I reveal I'm not sure what to do next," means that she and a student seeking an answer need to plan together how to find one, and modeling that planning is another opportunity to demonstrate what biologists face in practice. Unfortunately, the current selection of textbooks does not aid in that pursuit. Or, as she says, "Even when they have that little page [inset or sidebar; exactly the point made by Cantrell and Robinson (2002)], they don't go into detail."

Her participation in BLOOM was thus part of her active search to find teacher materials to support this effort. She has always gone beyond the limitations of the Punnett square in her classes, drawing sperm and eggs and filling in allele letters rather than just labeling rows and columns in the grid, and her quest for an easier way to predict inheritance made the BLOOM module attractive. What she discovered, however, in reviewing student work for the module was that aversion to mathematics (in favor of a visually oriented technique) reinforced use of the Punnett square, rather than replaced it, when students were given a choice.

Dorothy is facing a dilemma similar to that of undergraduate engineering education in the 1980s, when the engineering professions instigated NSF studies into the absence of design in curricula, and now Dorothy's administration is getting pressure from alumni of her school about incorporating more problem solving. But, while Dorothy currently intends to implement the BLOOM module or a modification of it with her classes next year, that is a decision attributable more to student preference than to administrative advice. The module is much more student directed than the way she normally teaches, and that novelty is an important consideration for her in developing student engagement.

Her candid evaluation is that the BLOOM materials' inadequacy (to make a mathematical alternative to a Punnett square seem attractive) is in part due to the confusing quality of the directions and examples. She herself was sometimes unsure about when an example was being presented. However, by the second professional development session, she felt confident that she was in command of the module.

Emma Is Concerned that Introducing Mathematics Reduces the Focus on Biology

Emma's parochial school sits squarely in a residential suburban area. Considering all of the participants' classrooms, hers is the most densely packed with models and living animals and relics of bygone projects. Whether feigning or sincere, her students are consistently vocal about their disdain for the BLOOM module, yet some of them have demonstrated dramatically beneficial effects.

This was the case with Melissa (not her real name). As Emma relates, "In the beginning they were shutting down ... and Melissa, that's her nature. She likes things the way they've always been." Indeed, Melissa had not been participating much at all throughout the first part of the module's implementation, when at the start of class 1 day, she started to weep to the extent that Emma was obliged to remove her from the classroom, requesting another teacher to monitor the class in the meantime. "She really struggles with things that are uncertain and not secure," was Emma's comment in the post-class interview that day. Melissa kept trying, though, and when the class was reviewing the charts that they were preparing to post about two-gene system combinations, she actually spoke up in her group; another student asked about how many combinations they should get, and Melissa set him straight on how many he had found and how many she had found.

Melissa then worked through a Punnett square that she had modified from a four-by-four grid in response to there being fewer combinations from homozygous parents than from a dihybrid cross. The discovery that she could change the square led her to interrogate how it could be made less complicated in order to serve a three-gene system. First it occurred to her that a three-gene Punnett square was going to have $(2 \times 2 \times 2) \times (2 \times 2 \times 2) = 64$ boxes to fill in, and that was something her group had not considered. When another student asked if they had made an accounting mistake, Melissa was ready to take charge. "Yes, I'll work on that," was soon followed by, "There! I fixed everything," an entirely reversed role from the student who had been led in tears from the room only a week or so before.

It may be that Emma's approach to pedagogical content knowledge has enabled her to adopt novel strategies more easily than other teachers can, leading to this sort of result. Although well grounded in content knowledge from her previous practical experience in a biology laboratory setting, she never formally learned to teach biology. In determining the details for her curriculum, she went through the textbook provided from her school and focused on topics that she thought would be interesting, at no time beholden to the reification of a standardized testing agenda that plagues teachers in the public schools. If not altogether laissez-faire, her philosophy is robust enough to tolerate significant change from 1 year to the next.

Of course, if mathematics was never an emphasis in her secondary biology or university courses (as was the case), and if the biology textbook she was perusing did not discuss mathematics applications (as was also the case), then there was no reason for her to introduce mathematics merely for the sake of novelty either. Thus, until the BLOOM module implementation, she did not plan to use any mathematics in her classes other than the Hardy-Weinberg equilibrium. She scheduled the Punnett square's annual appearance, demonstrated its traditional service as a widget for the presentation of offspring possibilities resulting from a dihybrid cross, and then ushered it from the stage without students explicitly examining its mathematical aspects. Emma even stopped offering extra credit for extending a Punnett square to four genes because only those students who were mathematically adept attempted it, apparently without beneficial effect to their understanding of inheritance as a biological process.

This does not mean that the BLOOM module was an unmixed blessing as far as clarifying advantages from mathematical applications. Instead, Emma felt that the amount and concentration of effort toward developing a mathematical expression for genotype proportions obscured the underlying reason for expending that effort to begin with. While using the Punnett square is simulative without being emulative of inheritance processes (i.e., simulative in that Punnett squares imitate content by generating allele permutations, but not emulative in that they do not imitate the processes of gametes combining that produce these permutations, per Moulton & Kosslyn, 2009; Stewart, 1982 makes a similar case without using that particular terminology), in essence ignoring the biology, it is also just plain easier to memorize for a couple of genes than deriving and grounding a generalized process, diverting attention from the biology. Her suspicion in this regard was borne out in an assessment item following the implementation, wherein she asked students for results that could have been determined from either a Punnett square or the mathematical expression they had recently derived; the majority chose the Punnett square.

Coding and Themes

Clearly, one code that all the participants shared explicitly was the absence from their backgrounds of mathematical expressions for biology. Had the participants been entirely middle school teachers this would not have been unexpected; Kuenzi

(2008, p. 10), citing a Congressional Research Service analysis of the School and Staffing Survey, reports, "Among middle-school teachers, 51.5 % of those who taught math and 40.0 % of those who taught science did not have a major or minor in these subjects." But there were high school teachers who participated in BLOOM, as well, and their mathematics backgrounds were similarly thin, and that does not correspond with the national survey data.

It is not that the mathematics was unavailable to the teachers, but that it was not stressed as an application to biology for them either at the secondary level or university, unless in the service of a capstone research project or the overlap between biology and chemistry. This appears to be a contextual code, describing a situation that participants accept as historically emic for themselves as biology teachers, but to which they now can only react, rather than affect, due to its nature as a fait accompli.

Two major kinds of these reactions needed to be coded. One appears in Carol's reflection about the difficulties of taking on the mathematics integration by herself:

> It's very hard for an individual teacher to get where [BLOOM] got me, and I think that's a frustrating aspect from teachers. It's not so much that we don't want to put in the math, but kind of like what you said, I seemed like I had to be convinced this would work. In a sense I do, because I need someone to provide that with me, 'cause it's not something that's been given to me ever before, and I need to see it happen in the classroom. But I can't create something that I don't even know about. …I was very amazed at the [previous version of the module], and thought, "Oh my gosh, I wish I had a team of five people working on a project for me for next month." Like, that's how much needs to go into something like that, and I think that's another reason why teachers are hesitant … y'know, you're gonna' get a biology textbook with worksheets and transparencies, that's what you're gonna' get. I don't have any way to incorporate math, unless it's my own creation with no background, and no [professional development], y'know, nothing.

Emma provides a similar view:

> In the science classroom, having to teach so much content in a short amount of time, I think trying to find a way to incorporate the math – on my own – is just a bigger challenge for me at this point. I've only been teaching for five years, so maybe I'm still trying to learn how to teach the science content? … This is probably my first year that I actually feel comfortable that I don't have to keep developing things. Y'know, the things that I've already developed have been working. I'm just kind of tweaking things here and there. I might be able to incorporate math here or there in something in my tweaking for future use, but to start from scratch and develop a whole unit or a whole lesson that does incorporate the math? I'd probably say, no, I wouldn't.

Without BLOOM, the integration of mathematics and biology would rest on participants' shoulders alone, because they couldn't expect any buy-in from the mathematics departments or administrations at their schools. Collaboration might still get plenty of lip service, but its implementation rarely occurs. As a result, coding also involved participants' recognition of limitations.

Theme: Interpreting an Unprecedented Emphasis on Mathematics as Instructional Improvement

The other major reaction is, of course, taking the next step after one decides that something needs to be done, but that the something is beyond one's individual resources. Less verbally explicit, it is instead operationalized by partaking in research that doesn't just passively recognize the absence of mathematics, but actively promotes mathematical expression as description and application of biological processes. While there lingered a hesitance among some participants to implicate mathematics directly, problem solving seemed an acceptable way to frame modifications to biology content. From their point of view, even those of their students who were visually-oriented could solve a problem, but only those who were mathematically- oriented could solve a problem mathematically.

The qualities of integrating mathematics and biology dealt with so far run along a continuum of participants' beliefs that mathematics can be beneficial to students in their biology classes: recall of a mathematical absence in their own biological education, perception of that continued absence (to varying extents) in their own teaching, their realization of an inability to address the absence individually, and their decision to participate in research that provides them with one example of how to deal with that absence.

The theme of becoming an expert teacher is evident, of course, but this is tempered somewhat for the public school teachers. Even though increasing student sophistication in problem solving is an acceptable enough goal for the parochial school teachers, there is further conflict to be resolved in the public schools with how problem solving applies to standardized testing.

Both parochial and public school teacher participants saw their higher level students struggle with having to work through the perceived ambiguities of the biology content. Likewise, most of these teachers also observed noticeably more participation than was typical from their lower level students, as a mathematical expression was derived through whole class negotiation. Carol described it as students leaning on each other for details during the implementation, but walking away with the big ideas. This is the basis for lifelong learning, in that recalling the specific content is not as important as the belief in one's ability to purposefully construct the content from available information when needed. She had also previously noted:

> I think too many times we do stuff with the kids, and don't reflect on it; and then they don't really know why they did it ... I'm taking advantage of the structure that you have already provided to me and using the time to reflect and hear how people are thinking ...' cause I think the ultimate goal anymore is, "learn how you think," evaluate your thinking, and I heard you guys say that someone was stressing that even; I mean, that's, that's what's kind of been pounded in teachers lately; the twenty-first century…

This merely highlights the conflict that public school teachers expect to arise from BLOOM versus standardized testing's format emphasizing the recall of details, to which parochial schools are not beholden (even though many actually do participate in standardized testing in order to provide a benchmark for their students' academic performance).

That noted, the National Board for Professional Teaching Standards (2012) does report a correlation for students having National Board Certified teachers (i.e., teachers seeking to increase their expertise, as the participants here are doing) and those students' higher scoring performance on standardized tests. Therefore, the question of how that conflict actually affects students is in need of further scrutiny.

Theme: Teachers Resolving Ambiguity for Themselves and in Preparation for Helping Their Students

A second theme was one of recognizing and dealing with contestability in biology content, at first apprehended as ambiguity by both participating teachers and their students. An example is that of the monohybrid cross discussed previously, where both parents are heterozygous, having a dominant and recessive allele, say *Aa*. Algebraically, there are four possible offspring genotypes: *AA, Aa, aA*, and *aa*. Genotypically, there are three: *AA, Aa*, and *aa*. This is so because *Aa* is indistinguishable from *aA* (most of the time, with this disclaimer: considering only one gene, an offspring having an A from mom and an *a* from dad is exactly the same as another having an *A* from dad and an *a* from mom, unless imprinting is involved, and the topic of imprinting, while correct, introduces complexity of limited utility at this level).

However, because there are twice as many *Aa* as either *AA* or *aa* the difference in relative ratios is not trivial, whether in a mathematical expression dealing with allelic permutations in the offspring from a single parental pair (especially when considering multiple genes) or in the subsequent generations for a much larger population.

The point is: there are reasons to consider both views, and it depends on the circumstances as to which of those views is relevant. Furthermore, this point is not peculiar to genotypes (e.g., *Aa* versus *aA*), or even to genotypes as those relate to phenotypes (because genotypically similar *Aa* and *aA* not only express the same trait phenotypically, but also share that phenotype with genotypically dissimilar *AA*), extending throughout biology and beyond. As Bruner, Goodnow, and Austin (1956), state:

> Do such categories as tomatoes, lions, snobs, atoms, and mammalia exist? In so far as they have been invented and found applicable to instances of nature, they do. They exist as inventions, not as discoveries. (p. 7)

It is easy to see that there are any number of biological constructs, such as taxonomy and speciation, that are not universally settled, making much of biology a wicked problem (Rittel & Webber, 1973) in the truest sense of the term.

Indeed, the BLOOM design challenge was purposefully contrived in order to require students to confront and define a concept of rarity and what that meant in various contexts. For example, what rarity means to a zoo administration capable of breeding a special gecko that is genetically consistent generation after generation, must be very different from what rarity means to the rest of the world in which such a special gecko does not even appear. For the zoo administration the special gecko is disproportionately small with respect to the population of geckos overall, while at the same time being incapable of reproducing any other genotype. To the rest of the world, the artificially selective breeding that resulted in this special gecko could not have occurred otherwise, thus confining this special gecko to a singularity that excites disproportionate curiosity. To gecko collectors and breeders who recognize the time and expense involved in the breeding sequence, the special gecko represents a disproportionate value exceeding that of geckos they have already encountered.

This theme of recognizing and dealing with contestability in biology content indicated teachers' efforts at coming to grips with how to support their students' comparisons of contexts in which to locate and develop views regarding biological constructs. When teachers can impel students to deal with problems for which the context is not immutable and thus must be established by the student, it stands to reason that students' self-efficacy improves as a result (at least with respect to these kinds of problems).

This has a profound effect on the models that students use in order to understand biological processes, and what teachers should expect when eliciting those models. In order for students to self-assess their models, Lesh, Hoover, Hole, Kelly, and Post (2000, p. 619) posit that those students should be able to judge when their responses need to be improved, or when responses need to be refined or extended for a given purpose, so they can determine when they have finished. The alternative is to continually ask, "Is this good enough?" (known as "satisficing," as coined by Simon, 1957), being the circumstance that they actually do face in professional practice. That also means an encounter with the phenomenon of mathematics-in- biology might not have been entirely satisfying for participants as far as sense making. Carol was not alone in saying:

> For me it was the, the math [in professional development] ... 'cause I didn't know I was, I mean, I knew because I know what a dihybrid is, I was supposed to get up, get sixteen possibilities. I knew that. But I didn't know mathematically how to show that, so that's why we just kept doing it randomly, to see if we, what number we ended up with. So the math is what held me up, like, how do they know when to stop?

To summarize, as shown in Table 14.1, two themes emerged from the interviews and observations, the first oriented toward increasing teaching expertise in one discipline by having to address another. That is, since mathematics was neglected in their own secondary education, these participants had had to deal with both the perception and perpetuation of biology as the math-less science, which was not beneficial for preparing their students who had any interest in biology for university study or a career in the field. For some participants this theme played out as a continuation of their efforts to include more mathematics, while for others (e.g., Carol and to some extent Emma) it was recognizing that mathematics needed to be addressed,

Table 14.1 Summary of participants and themes

Theme	Alice	Betty	Carol	Dorothy	Emma
Increasing Biology Teaching Expertise by Use of Another Discipline	Never in doubt, BLOOM materials could not have happened soon enough to please her	Similar to Alice, but willing to sacrifice some students' progress in order to maintain the entire class at about the same level	A repeat implementer, like Alice and Betty; out of a strongly felt duty to her students she at first made charts for herself in order to understand the BLOOM material, and weaning herself from reliance on the Punnett square	Most critical of the BLOOM materials, but also most willing to experiment with them alongside her students, without being entirely sure of the outcome beforehand	Most willing to implement BLOOM materials with fidelity, based on her practice of looking for content that she felt would interest her students
Perception of Emergent Student Self-Efficacy	BLOOM materials contributed, but would have happened anyway	BLOOM materials contributed, but absenteeism prevented optimal progress	Wary at first of her students' abilities to handle BLOOM materials, but progressively convinced by results	Similar to Carol, except that her typically higher level students faltered until they got used to the ambiguity of design challenges having multiple solutions	Similar to Dorothy

and if that meant biting the bullet in order for them to improve as biology teachers, then so be it.

The other theme remained more implicit than the first, perhaps because it was difficult to resolve and thus required some effort to follow. In any event, teachers' uneasiness from having to keep track of students' multiple solutions was assuaged somewhat when students who previously had been on the periphery of class discussions were able to assert their findings with confidence, having discovered their own abilities while the usual leaders in class were faltering without detailed direction. Clearly, exchanging the confusion of one student for that of another is not an end unto itself, but introducing one set of students to improved self-efficacy while another learns to deal with unprecedented yet desirable difficulty (Bjork & Bjork, 2006) needs to be pursued with additional study until those conditions can be replicated consistently.

Discussion and Conclusions

Three of the participant teachers in this study had implemented earlier versions of the BLOOM module, and to describe their first impressions of this implementation as naïve (Moustakas, 1994, pp. 14–16) is probably not as accurate as it would be for the others. In reading about Alice, Betty, and Carol, and what each had to say the reader should keep in mind that these teachers' familiarity with the materials and day-to-day expectations are likely to be grounded in typification (Gubrium & Holstein, 2000, p. 489) already. In fact, none of our participants was a novice in the classroom, with the least experience at 5 years or more, and some cultural and systems reifications of practice (Berger & Luckmann, 1967) may have inured them from, or impelled them toward, testing their own models of inheritance and reforming their own curriculum.

Keeping that in mind, it must be attended to that participants were not averse to introducing socio-mathematical norms for student class negotiation of mathematical expressions (a consideration of importance to professional development suggested by Elliott et al., 2009) and concepts of rarity, which probably would have been foreign to their own mathematical backgrounds. And they did try to embrace this attitude themselves in professional development. Yet, had the BLOOM developers addressed one particular limitation of professional development, then participant effort and effect might have increased substantially: facilitating continuous online contact among participants by providing a shared space for them to post questions, ask for help, and display big ideas they came up with themselves.

What is fairly certain is that the engineering practices (described in NRC, 2012, pp. 41–82 and Appendix F) that informed the BLOOM design challenge and required the student derivation of a mathematical expression in order to detail a solution, remained foreign to even the repeat participant teachers. This was apparent from the participants reflecting as a group at the professional development session after the module implementation had concluded; participant were asked about which aspects of the module they thought were directly related to mathematics and which to engineering. While it was easy for participants to flag the mathematics, their further responses indicated no distinction on their part between their typical classroom procedures and what they took engineering to be at the time the question was asked. In other words, if engineering had occurred during implementation, it was not purposeful engineering of which participants were aware or that they had intended or planned as such.

This should not be surprising when one considers two aspects of the implementation. The first is the NRC's eight categories of engineering practice one of the foundations for developing the BLOOM module:

- Asking questions (for science) and defining problems (for engineering)
- Developing and using models
- Planning and carrying out investigations
- Analyzing and interpreting data
- Using mathematics and computational thinking

- Constructing explanations (for science) and designing solutions (for engineering)
- Engaging in argument from evidence
- Obtaining, evaluating, and communicating information

Given these descriptions alone, it would be expected of participating teachers to read down the list, and check check check each of these items off in turn, because the headings appear familiar, and participants want to head off to the biology and mathematics content anyway. Those are the entries to the module for which their experiences have prepared them, after all, and of course they do all the activities on this list.

Yet it is not until one parses the items, as the NRC does (pp. 41–82) when pitting theoretical explanation versus useful enactment (rather than as the perpendicular axes of Pasteur's quadrant in Stokes, 1997), that what scientists do becomes distinct from what engineers do under each item. While this distinction is handy in promoting a variety of directions for classroom activities under each heading, it is not clear that raising awareness of engineering in apposition to science (thus maintaining the linear hierarchy of the results from primary basic research being transferred to secondary applied research; the analysis dominance over design that prompted all the NSF funded research as previously noted) is the most beneficial for teachers or students.

If Mathematics Was Something Daunting to Be Encountered, What Will Engineering be?

A second aspect of the implementation that might have restricted participant attention to engineering is the relative emphasis on mathematics day in and day out, versus the fewer periods of class time spent on the design challenge, leaving correspondingly fewer opportunities for participating teachers to define an engineering design process for themselves and then refine that with class discussions. There is not only a lot to do for design, but there is a lot to accept about it before doing can occur. For example, Carr et al. (2012, p. 18) provide an apparently comprehensive list of what engineers do (as currently being taught in P-12 curricula), from identifying criteria, constraints, and problems to describing the reasoning to designs and solutions to producing flow charts, system plans, solution designs, blue prints, and production procedures. And every single one is true, but those are activities that experienced engineers do, once they have already encountered and internalized the fundamental property that one enters a design process without any idea about what the problem is, much less what all the solutions might entail. All of that has to be determined, sometimes over and over again until clear enough to make progress. One participating teacher, Carol, displayed substantial anxiety about letting her students leave at the end of a class without a clear resolution, as if she were holding out on her end of a student-teacher contract that guaranteed a singular correct answer to every question she raised.

On the other hand, it did not take much convincing at the BLOOM professional development sessions to get these participating teachers to pose the design challenge as a student driven effort. While not a trivial achievement, this went much more smoothly than the developers anticipated, because there were teachers in previous implementations who were not at all convinced that their students could handle the challenge and thus saw fit to supplement and modify the BLOOM instructional materials at their discretion, thereby reducing the student- driven nature of the materials. In fact, the BLOOM developers were careful to iteratively prototype what Hashweh (2005) refers to as "Teacher Pedagogical Constructions" in order for teachers to have ready-made routines at hand for identifying and discussing naïve concepts with their students.

Finally, there is no dearth of research on either biology teachers learning to teach biology or mathematics teachers learning to teach mathematics, but studies of a teacher in one discipline making sense of what familiar subject matter looks like through the lens of another are somewhat more rare. For that teacher further to touch upon subject matter altogether foreign to P-12, as engineering is for the most part, has seemed up to now out of the question. Yet, standards related to engineering are headed straight for those classrooms, as previously noted, and, for good or ill, it is no longer desirable for biology to offer a refuge for those students who enjoy science without mathematics. That renders the implications of this study (i.e., that biology teachers motivated to improve their understanding and teaching of biology will take the risk of exposing students to novel ways of mapping biology onto other disciplines) of great interest to immediate impending instructional practice.

Key Insights: The Take-Aways

One very important observation to be communicated here is that, during this study, the teachers who entered with anxiety about mathematics and engineering came to terms not only with what they perceived as their personal or historical deficiencies regarding those fields, but also with their apprehensions about incorporating those unfamiliar approaches in their day-to- day instructional methodology. It is no mystery that a large part of this achievement was due to the exposure of all of the participating teachers to one another in the reflective portions of professional development as the implementation was taking place and then afterward. Certainly, those who had more confidence in their own abilities to handle the design challenge displayed and transferred some of that self-efficacy to their colleagues as the implementation ran its course. When other teachers who had been hesitant returned for additional rounds of implementation it was likely due to both previous instructional results and encouragement of the will to persist (itself engendered from friendships that had been struck up) that had produced those results.

Likewise, there were participating teachers who felt at first that the design challenge would prove beyond their students' abilities. Their expectation was that an encounter with the ambiguity of apparently wicked (albeit genuinely well-defined) problems that needed to be deconstructed and attacked without explicit step-by-step direction would inhibit their accustomed low performers into silence. As it turned out, because their otherwise already self-assured high performers needed to collect and regroup in the face of a strangely presented problem, the door was left open for actual collaborative input from those who had shied away from that before.

In closing, one notes that the participants brought a previously reified convention under scrutiny, betraying the haven against mathematics that biology had become at the secondary level. While it is not overreaching to declare this as courageously critical reflection for some of them, it is certainly overdue for them to correct this disservice to secondary level biology students and the sciences in general. That is, providing students with a clearer picture of what professional biologists (and, to some extent, engineers) can do and are expected to do enables them to make better informed choices about their career paths and interests than was possible before.

Appendix A

Protocol for first interview regarding experimental biology unit questions: teachers reflecting on mathematics proposed for inheritance instruction. We realize that experimental content might work for some students and not others. Please tell us the weak points as well as the strong ones.

Category 1: Personal Justification for Increasing Mathematical Exposure/Awareness/Mastery in General Studies, and in Biology Specifically

What math are you comfortable using off the cuff? Is the math you're using for the unit inside or outside your zone of comfort? [prompts: algebra and variables; geometry and progressions]

In your opinion, what place **does** math have in biology instruction? [prompts: on a continuum from good to neutral to bad, say, or with good being an important tool for understanding biological processes and their range and limitations]

In your opinion, what place **should** math have in biology instruction?

You can think of these next questions as ones of did you: learn and then retain the math through reuse; learn and then forget from disuse (certainly my case); or were you never exposed to it?

How was math used to define inheritance concepts when you were:

- A student in secondary school and university
- Learning to teach
- Since you've been at the present school [prompt: depending on who sets policy, well-defined administrative or departmental item?]

The National Research Council says this as part of its framework: **Mathematics serves pragmatic functions as a tool – both a communicative function, as one of the languages of science, and a structural function, which allows for logical deduction. Mathematics enables ideas to be expressed in a precise form and enables the identification of new ideas about the physical world.** Does that support how you feel about introducing math into biology? (2012, p. 64) [National Research Council of the National Academies (2012). *A framework for K-12 science education: Practices, crosscutting concepts, and core ideas.* Washington, DC: National Academies Press.]

Does this support what textbooks show or say about use of math in biology?

How did you use math in inheritance instruction before BLOOM? For example, did you use math to explain, calculate, or verify inheritance concepts for yourself before BLOOM?

Was it necessary for you to relate the math you used then to actual biological concepts and processes, or was it sufficient to find a reliable widget for calculation (e.g., a Punnett square) without investigating its limitations as a representation of a biological processes such as independent segregation, independent assortment, gamete formation?

Did you use math on any assessments when teaching inheritance before BLOOM implementation?

Category 2: Reflection on Interaction with Unit Content

When did you need to rely on math during the implementation: can you remember when math was helpful or any times when it was harmful to students' progress or understanding? [prompts: defining combinations; making combinations; counting combinations; predicting combinations, comparing combinations expected theoretically versus observed empirically]

Did you recognize any difficulty that the materials introduced or made worse, that might have gotten in the way of student understanding?

Did you include any items related to math on assessments subsequent to the implementation, and why?

Do you anticipate any circumstances that would cause you to include such items or revise the structure of your exam? [prompts: response to standardized testing of science, administrative or departmental directive]

How do you make sense of the concepts and the sequence of presenting rules in the BLOOM materials? [prompt: inheritance, combinations, expression, design as plan with scientific explanation]

Appendix B

Protocol for second interview regarding experimental biology unit questions: teachers reflecting on math proposed for inheritance instruction. We realize that experimental content might work for some students and not others. Please tell us the weak points as well as the strong ones.

Category 1: Triangulation of Data Analysis

Please look over the section for which your pseudonym is indicated. What do you think is inaccurate?
How would you change that to be accurate?

Category 2: Self-assessment Using the Design Challenge

At what stage of the implementation did you understand what the design challenge was asking students to do? [prompts: professional development, review on my own, while helping students, never really sure]
At what stage of the implementation did you feel confident in answering the design challenge yourself?

References

Ball, D., Thames, M., & Phelps, G. (2008). Content knowledge for teaching: What makes it special? *Journal of Teacher Education, 59*(5), 389–407.
Berger, P., & Luckmann, T. (1967). *The social construction of reality: A treatise in the sociology of knowledge*. New York: Anchor Books.
Bjork, R., & Bjork, E. (2006). Optimizing treatment and instruction: Implications of a new theory of disuse. In L.-G. Nilsson & N. Ohta (Eds.), *Memory and society: Psychological perspectives* (pp. 109–133). Hove/New York: Psychology Press.
Bruner, J., Goodnow, J., & Austin, G. (1956). *A study of thinking*. New York: Wiley & Sons.
Cakir, M., & Crawford, B. (2001, January). *Prospective biology teachers' understandings of genetics concepts*. Paper presented at the 2001 annual international conference of the association for the education of teachers in science, Costa Mesa, CA (ERIC Document Reproduction Service No. 463596), Retrieved May 24, 2012, from http://www.eric.ed.gov/PDFS/ED463956.pdf
Cantrell, P., & Robinson, M. (2002). How do 4th through 12th grade science textbooks address applications in engineering and technology? *Bulletin of Science, Technology & Society, 22*(1), 31–41.
Carr, R. L., Bennett, L. D., & Strobel, J. (2012). Engineering in the K-12 STEM standards of the 50 U.S. states: An analysis of presence and extent. *Journal of Engineering Education, 101*(3), 1–26.

Common Core State Standards Initiative. (2011). Retrieved March 31, 2012, from http://www.corestandards.org

Cox, C. (2009). *Legitimization of subject matter in an undergraduate architectural design program: A cultural and systems theory analysis*. Doctoral dissertation, Retrieved May 22, 2012, from Proquest dissertations and theses (AAT 3374470).

Davis, E., & Krajcik, J. (2005). Designing educative curriculum materials to promote teacher learning. *Educational Researcher, 34*(3), 3–14.

Drake, C., & Sherin, M. (2009). Developing curriculum vision and trust: Changes in teachers' curriculum strategies. In J. Remillard, B. Herbel-Eisenmann, & G. Lloyd (Eds.), *Mathematics teachers at work: Connecting curriculum materials and classroom instruction* (pp. 321–337). New York: Routledge.

Elliott, R., Kazemi, E., Lesseig, K., Mumme, J., Carroll, C., & Kelley-Petersen, M. (2009). Conceptualizing the work of leading mathematical tasks in professional development. *Journal of Teacher Education, 60*(4), 364–379.

Elmore, R. (1996). Getting to scale with good educational practice. *Harvard Educational Review, 66*(1), 1–26.

Gubrium, J., & Holstein, J. (2000). Analyzing interpretive practice. In N. Denzin & Y. Lincoln (Eds.), *Handbook of qualitative research* (2nd ed., pp. 487–508). Thousand Oaks: Sage.

Hashweh, M. (2005). Teacher pedagogical constructions: A reconfiguration of pedagogical content knowledge. *Teachers and Teaching, 11*(3), 273–292.

Kleickmann, T., Richter, D., Kunter, M., Elsner, J., Besser, M., Krauss, S., & Bumert, J. (2013). Teachers' content knowledge and pedagogical content knowledge: The role of structural differences in teacher education. *Journal of Teacher Education, 64*(1), 90–106.

Kuenzi, J. (2008). *Science, technology, engineering, and mathematics (STEM) education: Background, federal policy, and legislative action*. Congressional research service reports, Paper 35. Digital Commons at University of Nebraska, Lincoln, NE. Retrieved June 20, 2013, from http://digitalcommons.unl.edu/crsdocs/35/

Lesh, R., Hoover, M., Hole, B., Kelly, A., & Post, T. (2000). Principles for developing thought-revealing activities for students and teachers. In A. Kelly & R. Lesh (Eds.), *Research design in mathematics and science education* (pp. 591–646). Mahwah: Erlbaum.

Lincoln, Y., & Guba, E. (1985). *Naturalistic inquiry*. Newbury Park: Sage.

Moll, M., & Allen, R. (1987). Student difficulties with Mendelian genetics problems. *American Biology Teacher, 49*(4), 229–233.

Moulton, S., & Kosslyn, S. (2009). Imagining predictions: Mental imagery as mental emulation. *Philosophical Transactions of the Royal Society B, 364*, 1273–1280.

Moustakas, C. (1994). *Phenomenological research methods*. Thousand Oaks: Sage.

National Academy of Engineering and National Research Council of the National Academies, Committee on K-12 Engineering Education. (2009). Engineering in K-12 education. In L. Katehi, G. Pearson, & M. Feder (Eds.). Washington, DC: National Academies Press.

National Board for Professional Teaching Standards. (2012). Retrieved June 11, 2012 from, http://www.nbpts.org/

National Research Council of the National Academies. (2012). *A framework for K-12 science education: Practices, crosscutting concepts, and core ideas*. Washington, DC: National Academies Press.

Next Generation Science Standards. (2013). Retrieved June 18, 2013, from http://www.nextgenscience.org

Ralston, P., Hieb, J., & Rivoli, G. (2013). Partneerships and experience in building STEM pipelines. *Journal of Professional Issues in Engineering Education and Practice, 139*(2), 156–162.

Rittel, H., & Webber, M. (1973). Dilemmas in a general theory of planning. *Policy Sciences, 4*(2), 155–169.

Robinson, M., & Kenny, B. (2003). Engineering literacy in high school students. *Bulletin of Science, Technology & Society, 23*(2), 95–101.

Rossman, G., & Rallis, S. (2003). *Learning in the field: An introduction to qualitative research* (2nd ed.). Thousand Oaks: Sage.

Shulman, L. (1986). Those who understand: Knowledge growth in teaching. *Educational Researcher, 15*(2), 4–14.

Simon, H. (1957). *Models of man: Social and rational*. New York: Wiley.

Stewart, J. (1982). Difficulties experienced by high school students when learning basic Mendelian genetics. *The American Biology Teacher, 44*(2), 80–82, 84, 89.

Stokes, D. (1997). *Pasteur's quadrant: Basic science and technological innovation*. Washington, DC: Brookings Institution Press.

Sweller, J. (2011). Cognitive load theory. In J. Mestre & B. Ross (Eds.), *Cognition in education* (The psychology of learning and motivation, Vol. 55, pp. 37–76). Oxford: Academic.

Tolman, R. (1982). Difficulties in genetics problem solving. *American Biology Teacher, 44*(9), 525–527.

van Manen, M. (1997). *Researching lived experience: Human science for an action sensitive pedagogy* (2nd ed.). London: Althouse Press.

Chapter 15
Final Commentary: Connecting Science and Engineering Practices: A Cautionary Perspective

Michael P. Clough and Joanne K. Olson

Introduction

Ask most any science teacher what overarching aims they have for student learning in science and their responses will include goals like those appearing in Table 15.1. Many of these goals have a long history in science education (DeBoer 1991). Unfortunately, the well-documented failure of common science teaching practices to promote these goals has an equally long history. The *Next Generation Science Standards* (NGSS Lead States 2013) is the latest of several science education reform efforts in the United States to promote goals like those in Table 15.1, preceded by the *National Science Education Standards* (NRC 1996), *Benchmarks for Scientific Literacy* (AAAS 1993) and *Project 2061* (AAAS 1989). A prominent aspect of the *Next Generation Science Standards* is their emphasis on connecting science and engineering practices.

Appropriately connecting science and engineering practices clearly has the potential to aid in achieving often-stated goals for science education. For instance, robust and long-term learning of science concepts and practices is enhanced when they are meaningfully and repeatedly linked to other concepts and practices in various contexts. This is why the learning cycle (Abraham 1997) and subsequent variations of that instructional model all emphasize the importance of application, both for bolstering understanding of previously addressed science concepts and practices and also for setting a stage for introducing new concepts and practices. Including engineering concepts and practices provides new and potentially valuable contexts that may be used at times for applying science concepts and for introducing new science concepts.

M.P. Clough (✉) • J.K. Olson
Center for Excellence in Science, Mathematics and Engineering Education, Iowa State University, Ames, IA, USA
e-mail: mclough@iastate.edu

Table 15.1 Goals for students in science education (Modified from Clough 2015)

Demonstrate robust understanding of fundamental science and engineering ideas and practices
Exhibit an accurate understanding of the nature of science, technology and engineering
Effectively identify and solve problems
Be creative and curious
Use critical thinking skills
Effectively use communication and cooperative skills
Participate in working towards solutions to local, national, and global problems
Set laudable goals, make decisions, and accurately self-evaluate
Access, retrieve, and use credible scientific knowledge in socio-scientific decision-making
Convey self-confidence and a positive self-image
Express how a robust science education can promote personal and societal well-being

Unfortunately, policymakers and science educators have too often wrongly emphasized particular science education goals at the expense of others (DeBoer 1991) and largely ignored their interconnected and synergistic nature (Clough et al. 2009). For instance, developing a robust understanding of science and engineering content and practices *requires* attention to other science education goals such as critical thinking, problem solving, communication skills, the nature of science and others. And because what a learner understands impacts thinking, achieving any of the goals in Table 15.1 is impacted by the depth of understanding regarding content and practices. Thoughtful and overt attention to the nature of science and engineering can assist in helping students understand the similarities and differences between science and engineering, the importance of both, how they are intricately intertwined, and their respective strengths and limitations. This would then provide context for addressing the crucial, but widely misunderstood, nature of technology (Clough et al. 2013). When teachers overtly and effectively engage students in thinking about and linking science and engineering concepts and practices in appropriate and meaningful ways, learners are in a better position to exhibit actions that promote and reflect the student goals appearing in Table 15.1.

While including the teaching and learning of engineering concepts and practices in the science curriculum have merit, significant and legitimate concerns do exist with the kind and level of emphasis being placed on engineering practices. Generally speaking, the science education community has been remiss in its uncritical adoration of engineering and the inclusion of engineering concepts and practices in the science curriculum. Important concerns exist about K-12 engineering education in general and its inclusion in the science curriculum in particular. Raising these concerns is not an effort to maintain the status quo or a negative view of engineering and technology, but rather a thoughtful and scholarly effort to ensure students receive the best possible science and engineering education. Considerable thought and caution ought to occur in light of the marked changes being proposed regarding the content of the *science* curriculum in order to infuse engineering concepts and practices.

Connecting Science and Engineering Practices: Concerns and Issues

Overemphasizing Job Preparation as the Primary Purpose for Schooling and Science Education

The impetus behind emphasizing STEM in schools, including connecting science and engineering practices, is driven to a large extent by policymakers' and business leaders' desire to produce more engineers and grow the technical workforce in hopes of spurring economic growth and maintaining national security. While prosperity and security are undeniably important, casting the primary purpose of schooling and school science in terms of job preparation, economic growth and national security is ill-conceived. Economic growth and national security are, at best, only very loosely tied to the general state of schooling, and the need for a technical workforce hardly provides an impetus for most students to value STEM learning. Job preparation, when puffed up as the primary reason for schooling and STEM coursework, is equally bankrupt. Most students will not choose STEM careers, nor should they. Moreover, STEM education efforts are increasingly marginalizing the value of the humanities. Postman (1995), arguing against economic utility as a satisfactory reason for schooling, wrote:

> Putting aside its assumption that education and productivity go hand in hand, its promise of providing interesting employment is, like the rest of it, overdrawn. …If we knew, for example, that all our students wished to be corporate executives, would we train them to be good readers of memos, quarterly reports, and stock quotations, and not bother their heads with poetry, science, history? I think not. Everyone who thinks, thinks not. Specialized competence can come only through a more generalized competence, which is to say that economic utility is a by-product of a good education. (pp. 30–31)

STEM *education*, as opposed to a mere training, ought to be a concern of all educators. That first demands thoughtful examination of the exaggerated and ballyhooed rationales for connecting science and engineering practices. Working with and ennobling those in the humanities would certainly result in far more noble and ethical rationales for STEM *education*.

In *The End of Education*, Postman (1995) argued that without convincing transcendental ends, "schools are houses of detention, not attention". Perhaps the lack of a compelling purpose is why schools and prisons share so many similarities in how they are structured and administered. We often forget that:

> ...compulsory schooling *is* a sustained exercise in force in which individual freedom of action and freedom of thought are interfered with. Individuals are incarcerated for years on end and made to act in ways they would not freely choose to and to acquire beliefs and skills they would not freely choose to acquire. (Davson-Galle 2008, p. 684)

Within compulsory schooling is mandatory course work, including science which increases significantly in middle school and high school. Even within science courses, high stakes testing and national science standards cast an obligatory shadow over what should be taught and learned. As with schooling in general, compelling reasons ought to exist for privileging particular purposes for science teaching and learning because such decisions tend to marginalize other purposes.

A meaningful schooling, one worth requiring *all* students to complete, ought to "persistently and earnestly engage students in a manner that models and promotes action resulting in attitudes, understandings, and skills that make for a well-educated (as opposed to trained), self-actualized, caring, curious, motivated, responsible and reflective human being" (Clough 2015, p. 25). A science education that promotes the goals appearing in Table 15.1 would be an indispensable part of achieving those noble ends of schooling.

Marginalizing Science Content

While the possibilities noted in the introduction to this chapter support addressing engineering concepts and practices to some degree in the science curriculum, at what level such instruction should occur is deserving of far more discussion and analysis than has thus far occurred. While engineering concepts and practices should be infused (i.e., tightly linked to fundamental science ideas) rather than merely added into an already overstuffed science curriculum, the reality is that regardless of how engineering practices are integrated, they will take much time if done well and some science content will have to go. History is not on our side regarding depth replacing coverage, particularly in the United States. The "mile wide and inch deep" science curriculum—both formal and enacted—persists despite almost three decades of effort to promote depth of understanding of fundamental science ideas (Banilower et al. 2013; Goodlad 1983; Schmidt et al. 1997; Weiss et al. 2003). Moreover, adding engineering into the science curriculum is not merely a matter of depth replacing coverage, but adding non-science concepts and practices. Science educators ought to be gravely concerned about what science content might be downplayed or sacrificed, particularly when not all science content is equally amendable to engineering connections. For example, to what extent might particular fundamental ideas in fields of study such as biological diversity, biological evolution, animal behavior, ecology and others be marginalized in efforts to incorporate engineering in the biology curriculum?

Devaluing Basic Science

While both science and engineering produce knowledge, generally speaking, science is primarily directed toward developing knowledge regarding the natural world while engineering is directed toward developing knowledge regarding the development of technology (e.g., artifacts, processes, and procedures) that extend human capacities to achieve a desired end (Clough 2015). Despite these fundamental differences, science, engineering, and technology are so intricately linked together that people often judge the value of science research by how clearly and quickly its knowledge may be useful for supporting engineering design and technology development. People readily grasp the value of engineering, and often think that all science ought to be in some way targeted toward understanding aspects of the natural world that will likely be useful in technology development. Most science research is directed toward that end (often called applied science research). However, many people are surprised and dismayed to learn that many other scientists conduct what is called basic science research (sometimes referred to as pure or fundamental science) that appears to have little if any possible application to human wants or needs.

When engineering application is emphasized, applied science research is privileged over basic science research—that is, science research that is done for the sole purpose of understanding the natural world. Thus, emphasizing engineering design in the science curriculum could easily exacerbate the already prevalent problem regarding the lack of support for basic science research. Not understanding the crucial role basic science plays in generating knowledge about nature that no one could foresee as essential in technological advancement, objections are made regarding supporting such research. Astrophysicist Neil deGrasse Tyson (2011), challenges these mistaken notions regarding basic science:

> This notion that science is the path to solve your problems; I think that misrepresents what drives scientists. Do you think when you speak with Brian Green he's going to say, "I am trying to come up with a coherent understanding of the nature of reality so that I can solve people's problems?" Do you think that's what driving him? Do you think I'm being driven when I look at the early universe or study the rotation of galaxies or the consumption of matter by black holes, do you think I'm being driven by the lessening of the suffering of people on Earth? Most research on the frontier of science is not driven by that goal—period! Now, that being said, most of the greatest applications of science that *do* improve the human condition *come* from just that kind of research. Therein is the intellectual link that needs to be established in an elective democracy where tax-based monies pay for the research on the frontier. …The purpose of science is to understand the natural world. And the natural world has, interestingly enough, built within it forces and phenomena and materials that a whole other round of clever people—engineers, in the case of the magnetic reso-

nance imager—these are biomedical engineers basing their patents and their machine principles on physics discovered by a physicist, an astrophysicist at that. So I take issue with the assumption that science is simply to make life better. Science is to understand the world. Now you have a utility belt of understanding. Now you access your tools out of that, and use those, that ever increasing assortment of power over nature, to use that power in the greater good of our species. You need it all.

When teachers are seeking to integrate engineering design and science practices, science may appear to students as existing for the purpose of technology advancement, and science topics that have no clear link to technology may be downplayed or neglected.

Overemphasizing Personally Relevant Teaching and Learning

A popular rationale for the inclusion of engineering in the science classroom is that engineering provides relevance and application of science concepts for students. Inherent in this view is the assumption that students should study those things that they find personally meaningful, and some even assert that students should *only* study what is relevant to them. Dewey noted in 1902 that this perspective causes harm because it assumes "…that a child of a given age has a positive equipment of purposes and interests to be cultivated just as they stand" (p. 193), creating the potential for well-meaning educators to "arrest development upon a lower level" (p. 192). Fixating our attention on that which is "relevant" to children's current interests is also impossible given children's diverse and ever-shifting abilities and interests. A crucial role of education is to engage students in the experiences and knowledge of humankind that transcends students' limited personal experiences. Dewey argued that the curriculum's "genuine meaning is in the propulsion it affords toward a higher level" (p. 193), which means that we must carefully consider what an educated person needs to experience, which may be outside of immediate relevance to students, but broaden their thinking and expand their world. For example, consider that many non-scientist members of the public own small telescopes or subscribe to astronomy-related magazines. Astronomy could be considered irrelevant to the lives of most of the public, yet peering into the night sky is awe-inspiring and enhances our perspectives and lives. Shall we eliminate or reduce the teaching of astronomy, the history of Earth, biological evolution, ecology or myriad other topics because students do not find them particularly relevant? Of course, great teachers do things that make far more likely students will find subject matter interesting, but limiting ourselves to only that which is relevant unnecessarily narrows the curriculum and consequently, narrows children's experience. Thus, engineering need not, perhaps at times cannot, and often should not serve as a gateway for learning science. Much science learning should occur to enhance our understanding and appreciation of the natural world, and make more wise personal choices, not merely because it has an engineering application.

Naïve Adoration of Engineering and Technology

The uncritical adoration of engineering and technology in schooling largely paints a picture that misportrays the impetus for much engineering and the nature of technology. For instance, romanticizing engineering as primarily an empathic endeavor focused on addressing human needs ignores that new technologies are often developed solely for business profit motives (Bunge 2003). The perpetual upgrades of phones, computers, software and countless other technologies illustrate that meeting human "needs" is not the motive for much engineering. This is not cynical, but rather a crucial and more balanced view of what initiates a design process in the real world. Furthermore, much engineering creates unnecessary, unhealthy, and previously undesirable "wants" (Marcuse 1964). Sometimes the impetus for designing new technologies is improving human welfare, but just as often it is not. Some engineers *are* interested in the design process for noble reasons, while others are not. As with most jobs, most are merely doing what their employer tasks them to accomplish.

Moreover, engineering and the resulting technologies are not going to solve the most significant problems that have pervasively plagued humanity. Engineering education objectives in the science curriculum largely ignore and certainly marginalize individual and collective responsibility to make decisions and behave differently in ways that would go much further in mitigating personal, community and world-wide problems (Olson 2013). For example:

> …the unintended consequence of [many] drugs (a technology) is to diminish in many individuals their personal responsibility for adopting healthier habits. That impact extends beyond individual responsibility to societal health care costs that, to a large extent, reflect the eschewing of prudent health decisions in favor of relying on current and possible future medical technology. (Clough 2013, p. 374)

Education regarding science and engineering must convey a more realistic view regarding why engineering efforts alone will not and cannot solve the most pressing human problems just as science alone cannot answer the most meaningful questions that humans ask (Olson 2013).

Because all technologies have pros and cons, at the forefront of engineering and technology education objectives ought to be inculcating among students the habit of examining what is gained and lost with any particular technology. Engineering new solutions to the unanticipated effects of yesterday's technologies does appear in engineering education literature, but those newly developed technologies will, of course, have their own unanticipated pros and cons. This is the case with any technology. For example, communication technologies have undeniable benefits, but also significant drawbacks that largely go unexamined and thus are downplayed or unnoticed. In our "highly advanced technological society", we work more than ever before; aided, encouraged, and at times required by communication technologies such as Microsoft's aptly named *Office 365* e-mail program that ensure we are tethered to the demands of work each day of the year. People now willingly purchase wearable technologies that immediately alert them when a new e-mail, text message, or call has arrived, further tying every waking moment to work and creating a life punctuated by unending interruption and scattered demands on our attention. Intrusions of work in our personal lives that would have only a decade or two ago demanded overtime pay are now accepted as a matter of course along with the gadgets we adopt. Perhaps the emerging wearable communication technology may soon include software titled *Office 24/7* to reflect that every moment of our lives must be available for work.

How communication technologies have, for many people, destroyed any meaningful separation between a job and personal life illustrates another concern regarding the naïve adoration of engineering and the resulting technology—the mistaken view that technology is merely a tool and that the user bears all responsibility for its outcomes (Huesemann and Huesemann 2011; Proctor 1991). Technology *is* a tool for accomplishing some end, but tools are not neutral; they have affordances and limitations that bias and thus change behavior, often in ways we would not knowingly have chosen. We often forget that the communication technologies that make us constantly available are fairly recent. Adopted at first largely for their novelty and for the convenience they provide at times, few foresaw how communication technologies would change our work habits, views regarding the importance of face-to-face interaction, our psyche (e.g., anxieties associated when instant communication technology is inaccessible), and countless other transformations we would not have knowingly have chosen (Brende 2004). Efforts to link science and engineering in school science ought to have at the forefront the need for students to understand the nature of technology (alongside understanding the nature of science) and make clear that without overt attention to examining the purposes and end products of engineering, we may unwittingly have our behavior shaped and changed in ways which we would have never consciously assented (Huesemann and Huesemann 2011; Hull 2013; Postman 1992).

Pedagogical Issues

Dissatisfaction with science education has a long history in the United States. Mind-numbing science teaching practices that fail to mentally engage students and demand little more than superficial recall of information has been pervasive and persistent. Reform efforts emerge, wane, and re-emerge, as do educational fads that policymakers and administrators adopt, seeking magic bullets for complex problems. Unsurprisingly, student learning falls far short of the goals in Table 15.1 that both educators and policymakers champion. At the root of these unending disappointments are superficial considerations regarding what deep learning entails and the complexity of teaching that effectively promotes such learning.

For instance, efforts to have science taught *through* inquiry (pedagogical practices that require and support student learning of science ideas via inquiry) and *as* inquiry (pedagogical practices that help students understand scientific practices and how scientific knowledge is developed) has a rich and long history extending back

Table 15.2 Fundamental requirements for effectively teaching science through and as inquiry

Understanding how students learn
The unnatural nature of much scientific thinking and the counter-intuitive nature of many science ideas (Cromer 1993; Matthews 2015; Wolpert 1992)
Commonly held misconceptions students possess, why they make sense, and their tenacious nature
Content understanding of teachers
Deep and robust understanding of relevant science content
Deep and robust understanding of the history and nature of science
Pedagogical content knowledge related to inquiry in general and specific inquiry lessons
Selection of content, activities and materials within students' zone of proximal development
Teacher behaviors
Asking thought-provoking questions
Asking questions that overtly draw students' attention to the nature of science and science practices
Wait-time I and II
Encouraging non-verbal behaviors
Responding to students' ideas with questions that effectively support and scaffold thinking

at least 150 years (DeBoer 2006). But such teaching is uncommon because doing so demands, at the very least, what appears in Table 15.2. Teaching science through and as inquiry is incredibly complex, and now another field of inquiry, engineering, is expected to be integrated. The desire to have science teachers connect science and engineering practices further complicates matters and places additional demands on science teachers. To what extent science teachers can be reasonably expected to effectively integrate engineering concepts and practices is a formidable, and possibly unreasonable, challenge. The following are just a few of the many pedagogical concerns that must be acknowledged and addressed in order for engineering practices and concepts to be meaningfully integrated in the science classroom:

1. Science teachers, generally speaking, do not exhibit the pedagogical understanding and/or skills appearing in Table 15.2 that are crucial for teaching science and engineering through and as inquiry (Banilower et al. 2013; Schmidt et al. 1997; Weiss et al. 2003).
2. What fundamental engineering concepts should be taught to students in science courses is unclear as well as how such concepts may be meaningfully connected to science ideas in a manner that bolsters science content understanding. Absent this connection, science teachers, particularly at the secondary school level, will unlikely integrate engineering concepts and practices to any appreciable level.
3. Because efforts to integrate engineering in the science curriculum too often fail to meaningfully incorporate science and mathematics concepts, engineering is often misrepresented as merely an iterative trial-and-error tinkering process. It may thus appear as yet another add-on to the science curriculum, exacerbating the mile wide, inch deep problem that has persistently plagued science education.

Moreover, students who learn engineering as primarily a trial-and-error tinkering process will be quite surprised to learn that becoming an engineer has little to do with the tinkering they did in their K-12 schooling, and that engineers must often employ a great deal of higher mathematics and an understanding of science concepts in their work.
4. Maintaining that engineering activities will lead naturally to science inquiry is overly optimistic and ignores what critics of the Science/Technology/Society (STS) curriculum reform effort accurately noted move than three decades ago—that rarely are students motivated to learn difficult science ideas via practical application. What DeBoer (2000) wrote regarding criticisms of the STS approach is relevant to incorporating engineering in the science curriculum:

> The major concern of STS critics was that science would lose out to technological issues and social analysis since technology would become the starting point for virtually all problems that had contemporary interest at the science/society interface. As Kromhout and Good (1983) put it, under such an organization, social issues "do not convey any real understanding of the structural integrity of science" and "the basics simply do not get taught" (p. 649). ...Others were concerned that the goals of STS would not be attainable since most real-world issues involving science and technology are complex and require ...more knowledge of science than can be expected of school students... . (p. 589)

Science Teacher Education Issues

1. Science teachers, generally speaking, do not have sufficient understanding of science and engineering content, science and engineering practices, or the nature of science and technology necessary to effectively promote the goals in Table 15.1. Many states have reduced required science coursework requirements for teaching science to levels that assure teachers are ill-prepared to effectively teach particular science subjects for which they hold endorsements (Olson et al. 2015). Few science teachers have any formal education regarding engineering concepts and practices or the nature of science and technology (Backhus and Thompson 2006; Banilower et al. 2013; Clough et al. 2013). Emerging STEM teaching endorsement requirements are generally a mile wide and inch deep and are insufficient for preparing teachers to promote the goals in Table 15.1.
2. The mile-wide inch-deep criticism of school science curricula is equally relevant to many science teacher education programs, and is exacerbated with the expectation to also prepare science teachers to connect science and engineering. Too few science teacher education programs currently require the extensive pedagogical coursework necessary for preparing teachers to effectively teach science concepts through and as inquiry. While much about effectively teaching science is applicable to connecting science and engineering concepts and practices, additional time will be required to overtly help preservice teachers understand the similarities and differences and effectively integrate science and engineering. Moreover, few science teacher education programs require the necessary

coursework to accurately and effectively teach the nature of science (Backhus and Thompson 2006), and this along with additional coursework regarding the nature of engineering and technology will be crucial for preparing teachers to *educate*, as opposed to *train*, students about science, engineering and technology.
3. Teachers must be taught how to ask questions that overtly draw students' attention to targeted science and engineering concepts and practices in a manner that demands mental engagement with those concepts and practices. Many engineering activities, like many science activities, do not demand attention to underlying concepts and processes, resulting in what Moscovici and Nelson (1998) refer to as "activitymania". Merely having students take part in activities illustrating science and engineering concepts and practices rarely promotes mental engagement and reflection regarding those concepts and practices.

Final Thoughts

Simply because the NGSS call for engineering concepts and practices to be a significant part of the science curriculum does not mean they will be or that needed pedagogical reform will occur. Past reform documents such as *Project 2061* (AAAS 1993), *Benchmarks for Science Literacy* (AAAS 1993), and the *National Science Education Standards* (NRC 1996) were well-conceived, yet had little impact on what occurred in most science classroom. Merely providing curriculum and instructional models designed to integrate engineering concepts and practices, no matter how well conceived, will unlikely result in their being effectively taught.

As a parent and step-parent of a child with type-1 insulin-dependent diabetes, we understand very well the importance of engineering and the benefits of many technologies. And as we noted at the beginning of this chapter, appropriately connecting science and engineering practices has potential to assist in achieving important science education goals. However, our cautionary perspective challenges simplistic rationales and strategies for integrating engineering in the science curriculum, and raises issues that need considerable thought and action for reform efforts to successfully promote a meaningful STEM *education*. Children are far more than future cogs in an economic machine, and we owe them a meaningful and robust science education that is centered on the goals appearing in Table 15.1. College and career readiness would then be a byproduct of rather than the purpose for, science teaching and learning.

References

Abraham, M. R. (1997). *The learning cycle approach to science instruction* (Research matters – To the science teacher, No. 9701). National Association for Research in Science Teaching (NARST). http://www.narst.org/publications/research/cycle.cfm. Retrieved 30 May 2015.

American Association for the Advancement of Science. (1989). *Project 2061: Science for all Americans*. Washington, DC: American Association for the Advancement of Science.
American Association for the Advancement of Science. (1993). *Benchmarks for science literacy*. New York: Oxford University Press.
Backhus, D. A., & Thompson, K. W. (2006). Addressing the nature of science in preservice science teacher preparation programs: Science educator perceptions. *Journal of Science Teacher Education, 17*(1), 65–81.
Banilower, E., Smith, P. S., Weiss, I., Malzahn, K., Campbell, K., & Weis, A. (2013). *Report of the 2012 national survey of science and mathematics education*. Chapel Hill: Horizon Research.
Brende, E. (2004). *Better off: Flipping the switch on technology*. New York: HarperCollins.
Bunge, M. (2003). Philosophical inputs and outputs of technology. In R. C. Scharff & V. Dusek (Eds.), *Philosophy of technology: The technological condition* (pp. 172–181). Malden: Blackwell.
Clough, M. P. (2013). Teaching about the nature of technology: Issues and pedagogical practices. Chapter 18. In M. P. Clough, J. K. Olson, & D. S. Niederhauser (Eds.), *The nature of technology: Implications for learning and teaching*. Rotterdam: Sense Publishers.
Clough, M. P. (2015). A science education that promotes the characteristics of science and scientists. *K-12 STEM Education, 1*(1), 23–29.
Clough, M. P., Berg, C. A., & Olson, J. K. (2009). Promoting effective science teacher education and science teaching: A framework for teacher decision-making. *International Journal of Science and Mathematics Education, 7*(4), 821–847.
Clough, M. P., Olson, J. K., & Niederhauser, D. S. (Eds.). (2013). *The nature of technology: Implications for learning and teaching*. Rotterdam: Sense Publishers.
Cromer, A. (1993). *Uncommon sense: The heretical nature of science*. New York: Oxford University Press.
Davson-Galle, P. (2008). Why compulsory science education should *not* include philosophy of science. *Science & Education, 17*(7), 677–716.
DeBoer, G. E. (1991). *A history of ideas in science education: Implications for practice*. New York: Teachers College Press.
DeBoer, G. E. (2000). Scientific literacy: Another look at its historical and contemporary meanings and its relationship to science education reform. *Journal of Research in Science Teaching, 37*, 582–601.
DeBoer, G. E. (2006). Historical perspectives on inquiry teaching in schools. Chapter 2. In L. B. Flick & N. G. Lederman (Eds.), *Scientific inquiry and nature of science: Implications for teaching, learning, and teacher education*. Dordrecht: Springer.
DeGrasse Tyson, N. (2011, January 20). The moon, the tides and why Neil DeGrasse Tyson is Colbert's god: A conversation about communicating science. *The Science Network*. http://the-sciencenetwork.org/programs/the-science-studio/neil-degrasse-tyson-2. Retrieved 30 May 2015.
Dewey, J. (1902). *The child and the curriculum*. Chicago: University of Chicago Press.
Goodlad, J. I. (1983). A summary of a study of schooling: Some findings and hypotheses. *Phi Delta Kappan, 64*, 465–470.
Huesemann, M., & Huesemann, J. (2011). *TechNo-Fix: Why technology won't save us or the environment*. Gabriola Island: New Society Publishers.
Hull, G. (2013). Know they cyborg-self: Thoughts on Socrates and technological literacy. Chapter 2. In M. P. Clough, J. K. Olson, & D. S. Niederhauser (Eds.), *The nature of technology: Implications for learning and teaching*. Rotterdam: Sense Publishers.
Kromhout, R., & Good, R. (1983). Beware of societal issues as organizers for science education. *School Science and Mathematics, 83*, 647–650.
Marcuse, H. (1964). *One-dimensional man: Studies in the ideology of advanced industrial society*. Boston: Beacon.
Matthews, M. R. (2015). *Science teaching: The contribution of history and philosophy of science. 20th anniversary revised and expanded edition*. New York: Routledge.

Moscovici, H., & Nelson, T. H. (1998). Shifting from activitymania to inquiry. *Science and Children, 35*(4), 14–17. 40.

National Research Council. (1996). *National science education standards*. Washington, DC: National Academies Press.

NGSS Lead States. (2013). *Next generation science standards: For states, by states*. Washington, DC: National Academies Press.

Olson, J. K. (2013). The purposes of schooling and the nature of technology: The end of education? Chapter 12. In M. P. Clough, J. K. Olson, & D. S. Niederhauser (Eds.), *The nature of technology: Implications for learning and teaching*. Rotterdam: Sense Publishers.

Olson, J. K., Tippett, C. D., Milford, T. M., Ohana, C., & Clough, M. P. (2015). Science teacher preparation in a North American context. *Journal of Science Teacher Education, 26*(1), 7–28.

Postman, N. (1992). *Technopoly: The surrender of culture to technology*. New York: Knopf.

Postman, N. (1995). *The end of education: Redefining the value of school*. New York: Vintage Books.

Proctor, R. N. (1991). *Value-free science? Purity and power in modern knowledge*. Cambridge: Harvard University Press.

Schmidt, W. H., McKnight, C. C., & Raizen, S. A. (1997). A splintered vision: An investigation of U.S. science and mathematics education—Executive summary. http://hub.mspnet.org/index.cfm/9109/?print_friendly=true Retrieved 30 May 2015.

Weiss, I. R., Pasley, J. D., Smith, P. S., Banilower, E. R., & Heck, D. J. (2003). *Looking inside the classroom: A study of K-12 mathematics and science education in the United States*. Chapel Hill: Horizon Research.

Wolpert, L. (1992). *The unnatural nature of science*. London: Faber & Faber.